RADIO HISTORY
AMATEUR RADIO

SPURGEON "SPUD" ROSCOE VE1BC

RADIO HISTORY
AMATEUR RADIO

SPURGEON "SPUD" ROSCOE VE1BC

COVER PHOTO: This is Sea Cadet Petty Officer Second Class Duncan Culliton, VA3FSY, of RCSCC *Quinte*.

The photograph is from Dona Neves @ primadonadesigns photography and belongs to RCSCC *Quinte*. It is on their web site as RT-WEB21.

My father, George S. Roscoe, C1BK, his brother William A. Roscoe, C1EM, and I, all served in the navy under the white ensign. My father was so "Gung Ho" navy he formed RCSCC *Fieldwood* at Canning, Nova Scotia, when he retired.

The Halifax Amateur Radio Club is in the process of forming a sea cadet amateur radio station in HMCS *SACKVILLE*, VE0CNM, in memory of Chief Petty Officer Wayne N. Catchpaugh, VE1WC.

SACKVILLE is the last of the famous World War II Corvettes. Wayne was a retired C1CD, a 1st Class Chief Clearance Diver. His last posting was as coxswain on HMCS *CORMORANT*, the RCN diving tender. Wayne was also a pilot on the submersible *SDL-1*, and the remote-control submersible.

Wayne's friend was another retired C1CD, Sam Semple, VE1 YVN. Both Wayne and Sam were active members of the Halifax Amateur Radio Club. Wayne and Sam made several dives to over 2,000 feet in the Gulf of Mexico. Sam started his naval career as a radioman. He was sent up on the St. Lawrence River as the radioman in a diving tender. He enjoyed that experience so much he transferred to clearance diver. His first ship, as radioman was HMCS *MICMAC* with international call sign CYVN, the reason he has the VE1YVN call sign. Sam is donating an amateur radio station, he Dick Grantham, VE1AI, and Bruce Wade, VE1NB are the ones creating the sea cadet station in HMCS *SACKVILLE*. All three are active CW operators.

 FriesenPress

One Printers Way
Altona, MB R0G 0B0
Canada

www.friesenpress.com

ISBN
978-1-03-913321-1 (Hardcover)
978-1-03-913320-4 (Paperback)
978-1-03-913322-8 (eBook)

1. TECHNOLOGY & ENGINEERING, RADIO

Distributed to the trade by The Ingram Book Company

CONTENTS

RADIO HISTORY — AMATEUR RADIO

Radio Station
VE1FO
The Halifax Amateur Radio Club
HARC

VE1MK – VE1FO – VE1FQ

VE0MMA
CSS ACADIA

VA1MMA
The Wireless Room at the Maritime Museum of the Atlantic

VE0CNM
HMCS SACKVILLE

VE1ECT
Emergency Communication Trailer

VE1CPA
The call sign for the International Space Station contacts

VE1HNS
The 146.940- Repeater

VE1PSR
The 147.270+, 444.350+ and 53.550- (151.4) Repeaters

www.halifax-arc.org

THANK YOU

The American Radio Relay League, Inc., 225 Main Street, Newington, CT 06111-1494 U.S.A. owns the copyright to some of the material used in this exercise. We sincerely thank The American Radio Relay League for the use of this material.

Special thanks to Maty Weinberg, KB1EIB, at The American Radio Relay League Headquarters.

Special thanks to the Halifax Amateur Radio Club presidents Bill Elliott, VE1MR, and Murray MacDonald, VE1BB. We would like to thank HARC members Art Grant, VE1EP, John Doull, VE1WC, Ralph Fraser, VE1HJ, Don Watters, VE1BN, Dick Grantham, VE1AI, Sam Semple, VE1YVN, Wayne Catchpaugh, VE1WC, Howard Dickson, VE1DHD, Helen Archibald, VA1YL, Fred Archibald, VE1FA, Fraser McDougal, VE1WO, Brian Allen, VE1AZV, John Goodwin, VE1CDD, Murray Alary, VE1ALS, Elmer Naugler, VE1OD, and Rod Padmore, VE1BSK. We would like to thank Bill Gillis, VE1WG, Laval Desbiens, VE2QM, Roland Peddle, VO1BD, Mike Goldstein, VE3GFN, Dave McClafferty, VE1ADH, and a special thanks to Dave for holding Mike's old call for him.

A special thank you to granddaughter Josie C. Roscoe for the help with the computers.

ERRORS

When I first started to record the history of radio stations many years ago, I was convinced it was possible to record an accurate record. I now realize all these records have errors. I have no idea why one cannot put an exercise like this together and have it error free. This applies to all these records, whether it is radio stations, ships, and aircraft, even the history of a family or village and everything else one can think of. There will be errors in this exercise that we tried hard to avoid. We apologize for any and all errors. If you find any please notify us so we can make a note of the error.

Some will claim that all history is simply the opinion of the author. This exercise is simply my opinion. I believe that if anyone else had done the research and compiled this record it would be their opinion and quite different from this.

THE BEGINNING

When the news broke that Marconi had spanned the Atlantic with wireless on December 12th, 1901, those with an interest in electricity and telegraphy became very interested in wireless. This was the beginning of amateur radio. It has become tradition for amateur radio to create and the commercial radio interest to take it over. The Marconi Company claims this first signal was around 125 kilohertz in today's terminology although they did not realize these radio waves could be formed in frequencies or bands at the time. There are at least two other sources that claim this frequency was higher.

Einar Dessau of Denmark claims to be the first Amateur Radio operator from a transmission he made on March 18th, 1909.

Guglielmo Marconi (1874-1937) never held a radio license but one would have to agree that he was one of the first radio amateurs. He actually stated that he was only an amateur.

Hiram Stevens Maxim (1840-1916) was born in Maine. He invented the automatic gun and made a fortune from it. This gun was known as a Maxim gun. The 23-year-old North West Mounted Police used these Maxim guns to establish law and order in the Yukon during the gold rush of 1896 that became known as the Klondike Gold Rush. The North West Mounted Police became the Royal Canadian Mounted Police in 1920. Oh, to be twenty years old and whipped in shape by George Arthur French the first commissioner of the force. One would have the feeling they could run down a freight train; I am sure. He drove the recruits hard on their march west in 1874. The RCMP Historian, S. W. Horrall, in 1973, published "The Pictorial History of the Royal Canadian Mounted Police," ISBN 0-07-077366-1 if one would like a description of the history of the Force. Page 63 has a photograph of a .303 caliber Maxim machine gun held at the RCMP Museum, Regina, Saskatchewan. This gun was acquired in 1898 for service on the summit of the White Pass during the Yukon Gold Rush.

They built 2,710 merchant ships during World War II known as Liberty ships. The ***HIRAM S. MAXIM*** was named for Hiram Stevens Maxim. This ship was built in Portland, Oregon, in 1943 as the 468th Liberty ship of the 2,710 built. Once Hiram Stevens Maxim had made his fortune, he decided to build himself a twin-engine aircraft powered by two steam engines. This meant he more or less had money to burn. This also meant his son **Hiram Percy Maxim** (1869-1936) had the political connections and the power to help create this fascinating hobby known as Amateur Radio.

Hiram Percy Maxim became interested in amateur radio when his son Hiram Hamilton Maxim (1900-1992) became interested in the hobby. This was in 1911 when Hiram Hamilton Maxim was 11 years old. Hiram Percy Maxim did not learn Morse code until he was 40 years old. **Hiram Percy Maxim** and **Clarence Tuska** organized the **American Radio Relay League (ARRL)** on May 18th, 1914. Hiram Percy Maxim was the first president of the ARRL.

The ARRL published the first issue of their well-known and excellent magazine titled **QST** in December 1915. Clarence Tuska was the editor and at the time each monthly issue was 10 cents. This had crept up to 25 cents per issue by 1925. The price kept creeping up and was $2.00 per issue by 1978 and in 2008 was $4.99 U.S. or $6.99 Canadian. Mr. Maxim often wrote articles in this publication. He often used the pseudonym The Old Man – T.O.M. From reading his writing I feel he is someone I would have enjoyed meeting. The full story of when Hiram Percy Maxim became interested in amateur radio starts on page 9 of the May, 1979, issue of QST.

Marconi was very interested in giving ships a voice. The ship owners and crews felt that if they humoured this interest along it would soon disappear. When the two passenger liners *REPUBLIC* and *FLORIDA* collided off New York on January 23rd, 1909, it was evident that this passing fad was not going to disappear. Jack Binns was the wireless operator in *REPUBLIC*. *FLORIDA* was not fitted with wireless. Jack managed to talk a sister ship, the *BALTIC*, in alongside and rescue the passengers in both ships. *FLORIDA* limped into New York for repairs but *REPUBLIC* sank. Jack was given credit for the saving of 1600 lives. The distress signal at the time was CQD and Jack was known as "CQD" Binns until the day he died. John Robinson "Jack" Binns died in December 1959 in a New York hospital.

Marconi had little trouble in fitting more ships with wireless after that incident. By the time the *TITANIC* hit the iceberg on April 15th, 1912, the wireless world was getting "out of hand". There were no licenses and only a few company rules. The wireless operators were more or less doing as they pleased. This also applied to amateur radio.

The London Radio Convention of 1912 was just getting started when *TITANIC* hit the iceberg and this brought about the international licenses, for both the stations and operators, and also the assignment of call signs by the various governments on an international level. This included all wireless stations; ships, coast stations and amateur stations. They were the only stations at that time.

The only decision relative to Experimenters and Amateur Radio made at the 1912 London Conference was to ban such activity to 200 metres and below.

200 metres is 1500 kHz in today's terminology and therefore everything above today's present AM broadcast band was considered to be of no use. It was the amateur radio community that proved this theory wrong. In Canada these licenses were issued by the Naval Service of the federal government in 1912. This radio regulation service is provided by Industry Canada in 2009. The amateur radio portion is under the supervision of Radio Amateurs of Canada with head office in Ottawa.

There is a record that the first amateur radio license issued in Canada was to Frank Vaughan of Saint John, New Brunswick, in the fiscal year ending March 31st, 1911. Canada must have had some kind of amateur license before the London Radio Convention. This license was issued one year before the London Radio Convention.

An American Amateur Operator in Chicago made this statement in QST "In the good old days of 1913 – 1914, when there were more unlicensed amateurs than otherwise, I operated a "bootleg" ham station "CH" in Chicago and returned to the air after the war as 9ASR." There was a lot of illegal activity back then, but according to the 1916 issues of QST they were rounding them up and putting them through the legal system. Many were operators from the telegraph lines. Switzerland only had one officially licensed station in 1925 but had 30 transmitters on the air. There must have been some weird and wonderful stuff to listen to at least back then.

The first amateur radio operators used whatever came to mind as a call sign. Telegraph operators for fifty years had often used the initials of the first operator for the call sign or call code of a station. They had assigned a few two letter codes as special calls. One, CQ, meant all stations. This same system of call signs was popular with our first amateur operators.

The amateur and commercial operators were more or less QRM (manmade interference) for each other during that time. The primitive nature of the equipment in use by both the amateur and commercial operator meant that the frequency in use at any time by either operator was more or less all the frequencies then known. The majority of these early stations transmitted on a wide band of frequencies, which was a large portion of that area known as 100 kilohertz in modern terminology. When an amateur operator interfered with a commercial operator he was to terminate when the commercial operator transmitted the signal "STP" and was not to commence until the commercial operator transmitted "CANCEL STP".

The Wireless Association of Ontario was formed in 1912 to provide some form of regulation to the amateur radio community. This association assigned call signs with the letter X as the prefix, and then the first letter of the holder's surname, and then a letter in alphabetical order. For example, we will break down call sign XHC. X is the prefix, then the H is the first letter of the holder's surname, D. Hall, then C indicates it is the third call sign assigned someone with the letter H as the prefix of their surname.

Soon after this the government of Canada saw the need to have all amateur radio activity under its jurisdiction and drew up a set of regulations. These came into force in 1913-1914. These regulations had over thirty items and item 25 are as follows:

"25. A distinctive call signal will be allotted to each station commencing with the letter "X", e.g., XAA, XAB, which signal must be sent not less than three times at the termination of every transmission."

When the government took over the assignment of call signs, they simply went down the alphabet from XAA. This has left one wondering why Frank Vaughn, as the first, was assigned XAO per the 1911-1912 lists as shown below. It would appear that they may have been given a choice and were able to choose from certain blocks of call signs.

THE 1912 CALL SIGNS

These Experimental Amateur Station call sign lists for the Maritimes are from official government records and go back to the first list published in 1912.

1911-1912

XAI	Donald Lawson	Yarmouth NS
XAJ	Carl O. Elderkin	Weymouth NS
XAK	Charles J. O'Hanley	Yarmouth NS
XAN	Militia & Defense	Charlottetown PEI
XAO	Frank P. Vaughn	Saint John NB
XAR	K. S. Rogers	Charlottetown PEI

1913-1914

XAK	Charles J. O'Hanley	Yarmouth NS
XAR	K. S. Rogers	Charlottetown PEI
XBB	Jack Oak Hum	Saint John NB
XBS	C. P. Logan	Saint John NB
XCA	G. D. Crowell	Sydney NS
XCJ	H. Reading	**Halifax NS**

It is hard to believe there were only one Halifax and no Dartmouth stations listed in the above lists.

Apparently in some areas at least the first call sign was known as an "Identifier". I have a 1915 list of marine radio stations and they are called a call sign in that.

HOMEWARD BOUND

Lloyd Manuel QST July 1916
We have all felt like this "Old Boy" on passing our certificates, I am sure.

This article was recorded in the "International Amateur Radio" news section of QST for September, 1922.

CANADIAN REGULATIONS

The Canadian Radiotelegraph Act of 1912 does not lay down any operating rules but gives the administrative department the power to make regulations from time to time, making revision of regulations red tape and delay. The department is very favourable to the amateur and is willing to let him have his fair share of the ether. They do not recognize any distinction between the various classes (telegraphers and broadcast listeners) and are very anxious that these two groups live at peace with each other without interference from headquarters.

The wave length limits favour C.W. transmission, amateurs being permitted up to 180 meters for spark and up to 200 for C.W., while experimental stations observe the same maximum for spark but may use 275 for C.W. The maximum power permitted is 500 watts, which is not all that might be desired for spark, but the spark is rapidly dying in Canada and nobody wants or can afford more than that in C.W.

One will note that this article lists CW as C.W. This has become CW and this had me confused for a while. CW stands for continuous wave and is the type of radio transmission that replaced the spark transmissions, the first means of radio communications. CW is also known as radiotelegraph (Morse code), but the amateur radio community is the only place one will hear it today.

The American Radio Relay League was created in 1914 and the first issue of their magazine, QST, was printed in December 1915 as stated. There had been quite a shake-up of this organization in April 1917 just before it shut down for World War I. Their system of divisions was created at this time. These divisions were much the same as they are today.

The Old Man wrote an article titled Rotten QRM (manmade interference) that appeared in the January, 1917, issue of QST. The Old Man was Hiram Percy Maxim who wrote a number of rotten articles for QST. In this one on QRM he was describing the gibberish he was receiving at his station one evening. A couple of operators were talking about something called a wouff hong. The Old Man felt it was probably some piece of apparatus used in the southern states to beat monkeys with. Whatever it was he felt it simply some form of uninteresting conversation which clutters up the air with QRM. This took off to the point that the Royal Order of the Wouff Hong was created.

This description appears on page 63 of the December, 1986, issue of QST:

> "The ROWH is a secret society of radio amateurs who are members of the ARRL. The Order of the Wouff Hong can be conferred only at a National,

Division, State or Section ARRL Convention. Each inductee receives a certificate of membership to be displayed prominently in his or her shack.

The ceremony is not conducted at every League convention, so you'll have to watch the convention write-ups in QST or publicity mailings to determine whether it is one of the scheduled events at a convention in your area. Then, with proof of League membership in hand, register to be one of the inductees into the great, secret fraternity of Amateur Radio, the Royal Order of the Wouff Hong."

There was also something called a Rettysnitch mentioned in this conversation the Old Man was listening to. Both the Wouff Hong and Rettysnitch became a major part of amateur radio tradition. L. R. Cebik, W4RNL, describes them both in an article on page 59 of the September, 1996, issue of QST.

The Radio Amateur's Handbook for 1930 describes them both as follows:

> "The Wouff-Hong is Amateur Radio's most sacred symbol and stands for the enforcement of law and order in amateur operations."

> "The Rettysnitch…is used to enforce the principles of decency in operating work."

The Old Man presented an actual Wouff Hong to ARRL Headquarters in June, 1919. The Headquarters staff did not know which end to use as a handle in order to beat off the QRM. There has been at least one yacht named *WOUFF HONG*. At least two members of HARC received this Royal Order of the Wouff Hong. Brit Fader, VE1FQ, received his in 1965 and Art Crowell, VE1DQ, had his in 1948 but I did not learn the date he received his.

The Old Man presented an actual Rettysnitch to the League's Traffic Manager in 1921. Both instruments are still held in the ARRL Museum.

QST was shut down during World War I for 20 months. The last issue was the September, 1917, issue. A notice of several pages was distributed during April, 1919 and the first post war issue was the June, 1919, issue. The United States Navy had authorized American amateur stations to receive only on April 15th, 1919, and transmitting or full station operation would be allowed on the signing of the Peace Treaty. Transmitting and the full use of the amateur radio stations were permitted on October 1st, 1919. The November, 1919, issue was labeled the Liberty Number and showed an amateur in a garbage can with the lid in the air stating "Lid Off". The lid was off and the restrictions on operating during World War I were removed.

Mr. Hiram Percy Maxim wrote an excellent article in the November, 1919, issue of QST that would probably apply to a number of things today. He stated it was organization that retained amateur radio. There had been a strong opposition to amateur radio in any form. Mr. Maxim praised the various amateur radio clubs and especially the Pennsylvania Wireless Association for their work at the 1912 international radio convention, stating it was these organizations that made amateur radio possible. He went on to state that organization brought amateur radio from the realm of Toyland to the dignity of a national asset and the reason they were able to supply a flood of operators for the Army and Navy when needed most at the outbreak of the war. This November issue went on to praise several who had been instrumental in making amateur radio possible again.

S. G. "Spud" Roscoe, VE1BC

Mr. Maxim went on to state that amateurs should join ARRL simply for protection. This makes a lot of sense. I have been a life member for years. I am recorded with the life member applicants for January 23rd, 1979, and appear in Moved and Seconded on page 63 of the October, 1979, issue of QST. The life membership program was created for ARRL in 1967. I do not remember when I first became a member of ARRL. I have a 1958 Handbook I purchased new and I know I was a member in 1962. Working shift work as a radio operator tended to dampen anyone's enthusiasm with radio as a hobby.

The American Radio Relay League, Inc.

A-1 Operator Club

This certifies that:

S. G. "Spud" Roscoe, VE1BC

is a member of the ARRL A-1 Operator Club and is authorized to nominate other deserving qualified radio amateurs for membership.

Membership in the A-1 Operator Club represents adherence to several principles of good operating: careful keying, good voice operating practice, correct procedure, copying ability, judgement and courtesy.

3/5/2014

Kay Craigie N3KN
PRESIDENT

These are two of the certificates I have received from ARRL.

THE HALIFAX EXPLOSION

On December 6th, 1917, the French munitions ship *MONT BLANC*, carrying benzine on her upper deck, pitric acid, gun cotton and nitro glycerine (T.N.T.) in her holds was proceeding into Halifax. *MONT BLANC* was a floating bomb. At the narrows she collided with the Norwegian Ship *IMO* that was carrying relief supplies for the war victims of Belgium. This caused the benzine on the *MONT BLANC* to catch fire and roughly twenty minutes later caused the TNT to explode. This explosion nearly wiped Halifax off the map and the total count of those injured or killed was impossible to calculate. The estimates were placed at 2,000 killed, 2,000 wounded, and 6,000 homeless.

The only radio or wireless involvement I have found for the Halifax Explosion is that George Harris was the Warrant Officer Telegraphist in *HMCS NIOBE*. He was going down an outside ladder at the time of the explosion and landed on his butt on the upper deck. He was not injured and ran to the wireless cabin and found the equipment in working order. He called Camperdown Radio at Portuguese Cove in the approaches to Halifax Harbour and this alerted the world to the disaster. *HMCS NIOBE* had wireless call sign VDA and Camperdown had call sign VCS. *HMCS NIOBE* received a lot of damage to her superstructure from this explosion and simply laid alongside here in Halifax as a Depot Ship until sold and scrapped in 1920.

My main reason for recording this is that later on that day there was a severe snowstorm. We can hear these storms coming for a day or so, especially on the lower frequencies that were in use during the CW world of marine radio. The October, 1946, issue of QST states that the static charge in a snowflake is the equivalent to 17,000 electrons. I would agree with that from the static I have heard over the years from these storms. This static was difficult to work through on December 6th, 1917, with the primitive equipment in use at that time. Camperdown VCS and Sable Island VCT were not able to hear each other quite often through this static they called X's and we call QRN. They worked each other around 143 kilohertz at that time.

HARC Held a Special Event for the 100th Anniversary of the Catastrophic Halifax Explosion of December 6, 1917

From 0000Z (midnight Greenwich Mean Time) December 2 to 2359Z (midnight Greenwich Mean Time) December 10, 2017, various members of HARC used the call sign CK100VDA to remember this terrible explosion. CK is a rarely used Canadian call sign prefix. 100, as in 100 years to the day of the explosion and VDA the call sign of *HMCS NIOBE*. Greenwich mean time is the time at Greenwich England and the standard time for all radio rooms so that every radio room is on the same time.

Call	Name	CW	SSB	Digital
VA1CM	Craig			43
VA1MMA	Museum	36		
VE1AI	Dick	122		
VE1ANU	Mike	67		
VE1AST	Eric	118		
VE1BAB	Wayne	164		
VE1FA & VA1YL	Fred & Helen		1669	
VE1JG	Jim	359		
VE1NB	Bruce	1136		
VE1QD	Scott		238	
VE1YVN	Sam	18	9	
VE1ZD	Howard		271	
Mode Totals		2020	2187	43
Grand Total (not counting dupes)		4250		

These are the results of the CK100VDA commemoration of the Great Halifax Explosion of 1917 as recorded by Fred VE1FA on page 35 of the March/April 2018 edition of The Canadian Amateur magazine.

The highest level of static is around 1600 kilohertz or just below the 160-meter amateur radio band. A continuous watch had to be kept in this area on 500 kilohertz CW and 2182 kilohertz AM and later USB at the marine radio stations. It was a noisy environment at times and 2182 was by far the noisiest of them all. 2182 should have been in a separate sound proof room away from the other circuits or positions.

The Halifax explosion occurred during World War I and there is one amateur radio station listed for Halifax in 1914 only. The amateur radio stations were shut down during the war. Until the atomic bomb, a period of nearly thirty years, the size of an explosion was compared to the size of the Halifax explosion.

THE 1919 CALL SIGNS

In 1919 the various provinces and territories of Canada were divided up and the provinces of Prince Edward Island, New Brunswick and Nova Scotia were assigned the digit one as the prefix of their amateur radio call sign.

This is the numerical prefix of the Canadian Amateur Radio Call Signs when they were first assigned a numerical prefix in 1919:

1 Nova Scotia, New Brunswick and Prince Edward Island
2 Quebec
3 Ontario
4 Alberta, Saskatchewan and Manitoba

5 British Columbia, North West Territories and Yukon Territory
6 Training Schools
7
8
9 Experimental
10 Amateur Broadcasting

This list of call signs is one of the reasons I became interested in the history of radio in Canada, because so much of my day-to-day activity with radio made little if any sense. The 1 through 5 call signs inclusive are self-explanatory. I have not found a training school assigned a 6-call sign. I have found stations at various schools but they simply used a call sign from the area of the school, such as 3 in Ontario. The 7 and 8 call signs have always been left blank on this list but one assumes they were for future use. One also assumes this had some effect on the Dominion of Newfoundland assigning the digit 8 as the prefix of their first amateur radio call signs. There were eight active amateur radio stations in Newfoundland in 1926.

The 9-call sign is hard to understand today. We will use Keith Rogers as an example to try and understand this call sign. Keith built his first wireless station in 1907 and served as the wireless operator in the ferry *MINTO* during the winter of 1910-1911. Keith was XAR but became 9AK in 1919. The 9-call sign had no regional assignment; 9AK in Charlottetown, Prince Edward Island, 9AL in Toronto, Ontario and 9BP in Prince Rupert, British Columbia and so on. Amateurs had to work below 200 meters from instructions created at the 1912 London Radio Convention. If a person held a commercial radio certificate, they could obtain a 9 amateur call sign. 200 meters was so close to 600 meters that an amateur station close to a marine station often created so much interference with the primitive equipment in use that the amateur station had to stop transmitting.

During the annual shipping season on the Great Lakes all amateur activity within a certain range of this shipping activity had to move down below 50 meters. But if one held a commercial radio certificate, they could operate on 200 meters with a 9-call sign. I do not believe this applied to the East and West Coast but Keith Rogers had the 9AK call sign. The 9-call sign participated in the regular amateur activity like any other amateur call sign. Our American neighbours had no such prefix and they did not have to move off 200 meters for

Department of the Naval Service,
Ottawa.

30th December, 1919.

Sir:—

I have the honour to advise that the Minister of the Naval Service has been pleased to authorize all amateurs on the Great Lakes and River St. Lawrence from Port Arthur, Ont. to Quebec, P. Q. to use a transmitting wave-length of 200 meters until the re-opening of navigation, approximately the 15th April 1920.

The Department is anxious to accord to amateur stations every possible latitude compatible with the proper protection of Naval and Commercial services against interference, and the authorization of the use of 200 meters this winter is in the nature of an experiment.

Should no interference result the Department is prepared to consider a permanent amendment to the regulations regarding wave-lengths, and it accordingly behooves every amateur to see that his transmitting apparatus is sharply tuned and its decrement reduced to a minimum.

Your attention is particularly called to Radio telegraphy regulation No. 25, reading as follows:

"A distinctive call signal will be allotted to each station commencing with a figure eg. 3 A A etc. which signal must be sent not less than three times at the termination of every transmission."

I am, Sir,

Your obedient Servant,
G. G. Desbarats,
Deputy Minister.

QST March 1920

This is the only letter I have found restoring amateur radio in Canada after World War I.

One can see it is for Ontario and Quebec and that the call signs are from the 1919 assignment.

The ARRL used this letter as though it were the only letter lifting the World War I restrictions of the amateur radio service in Canada.

any reason, unless told to do so by the nearest commercial station. This 50-meter 200-meter split made it difficult for Canadian amateurs to communicate with American amateurs.

The 10-call sign was recorded by one Canadian amateur in the ARRL publication QST as another national joke right out of Ottawa. I more or less agree with that along with many other similar records. Keith Rogers founded broadcast station CFCY Charlottetown but I do not know if he did that via his 9AK call sign or if he had a 10-call sign. He started this broadcast station in 1924 and it was known as "The Friendly Voice of the Maritimes". Keith became VE1HI and this call is the call of the Charlottetown VHF repeater and has been for years in memory of Keith. That was the Keith Rogers Memorial Amateur Radio Club call sign for several years. That club terminated years ago.

The following stations were on the 1919 list:

1AA	J. C. Hanley	St Stephen NB
1AB	C. W. Alexander	Campbellton NB
1AC	L. L. Smith	Yarmouth NS
1AE	H. Holden	Glace Bay NS
1AF	G. D. Davidson Jr.	Saint John NB
1AG	J. F. C. Wightman	Amherst NS
1AH	George Sandoz	**Halifax NS**
1AI	P. Whitman	**Halifax NS**
1AJ	J. E. A. Demers	Canso NS

The list of Canadian call signs on page 42 of the July, 1920, issue of QST simply confirms the above list. The Halifax Amateurs on the 1920 list were the same two as on the 1919 list:

1AH	George Sandoz
1AI	P. L. Whitman

THE 1922 CALL SIGNS

The 1922 list includes the following **twelve Halifax-Dartmouth** area amateurs:

1AH	G. A. Sandoz
1AI	P. L. Whitman
1BI	L. A. Martin
1BM	G. D. Crowell

1BQ	A. W. Greig
1CM	C. H. L. Baker
1DD	W. C. Borrett
1DE	C. A. Landry
1DF	C. C. Curran
1DJ	E. S. Campbell
1DQ	A. M. Crowell
1DT	H. Lardner

Joe Fassett, 1AR, Dartmouth first appears on the 1923 list. He set some early DX (distance) records that I have recorded elsewhere.

THE ARRL SECTIONS

This is the front cover of the April, 1920, issue of QST signifying hands across the border and welcoming us Canadians as members of the American Radio Relay League (ARRL). One can see that we in Nova Scotia did not make it on the map but we were represented none the less. One will note this map shows the ARRL divisions when Canada first became a part of ARRL. This close relationship with our fellow amateur radio operators south of the border lasted 67 years until January 1st, 1988, at which time some of us felt we were big enough to go on our own.

The January 1920 issue of QST contained over a half page with the heading Greetings Canadians and this described in detail how the ARRL could help Canadian amateurs. There were four Canadian divisions in 1920 and they were: St. Lawrence Division, comprising the provinces of Quebec, New Brunswick and Nova Scotia. Ontario Division, comprising the province of Ontario. Winnipeg Division, comprising the provinces of Saskatchewan and Manitoba. Vancouver Division, comprising the provinces of Alberta and British Columbia. There was no provision for Prince Edward Island, the Yukon or the North West Territories. The Yukon and the North West Territories had the same prefix in the call sign as British Columbia so was considered part of the Vancouver division.

The first report from a Canadian division was from Ontario in March, 1920. Each division had a manager and Albert Lorimer, 2BF, was the manager of the St. Lawrence Division. It was not until April 1923 that they elected a Canadian General Manager at ARRL Headquarters in Hartford in the person of a Mr. W. C. C. Duncan, 9AW, from Toronto. The first report from Albert, 2BF, was in April, 1920, and he had no contact with this region. He stated if anyone down here wanted to participate in the League to contact him. This may be the reason we in Nova Scotia were not included on the map of April 1920.

QST October 1923

The St. Lawrence report for January, 1922, stated that Keith Rogers, 9AK, Charlottetown had been heard in Hartford, Connecticut. It was not until March, 1922, that the St. Lawrence division splits into the Quebec and Maritime divisions and there are now five Canadian divisions; Quebec, Ontario, Winnipeg, Vancouver and Maritime. Albert J. Lorimer, 2BF, is manager of Quebec and Keith S. Rogers, 9AK Charlottetown, is manager of the Maritime division. The Maritime division finally shows up with a report in July 1923 but it is not Keith Rogers. C. C. Curran, 1DF, is the author of that report although Keith is listed as District Manager. Keith Rogers, 9AK, was the only amateur radio station in the province of Prince Edward Island at this time.

THE MARITIME DIVISION

The Maritime division was the province of New Brunswick, Nova Scotia, Prince Edward Island and the Dominion of Newfoundland and Labrador.

This is the first Maritime division news to appear in QST. This was on page 48 of the July, 1923, issue. Art Greig, 1BQ, managed to work 3GK in Toronto with 20 watts of CW on about 1500 kilohertz through all the QRM and what have you. One would have to agree that was pretty good and worth recording.

MARITIME DIVISION
Reported by C. C. Curran, 1DF

1BQ is off the air at present. He was using 20 watts and has been reported 2600 miles east of New York. He formed the Eastern leg of the successful Canadian transcontinental tests, and succeeded in working Toronto.
1DD maintains several schedules with American friends and is easily QSA at Montreal and at American 4FT. Cap. is using 16 watts. 1AR is reported QSA in the Yanks first district with his 10 watts.
By the time this is published 1BI will be on with two fifties. 1DT, 1DF, 1DJ, 1DG, 1DE, 1BV, 1EB, and 1EF already have, or will soon have, 5-watters splitting the air.

QST July 1923

Captain Bill Borrett, 1DD, had replaced Keith Rogers, 9AK, as the District Manager on the December, 1923, report and Bill gave a very good report. Bill Borrett, 1DD, organized the American Radio Relay League (ARRL) activities in the area of Halifax. Bits and pieces of his first report will be found throughout this exercise. The local Halifax gang called him

"Cap" from his rank as Captain in the army militia and when promoted Major, Bill claimed most still called him "Cap".

THE ARRL EMBLEM

 The ARRL Emblem first appears on page 23 of the July, 1920, issue of QST. ARRL had asked for suggestions for this emblem and the board of directors accepted the one we have become so familiar with and still in use today. Unfortunately, they did not state who designed the emblem.

AMATEUR RADIO HELPING LOCAL POLICE FORCES

The ARRL had been asking amateurs to contact their local police force and see if they could help in any way, especially with stolen automobiles. It is hard to believe but this appeared in the 1920 issues of QST. Amateur radio assisted in the capture of a stolen vehicle in early 1921 when the New York Police station KUVS transmitted a list of stolen vehicles each evening. Amateur radio station 2TK at Union Hill, New Jersey copied these broadcasts and alerted their police department on the location of a stolen 2-ton truck loaded with 97 boxes of oranges. Another stolen automobile was found in January, 1922. This one was stolen in Boston and found in New Hampshire from a broadcast made by the Boston police and copied by an amateur operator. These broadcasts were in spark and telegraph at that date and time of course.

THE CONFUSION BETWEEN CANADA AND USA STATIONS

The January, 1921, issue of QST stated that there was considerable confusion between Canadian and U.S. call signs now that they were starting to hear each other. The ARRL came out with the following scheme that was to be adopted at once:

U.S. stations were to continue to use **DE** as the separation signal between calls.
A U.S. station calling a Canadian station was to use **AA** as the separation signal.
A Canadian station calling a Canadian station was to use **V** as the separation signal.
A Canadian station calling a U.S station was to use **FM** as the separation signal.

In other words, if 1DD in Halifax called 1BQ in Halifax it would be: 1BQ V 1DD
If 2NJ in New York called 8BB in New York it would be: 8BB DE 2NJ
If 8BB in New York called 3AB in Toronto it would be: 3AB AA 8BB
If 3AB in Toronto called 8BB in New York it would be: 8BB FM 3AB

This was changed again in two years in 1923 when the amateur stations started to reach across the Atlantic Ocean.

THE FIRST CANADIAN BROADCAST STATIONS

In January, 1922, the Canadian Marconi Company, with headquarters in Montreal, was operating two broadcast stations in Canada. One station was in Montreal and the other in Toronto. Every Tuesday evening from 8 PM until 9:30 PM they would transmit news, market reports, musical entertainments, etc. from these two stations on 1200 meters. They were using a ½ kilowatt transmitter that gave a range of 200 miles. A third station of identical characteristics was nearing completion in Halifax but full arrangements for its operation had not been completed. 90 minutes a week on 249.9 kilohertz in today's terminology.

RENAMING THE WESTERN CANADIAN DIVISIONS

Those in Canada with the 4 and 5 call signs felt that the division names of Vancouver and Winnipeg did not properly describe their divisions so had this changed. They first appeared in the April, 1926, issue of QST as the Vanalta and Prairie divisions. Vanalta was British Columbia and Alberta and the Prairie division was Saskatchewan and Manitoba. This remained this way until the March, 1953, issue of QST. In that issue the names in the masthead at the front of the issue remained Vanalta and Prairie but the actual reports were divided into provincial divisions. Each division was labeled British Columbia, Alberta, Saskatchewan and Manitoba. Each province had a Section Communications Manager (SCM). This masthead was changed in the May, 1953, issue of QST to read the Canadian Division with each Canadian section listed; Maritime, Quebec, Ontario, Manitoba, Saskatchewan, Alberta, British Columbia and Yukon.

By 1926 these District Managers (DM) were helped with a Section Communication Manager (SCM) who issued a report for each section in their division. There were four in the Maritime division: Loyal Reid, 8AR for Newfoundland, Bill Borrett, 1DD, for Nova Scotia, Walter Hyndman, 1BZ, for Prince Edward Island and T. B. Lacey, 1EI, for New Brunswick.

OPERATOR CERTIFICATES 1924

This is a direct quote from the Canadian General Manager at ARRL, Keith Russell, 9AL, in the Canadian Section news for May, 1924.

"The latest Government regulations as regards amateur transmitters have caused considerable anguish among the brass pounding fraternity in so far as the boys are all required to have operator's certificates of proficiency prior to renewal of licenses on April 1st. This is expected to eliminate a considerable number of licensees who have licenses but no stations as well as those who are interested solely in radiophone and who have never taken up the code sufficiently to learn it up to an efficiency of ten words per minute. The license fee for amateur transmitters has also been raised from $1.00 to $2.50, this also with the same idea in view of eliminating the dead-wood."

This makes it clear that one received a station license with call sign without an operator's certificate or a station. All one needed was an interest in amateur radio and this is the reason for the big difference in the list of stations for 1922 and the one for 1926-1927.

THE FIRST THREE NEWFOUNDLAND STATIONS

At this time there were 60 members in the Maritime division of ARRL and they were hoping to have 100 by the fall of 1924. They still had no one in Newfoundland but were hoping to have someone on soon. The first three Dominion of Newfoundland stations showed up in 1924; Loyal Reid, 8AR, St. John's, James Moore Jr., 8AW, Carbonear, and E. Taylor, 8AY, Carbonear showed up. These three stations agreed to act as relay stations for traffic with ARRL.

Newfoundland had people interested in amateur radio of sufficient number to form an amateur radio club in 1921. Mr. Ernest Ash was elected president of this club, the St. John's Radio Club at that time. Ernie Ash was the first licensed amateur in Newfoundland and received license # 1 when formal licensing of amateurs started on January 3rd, 1923, and identifier 8AA as it was known. In 1925 Loyal Reid showed up with call sign 8LR. Roland Peddle, VO1BD, stated that Loyal Reid's 8AR call was his "signaling" call and 8LR was his "broadcasting" call. The odd feature is that one could not tell what was amateur radio or broadcasting.

Roland Peddle, VO1BD, told me there were no examinations, and the only requirement to obtain a license and "identifier" was to have an "interest in radio communications". Apparently, up until 1929 there were three categories:

1. "Signaling stations" – engaging in amateur radio activities much as they do now.

2. "Broadcasting" music and voice.

3. "Listeners" with receiving sets and an "interest in radio".

This all changed in 1929 when the amateur and broadcasting were separated forever.

Canadian amateurs were given permission to operate on the same short-wave bands as the American stations with fewer restrictions than the American stations in 1924. The big excitement in 1924 was the Superheterodyne receiver with just the one control.

AMATEUR RADIO CALL BOOK

There soon became a need for a call book where one could look up the name and address of each call sign one heard or worked. The Canadian government produced lists of amateur radio operators as described. I have seen one of the Canadian lists of those holding a license as early as 1924. The ARRL complained of the money they had lost in producing various books on amateur radio in the June 1916 issue of QST. They announced then that they would have no part of producing a call book because it would be another money loser. They repeated this statement in the January 1917 issue of QST, stating that amateurs would not pay the 35-cent cost for such a publication and that they were going to continue to use the government lists.

The Radio Amateur Callbook Magazine was created in 1920 and was owned and operated by Radio Amateur Call Book, Inc., at Chicago, Illinois. This started as the Citizens Radio Call Book, from the Citizens Radio Service Bureau in Chicago. There were few if any amateur operators that did not buy a copy now and then. The more active amateurs actually subscribed to this Callbook that was published quarterly. This Callbook came with a wealth of information for the amateur radio community in addition to the name and address of the holder of each and every active call sign around the world.

The last issue of the Callbook was the 75th edition in 1997. It was available on CD only until the winter edition of 2003 when it was phased out completely. The availability of this information on the Internet made it impossible to sell sufficient copies in order to stay in business. 77 years was a good run but I must admit I miss the book. I find it hard to believe that those 77 years of the Callbook will ever be available on CD. I am probably the only one who would have an interest in such a CD. The Callbook moved to Baunatal, Germany, and they produced the winter 2003 edition. Their advertising gave their German address, an address in Florida and also their website. Their website on January 15th, 2009, states that their shop is temporarily closed. There are other call books advertised on CD in QST but it looks like "The Callbook" is history.

TELEVISION MAY 1925

There is an article on television complete with pictures that had been transmitted in May, 1925. It is amazing that it took nearly thirty years for this to get going and become a useful form of entertainment. The July, 1925, issue of QST actually calls it Television. According to QST colour reception was obtained in the April, 1944, issue. The word "television" was created by Hugo Gernsback according to an interesting article on Hugo by Gil McElroy, VE1PKD. This article is in the February, 1995, issue of QST. The February, 2009, issue of the National Geographic magazine claims that Russian scientist Constantin Perskyi coined the word "television" at the Paris World's Fair to describe the transmission of still photos over electrical wires in 1900. This is another simple case of where the records provide two answers to one simple question.

FIRST AMATEUR CONTACT ACROSS THE ATLANTIC

Bill Borrett 1DD of Dartmouth was one of the first in this area to cross the Atlantic. He heard Leon Deloy, f8AB, of Nice, France working u1XW in the United States on a frequency of around 100 meters or 3 megahertz in today's terminology. The first to cross the Atlantic was on November 28th, 1923, when Leon, f8AB, worked Fred Schnell, u1MO, in West Hartford, Connecticut. Bill Borrett, 1DD, heard this crossing in December 1923 and all he could hear, in his mind, for the longest while was that 25-cycle fluttery note transmitted by f8AB. Leon Deloy, f8ab, became a silent key (died) on January 21st, 1969, and is recorded in the silent keys of May, 1969. Not long after these contacts took place across the Atlantic Frank Bell, z4AA, New Zealand, worked Cecil Goyder, g2SZ, in London, England on October 18th, 1924.

The main object of the amateurs in this the Halifax area in 1923 was to make contact with an Englishman, and Bill spread the news the next day after hearing f8AB to some ten amateurs in the Halifax area, who looked upon him as something out of the ordinary. Bill Borrett, 1DD, stated that Arthur Greig, 1BQ, had the better station in this area, and it was only a few days when c1BQ worked g2OD of Gerrards Cross, England. 1BQ and 2OD became very close friends after that initial contact.

The way in which they contacted each other back then is interesting.

They would transmit:

CQ CQ CQ gc 1DD 1DD 1DD

Bill stated it was a scheme that became very useful. All Canadian stations had the letter C in their call and all English stations had the letter G in their call, and this would indicate a Canadian station wanting contact with an English station.

This is explained in detail in the December, 1923, issue of QST. I always heard it called the separation signal but they called these "Intermediates" back then. The letters assigned in 1923 were as follow:

A – Australia
C – Canada
F – France
G – Great Britain
I – Italy
M – Mexico
N – Netherlands
O – South Africa
P – Portugal
Q – Cuba
R – Argentine
S – Spain
U – United States
Z – New Zealand

QST stated twelve letters are still unassigned for future developments and as the call arises, they will be allotted and the proper publicity given. In February, 1925, they added:

B – Belgium
C – Canada and Newfoundland
D – Denmark
CH – Chile
FN – Finland

If Keith Russell, 9AL, called Hiram Maxim, 1AW, it would be:

1AW 1AW 1AW uc 9AL 9AL 9AL

If Joe Fassett, 1AR called 3BP in Ontario it would simply be:

3BP 3BP 3BP c 1AR 1AR 1AR

And so on around the world.

These had been continually added and changed until the last list to become effective 0000 GMT, February 1st, 1927 was so long it took up a full page in the January, 1927, issue of QST. Canada, Newfoundland and Labrador ended up with **NC** explained further along in this exercise.

Courtesy Thomas Roscoe K8CX

This is a Joseph Fassett QSL dated November 16th, 1926.

Note the small "u" as a prefix in the 8ABW and the small "c" in the prefix of the 1AR call signs.

Note the Can. 1:AR "The Globe Trotter" across the top of the card and the question Do You Q.S.L.?

All good Ops. Do. across the bottom right corner.

Just below the Joseph Fassett Owner and Operator is the statement DX Worked: Continents of Europe, North and South America, Australasia and Africa.

I would say that is pretty good for 1926 with 50 watts input and a one wire aerial with counterpoise.

It was legal back then for an amateur radio station to transmit anywhere above 1500 kilohertz in today's terminology or as they knew it, 200 meters and below. It was all Morse code and if anyone wanted to experiment with radiotelephone, they had to go way down below 200 meters or above 1500 kilohertz in today's terminology. They were not allowed to play a gramophone over the air. Those who wanted to do that obtained a license with a call sign with a 10 prefix here in Canada. The majority with the call sign with a 10 prefix simply disappeared into the world of broadcast radio. They became the broadcast stations with a call sign like CHNS.

BROADCAST STATION CHNS ON THE AIR 1926

CHNS went on the air in 1926 with a borrowed 100-watt transmitter. A five-tube table model broadcast receiver was advertised in Halifax in 1925 for $175.00. The batteries and aerial were extra and batteries were an essential part of the radio in 1925. It was expensive because my mother received $300.00 for teaching school the first full year, she taught in the 1930's. In other words, that radio cost more than half a year's salary for a school teacher.

THE FORMATION OF THE IARU

Bill Borrett, 1DD, had done such an excellent job of organizing the American Radio Relay League (ARRL) in this area; he was selected as the Canadian representative at the formation of the International Amateur Radio Union (IARU) in Paris on April 17th, 1925. This organization commenced with a dinner on May 24th, 1924, in Paris, France. Mr. Hiram Percy Maxim the president of ARRL and representatives of all the world's amateur radio organizations were in attendance at this dinner in 1924.

There were 23 nations present the next year in 1925. A list of all 23 with each representative along with the complete constitution of IARU appeared in the June 1925 issue of QST. Keith Russell, 9AL, was disappointed in the response to his request for donations from all Canadian members but managed to get financial help for Bill Borrett, 1DD, via the League Headquarters in Hartford. Mr. Hiram Percy Maxim was so well liked by all the world's amateurs that he was elected the first president of the IARU. There were 698 members by late 1925 and 20 of them were Canadian.

Mr. Maxim sent Major Borrett some movie film taken at the IARU formation in April, 1925, and this was enjoyed by those attending a meeting of the local amateurs.

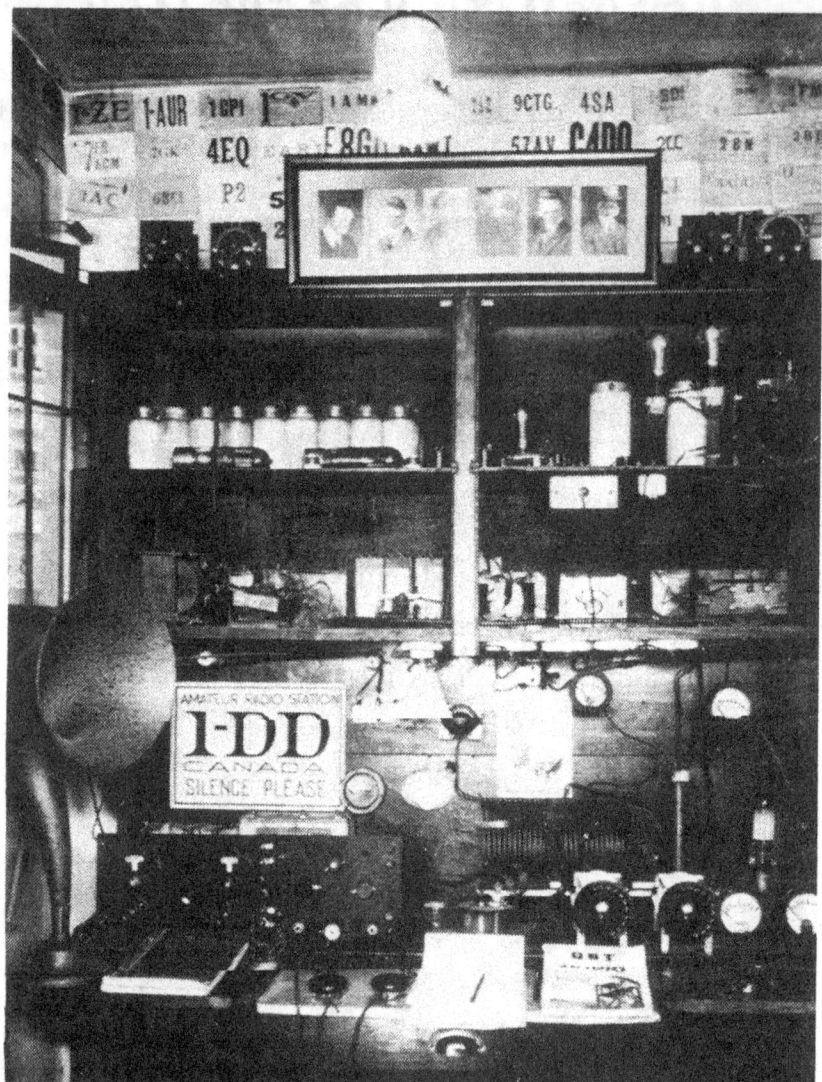

Major Borrett's station back in 1924—look at that loaded shelf, preserved plums.

28

This is page 28 in the book *"from Spark to Space"*, by the Saskatoon Amateur Radio Club, VE5AA.

This photograph is a treasure to HARC.

It is one of the first stations of that era to become part of the Maritime Amateur Radio Association.

This photograph is in the April 1959 issue of The Canadian Amateur.

Does anyone know the location of this photograph?

Thanks to Bill Elliott, VE1MR, for scanning this photograph.

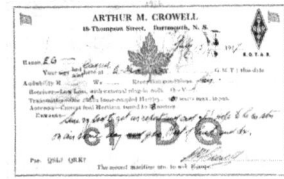

Courtesy Thomas Roscoe K8CX

Art Crowell, c1DQ, was the second station in this area to cross the Atlantic according to his QSL card.

Hams are doin' fine
They're heard across the sea - - -

They're getting' into England,
And into Germany - - -

And if their wave goes up too much,
They'll surely get into Dutch.
QST May 1923

THE ARRL TRUNK LINES

The ARRL had the United States laid out in trunk lines that were identified with a letter; Trunk A, Trunk B, Trunk C, and so on down the alphabet. These lines stretched across the United States from coast to coast with various amateur stations in each city on a trunk line that relayed traffic from one station to the other. This is the reason for the name the American Radio Relay League.

The ARRL encouraged Canadian amateurs to join them so that they could assist with a trans-Canada Trunk line via their border stations, until the Canadian amateurs could create one with all Canadian amateurs.

The Canadian amateurs managed this feat after three continuous days of tests. The following message was the longest, covering from Prince Rupert to Halifax and return on April 15th, 1923:

> From Prince Rupert, BC
> To 1BQ, Halifax, NS
> HOW IS THE WEATHER THERE NOW WE ARE ENJOYING SPRING HERE
> (Sig) Barnsley, 9BP

This message left 9BP at 8:53 PM PST. The routing was 9BP-4FN-3NI-2HG. Here it took a peculiar twist, as 2CG failed to raise 1BQ at Halifax and sent a reply which reached 9BP at

10:05 PM PST. In the meantime, 2CG raised 1BQ and gave him the message and got a reply, but got stuck with it and finally had to give it to 2HG who passed it along following his own. The real reply from 1BQ finally got back to 9BP at 1:14 AM PST April 16th. This was a dandy piece of work and might have made better time but for a slight slip-up at 2HG and 2CG, yet they both did their best. Total time of this message was 2 hours 14 minutes. This is from a report by F. H. Schnell, Traffic Manager at ARRL Headquarters.

Keith Russell, 9AL ex 3AB, terminated his Ontario division report with the statement "altogether a glorious three nights". One would have to agree when most of it took place plus or minus 200 meters or 1500 kilohertz in our terminology.

KEITH RUSSELL

When looking back at these guys I think of old men for some strange reason. They were not old at all. A lot of them were university students and quite a few carried on and obtained a commercial certificate after obtaining their amateur radio certificate. Some of them actually spent their summers sailing as wireless operators in the Great Lakes inland fleet.

9AL Toronto Canadian Manager ARRL.

QST March 1924

A.H. K. Russell was the first manager of the Ontario division. His first call sign was 3AB as noted and when he obtained his commercial certificate, he replaced this call with the 9AL call sign. Several of these early Canadian leaders in amateur radio received a biography complete with their photograph in QST. Keith had been a member of the Canadian Naval Volunteer Reserve stationed at the radio monitoring station at Newcastle, New Brunswick. When World War I broke out he transferred to naval air in the Royal Navy and became a pilot. After the war he went home to Toronto and back to university obtaining a law degree. He and two lawyer friends formed their own legal practice. Keith replaced W.C.C. Duncan as the Canadian General Manger at ARRL Headquarters and also became one of the Directors of ARRL. One would have to agree a great Canadian amateur. The world's amateur radio community is deeply indebted to Mr. Hiram Percy Maxim and we Canadians are deeply indebted to A. H. K. Russell. If Canada had amateur radio today it would definitely not be the same without those two.

BOWDOIN AND THE SHIPS NORTH

The front cover of the October, 1923, issue of QST had a sketch of the MacMillan Arctic Expedition schooner *BOWDOIN* frozen in up north for the winter. She was frozen in at

Refuge Harbour, Greenland. *BOWDOIN* is an 88-foot New England schooner. She had a New England fishing master in command. She is named for Bowdoin College in Maine. The total complement on board was 8 men. Don Mix, 1TS, from Bristol, Connecticut was her wireless operator and her call sign was WNP. Permission had been given for this vessel to communicate with amateur radio stations via her WNP marine station. This had been routine to give some of these vessels this permission back then and most vessels had a three-letter marine wireless call sign. **W**ireless **N**orth **P**ole, but I do not know if this vessel had been assigned that call sign for that reason.

Don Mix had not been heard from for a while but finally managed a QSO with R. B. Bourne, lANA at Chatham, Massachusetts. This is the cute part. It did not take the newspapers of the day long to state that Lana had been talking to Don on the *BOWDOIN*. This QSO took place at 11:35 PM EST on August 27th, 1923. This station was further north at the time than any amateur station had been. The distance between the two stations was 2,300 miles. There was heavy QRM on 200 meters but they managed this QSO on 180 meters or roughly 1666 kilohertz in today's terminology.

WNP was running a 500-cycle plate supply and worked about 180 or 220 meters. That would be roughly 1363 or 1666 kilohertz in today's terminology; within the present AM broadcast band. The equipment had been manufactured by the Zenith Corporation. The *BOWDOIN* went to her winter anchorage in Refuge Harbour in October 1923. Joe Fassett, 1AR, and Bill Borrett, 1DD, were the two stations Don worked in the Halifax area. When she went to anchor Don was unable to work the east coast but had good communication with the west coast. This I understand from experience in that area. Don's main contact was with Jack Barnsley, 9BP, at Prince Rupert, British Columbia. Jack was a former ship's wireless operator who became the Agent for Union Steamship Company at Prince Rupert.

QST January 1924

This drawing of Don Mix is a figment of someone's imagination. All the photographs of Don on this trip show him clean shaven and with a recent haircut.

BOWDOIN was held up before she sailed for compressed air. One assumes the compressed air was used to start the engine. The engine was a 60 horsepower Fairbanks Morse that could operate on crude oil, kerosene and even seal oil. Donald MacMillan the one in charge of the expedition, did not want to carry gasoline feeling it was too great a fire hazard. The World War I Submarine Chaser was started with compressed air. The engine was too big to crank. There were no electric starting motors and there were no Pup or Donkey motors. A Pup or Donkey motor was a small gasoline motor one started with a pull cord. Once it was going and warmed up, they clutched in the large engine it was used to start. The large motor would cough, sneeze, puff black smoke (unburned fuel) and finally start. Once the large motor was going the Pup or Donkey motor was stopped and one could enjoy what seemed to be the relative quiet of the large engine compared to the rather sharp bark of the Pup or Donkey engine. Some called these small starting engines a Pup and others called them a Donkey and that was the only difference in the terminology I have heard. The thing that has bothered me since learning this is how in the world did the engineer in *BOWDOIN* keep compressed air in order to start that engine after it had been frozen in for months? The engine would have had an air compressor and an auxiliary water pump. But when the compressed air was getting low, I am certain the engineer did not drain all the oil out of that large engine, heat it, replace it and start it just to make more compressed air. The answer to this question would be very interesting.

Don had not been heard from for a while again in early 1924 and ARRL made an all-out plea for all amateurs to try and make contact with him again. *BOWDOIN* sailed from Wiscasset, Maine on June 23rd, 1923, and arrived back at Wiscasset on September 20th, 1924. There was a large crowd gathered to welcome *BOWDOIN* back as one can imagine. Mr. Hiram Percy Maxim the president of ARRL was there with a speech along with several other dignitaries.

Don Mix, W1TS, continued on with the technical end of radio and eventually became a technical editor with ARRL. He has an excellent lengthy description titled "Operating the 807" in May, 1946, and we who knew World War II equipment will definitely remember that tube. Was there a tube more widely used than that one? Don retired as technical editor at ARRL and became a silent key at age 71 in 1973.

Canada gave the government vessel *ARCTIC* permission to carry an amateur radio operator and to communicate with amateur stations on her northern voyage during the summer of 1924. "Boisterous Bill" Choat, 3CO, from Toronto was her operator. "Boisterous Bill", 3CO, had a lengthy career in amateur radio. He went on to do some excellent and rewarding work with the blind amateur radio operators through the CNIB in Toronto, Ontario. He had been a four-year member of ARRL when he made this trip north having joined in 1920. Bill, and by this time was VE3CO, was in charge of the "White Cane" operations with the CNIB in Toronto. Bill Choat, VE3CO, resigned as manager of the amateur radio operator program with the CNIB Toronto in 1983.

Captain J. E. Bernier was captain of the *ARCTIC* on this trip north in 1924. There is a museum in Quebec and a coast guard ship was named for Captain Bernier. The *ARCTIC* was operated by the Department of the Interior and was sent north to establish and service police stations. She carried a Circuit Judge to hold court when required and was a typical Canadian government vessel serving the north.

This permission to carry an amateur radio operator was created by Keith Russell, 9AL, and Commander C. P. Edwards. Edwards was in charge of all radio activity in Canada at the time. I am inclined to believe the personal relationship between Charles Edwards and Keith Russell had a lot to do with the excellent arrangements given the Canadian amateur radio operator. This is my own personal feeling after many years of research.

The *ARCTIC* had wireless call sign VDM and communicated with Joe Fassett, 1AR, Bill Borrett, 1DD and Louisburg, VAS, in this area on this voyage. They found the 1AR station the most reliable of the three. VDM worked VAS on 2090 meters or 143 kilohertz in today's terminology and worked the amateurs on 125 and 140 meters, the most popular amateur bands. This is 2400 and 2143 kilohertz and not very high for we today. On occasion an amateur would use 80 and 68 meters in 1924 or 3750 and 4412 kilohertz. Bill complained of the shifting antenna capacity when the ship rolled while working short wave. Welcome to the effects of the wooden hull Bill. *ARCTIC* was fitted with two transmitters.

ARCTIC departed Quebec City on July 5th, 1924, and arrived back at Quebec City on September 24th, 1924. Bill paid a visit to the Halifax gang shortly after this with a most interesting lecture on "up nort" complete with over 200 photographs. He was made a member of the Royal Order of Trans-Atlantic Brass Pounders at this meeting. There were several of these voyages around the world carrying wireless, sponsored by and written up by ARRL at this time.

The *ARCTIC* and the *BOWDOIN* went "up nort" again the next year in 1925 along with a Navy vessel the *PEARY* named after another well-known arctic explorer. They also had three navy Loening Amphibian Aircraft complete with navy crews. Richard E. Byrd was in charge of this navy crew and the aircraft. He went on to command many expeditions into the Arctic and Antarctic. The aircraft were fitted with low power radiotelephones with CW capability for communication with the *BOWDOIN*, but it was felt they could be heard as far as the United States. They were using a trailing antenna on 40 meters.

A trailing antenna was a wire that was paid out behind the aircraft.

One of the main amateur radio stations that communicated with this expedition was that of 15-year-old Arthur Collins with call sign 9CXX at Cedar Rapids. Arthur Collins went on to create some of the best radio equipment made. What ham did not dream of one of his S-line stations?

This 1925 expedition was known as the Navy-MacMillan expedition and all members had naval rank from the reserve navy. The *BOWDOIN* and *PEARY* were loaded down with the management from Zenith Radio and all held reserve naval rank. All the radio equipment was Zenith built and supplied by them. Nearly all the equipment was donated to the expedition. The *BOWDOIN's* amateur station was operated by Lt. John L. Reinartz, USNRF, amateur call sign 1XAM but he used WNP as before. He was also the communications officer for the voyage and an employee of the Zenith organization. John became K6BJ and a silent key in 1964. The December, 1964, issue of QST has his obituary.

Paul J. McGee, 9AE, was the operator in the *SS PEARY*. Both vessels were fitted with 20-40-80-and-275-meter stations, designed, built and donated by Zenith. The *PEARY* was assigned call sign WAP. The stations in both vessels worked extremely well on this voyage. They used amateur radio stations for communications and they actually handled coded naval messages for the navy. The *BOWDOIN* and *PEARY* sailed from Wiscasset, Maine, on June 20th, 1925, with plenty of fanfare as one can imagine. And both returned safely on October 12th, 1925, with more fanfare but not the VIP treatment of the year before.

A Chronological History of the United States Navy communications makes this statement:

> "Reliable wireless communications are maintained by Donald B. MacMillan on the 1925 polar expedition to the North Pole with the U.S. Naval Communication Service on high frequencies. On the Voyage to and from the arctic, the naval radio station at Bar Harbor, Maine, is successful in communicating with the SS BOWDOIN and SS PEARY after the ships reached higher latitudes."

This reliable wireless communication was not that reliable with the record they created with the amateur radio stations.

The Army and the Navy must have been rather pleased with amateur radio via the ARRL because both set up various stations within their organization. The navy actually gave all the American officials of ARRL naval rank in the naval reserve. Mr. Maxim was made a Lieutenant Commander USNRF and one other ARRL official held that rank. The rest were officers of either Lieutenant or lower in rank. The Army made various amateur stations near each of their Army bases part of their communications and issued those stations with Army-Amateur Certificates. This was increased in 1926 to several auxiliary Army-Amateur nets with amateur stations not close to an army unit.

The National Geographic was involved with these voyages and the amateurs were advised to send any messages copied to their office in Washington collect. The pages of QST record many expeditions that used amateur radio for their communications. There were three into the Arctic alone during the summer of 1926. One was the famous Bob Bartlett in command of the *EFFIE M. MORRISSEY* with Newfoundland call sign VOQ and an American from Ohio, Edward Manley, u8JF, operating her station cross band on the marine and amateur bands. The amateur station that handled most of the message traffic on this expedition was J. R. Miller, u9CP. He was presented with a Narwhal tusk 6 feet 3 inches long and the twisters went around the tusk five times. He received this in appreciation of all the work he put into handling their messages. Quite often the captain of these vessels would have a write-up in QST followed by one from the radio operator in the vessel. I found them all most interesting. When the *EFFIE M. MORRISSEY* went to a four-letter call sign, she simply added the letter H, as in VOQH. Her Captain was still Bob Bartlett. The Canadian Coast Guard Ship *BARTLETT* is named for him. There is also a museum of his achievements in Newfoundland.

There was so much newspaper coverage on these voyages that it prompted the Duluth, Minnesota, Herald to note: "A professional radio operator seems to be one who connects with lost explorers after amateurs show him how". Well, described indeed. This record was recorded on page 16 of the January, 1929, issue of QST.

Robert M. Foster, 2AC, Montreal sailed as the radio operator in *ARCTIC* in 1925. "Bobbie" Foster had been a ham since 1917 and a graduate of the Marconi School in 1921. He worked for Marconi and held official positions within that organization. He held a first-class commercial radio operator certificate but for some unknown reason did not hold a 9-call sign. Bobbie checked into the Wednesday night schedule on 125 meters or 2400 kilohertz each week using the *ARCTIC's* marine station call sign.

This was an exclusive Canadian amateur radio frequency and it was changed in early 1926 to 52.51 meters or 5713 kilohertz in today's terminology. The reason for the change was that so many were working on 40 meters. Because their antennas were cut for this band, they were not checking into the Wednesday night schedule known as the "Prayer Meeting". This would make their antennas more compatible for the Wednesday night schedule.

The *ARCTIC* with call sign VDM also worked on 40 meters and the marine frequencies. *ARCTIC* sailed from Montreal on July 1st, 1925, but lay at anchor for nearly a week below Quebec City repairing several breakdowns on board. VDM arrived safely back on October 1st, 1925, but the voyage had been plagued with equipment trouble and was not near the voyage of the year before. Most of the problem was in the engine room and apparently her dynamos gave a lot of trouble. This would mean no power for the radio equipment. She had good communication during the latter part of the voyage and most of it was handled on 40-meters.

Many in the Maritime division worked these vessels with stations 1AR and 1AM the leaders in this communication. "Old Joe", 1AR, was one of the seven top stations to work the ships on this second trip. He was in contact with WNP nine times and handled 9 messages and 1055 words of press. He was in contact with WAP twice and this was all on 20-meters. *BOWDOIN* stopped at Sydney, Nova Scotia, and John Reinartz visited station 1AW at Glace Bay while there on the way north.

These guys sailing in these vessels from Zenith may have been paid by Zenith. MacMillan has a note in the April, 1948, QST for another amateur radio operator to sail in *BOWDOIN* and states these crews pay for this privilege and that it is $1,000.00 per person for the trip. The operator must hold a commercial certificate and that because of his duties as radio operator the fee is $500.00 for the trip, half the regular fee. If *BOWDOIN* had a New England Fishing Master in command on her first trip in 1923 it was the only trained seaman she carried. MacMillan carried college students without sailing experience from then on. They learned how to sail, stand a watch, steer, and whatever else was needed on the job.

It is a shame two other Canadian government ships were not given this permission at this time. Both have HARC connections. *ACADIA* held wireless call VDT at this time and is now a museum here in Halifax. The members of HARC operate her amateur station VE0MMA.

The *LADY LAURIER* was also here and serviced the buoys and lighthouses in the area. She held wireless call VDF at this time and her last Captain was Mel Lever an HARC member for many years with call sign VE1VX. It is impossible to name all the HARC members that served in all these ships.

There was someone sailing in the *CANADIAN FISHER*, a member of the Canadian Government Merchant Marine, operating as an amateur station and using call sign CR10 on about 42.5 meters. This was announced as a stray in QST April, 1927, and gave a Montreal address to send a QSL card. A stray is the name of a note used in QST. A QSL is a card mailed to an amateur operator confirming their radio contact. The *CANADIAN FISHER* was one of the early Canadian ships to receive a four-letter radio call sign before the radio and visual or flag call became one and the same. She had radio call sign VGBM and flag call TQDB. The flag and radio call sign became one and the same after the international agreements of 1927 held at Washington, D.C. The two-letter wireless or radio prefix was from Canada's block of call signs first issued in 1912. *CANADIAN FISHER* would have retained VGBM and simply dropped the TQDB call.

The first three Canadian ships fitted with wireless were fitted in early 1904 but I have not located the exact dates. The three were all government ships known as a Dominion Government Steamer (DGS). They were: *CANADA, MINTO* and *STANLEY. CANADA* was our first warship such as it was. It had a gun and did naval maneuvers until we managed to get a navy in 1910 complete with a couple of old British warships.

The other two *MINTO* and *STANLEY* were constructed as Prince Edward Island ferries. The reason I mention this is that in 1928 the *STANLEY* with call sign VDE and another government ship the *LARCH* with call sign VFW did another of these trips "up nort". They used the old standby amateur radio stations for communications. "Old Joe" 1AR was right in there as the top station of the bunch, with Bill Borrett, 1DD, and the Louisburg station VAS doing what they could. And of course, there were a number of other ships on the northern run that year along with all these I have mentioned and most were using amateur radio for communications. It was certainly a big plus to the amateur radio community.

The *LARCH* and *STANLEY* were loaded with aircraft, construction equipment, radio equipment and general supplies. They did a survey into Fort Churchill, Manitoba, and installed radio stations through Hudson Strait and Hudson Bay to assist shipping into the railroad at Fort Churchill. *STANLEY* carried 10 radio operators "up nort" so her radio station was on continuous. A few of these radio operators were dropped off as each radio station was constructed. The *LARCH* had long wave or low frequency only to communicate with *STANLEY*. Two of the operators involved in this mission were C. H. Starr, **NC**1AE, and R. L. Bunt, **NC**3MX.

The Royal Canadian Mounted Police schooner ST. ROCH with call VGSR was built for G Division of the RCMP. G Division is that section of the mounted police that patrols and governs the Yukon and back then the Northwest Territories. The ST. ROCH made her first voyage north in 1928 and made several voyages up to and including World War II, between 1928 and the outbreak of WWII when amateur radio terminated for the war, she used amateur radio for communication, but I found no record of this. I have found a record where ST. ROCH communicated with the Canadian Army Signal Station, VEF located at Aklavik, North West Territories' and radio station VBK at Coppermine, N.W.T. The reason for this is that I was concentrating on the Halifax area. The ST ROCH did communicate with amateur radio while up north like the rest of the ships up there, but she did it via west coast amateur operators and stations and appeared in the west coast news sections of QST. The International Telecommunication Union List of Ship Stations for 1933 lists ST. ROCH with two HF frequencies only. 6310 and 12820 kilocycles. In the List of Ship Stations for 1940, in addition to these two it lists 8220 kilocycles. The 6310 frequency would likely have been the main frequency for communications with the amateur radio stations.

NORTHERN RADIO STATIONS

PICTURE on top was taken during a leisure moment at Nottingham Island and shows Harold Vaughan, W. H. Williams, both of Halifax and Carl Baldwin, Toronto, left to right. Bottom picture was taken at Resolution Island and shows Alexander O'Grady, Francis Richards of Halifax in charge of the station, and C. A. Walker of Dartmouth.

By R. P. KEHOE
Staff Writer The Halifax Mail

"Calling VE5RA . . . ."
"VE1DQ calling VE5RA . . . ."
The amateurs are on the air!

THOUSANDS of miles separate them but the wives at home in Halifax, Nova Scotia, carry on regular conversations with their husbands on Resolution and Nottingham Islands in the Arctic Circle. Isolated, lonely islands with only three white men at the Canadian Government radio stations on each outpost to keep company with the Eskimos, speak from out of the frozen north; Resolution in the spluttering code of Morse; Nottingham by short wave transmitter. Nova Scotia answers back and the voices of the wives of the men are carried into the barren, desolate northern wastes to cheer lonely husbands.

HARC Files

This is 1936

The Carl Baldwin in the top photo is not correct. It should read Coll Baldwin – short for Collison.

Note three men at each station. This was typical of the Department of Transport back then. The simple fact is it is near impossible to find three people that can stand each other for a long period in isolation. Two will tend to ignore the third to the point the third did strange things. There is any number of cases of this on record up there. When one fellow came "un-glued" he simply sat staring at the other two sharpening a knife. You went in not knowing when you would be coming out. Those at Cambridge Bay had been requesting a transfer out for some time and were completely ignored. The next time the officials from down south came in for an inspection all on staff simply went about their daily routine in their "long John" underwear as though all was right and normal. They had their transfers out when the officials returned to their southern offices.

BOWDOIN NORTH AGAIN IN 1946

There was a note under Strays in the February, 1946, issue of QST that Captain Donald B. McMillan had visited Greenland in 1941 with the same rugged schooner *BOWDOIN*.

BOWDOIN was assigned WDDE when she was assigned the standard four-letter ship call sign for ships from the 1927 international agreements. *BOWDOIN* became *USS BOWDOIN* during World War II with Lieutenant Commander Donald MacMillan in command. She patrolled the Greenland area during this time. I have been unable to locate her four-letter naval call sign with the N prefix. Radioman Victor Paounoff was her radio operator as *USS BOWDOIN*.

BOWDOIN went "up nort" again in 1946 and she was assigned call sign KLPO on this run. She was working cross band; transmitting on marine frequencies 4150, 8250 or 12480 and listening on 80, 40 or 20 meters. Bill Matchett, W1KKS, was the operator on that run. Her radio room had been moved to a shack 5 feet high amidships capable of sleeping 2 men. She carried a total of 16 men north in 1946 and this was Donald Baxter MacMillan's 24th expedition north and the 18th for *BOWDOIN*.

If those marine frequencies are correct the KLPO station would have been capable of three frequencies in each marine radio band. Her three basic or bottom frequencies would have been 2062.5, 2075.0 and 2080.0 kilocycles. The FCC must have stated transmission on those three only. She would have been capable of:

2062.5	4125	6187.5	8250	12375
2075	4150	6225	8300	12450
2080	4160	6240	8320	12480

The fact those three frequencies are not harmonics of each other makes no sense. At least all three are working frequencies but all three are in the high traffic or passenger ship segment of the working frequencies. This transmitter would have been "Rock Bound" or crystal tuned and it would have had at least one more 2-megacycle frequency. It would have had one or two between 2089 and 2093 kilocycles. That frequency or frequencies would have permitted calling a coast station on a calling frequency within one of the bands; 2, 4, 6, 8 or 12 megacycles. I will take this no further but I find it unusual she was assigned those three frequencies and that they are within the passenger ship band and not with the low traffic freighters; QSC – I am a cargo vessel.

VE1MK, Sydney, kept a sked (radio abbreviation for schedule) with W2OXE/MM the schooner *BOWDOIN* during the summer of 1949. Brit Fader, VE1FQ, worked the *BOWDOIN* in 1950 and she was using the W2OXE/MM call sign. This is a clear indication she is using amateur radio and was not transmitting on the marine bands and simply listening to the amateur bands.

My copy of the August, 2008, issue of QST arrived on July 10th. For a minute I thought John Dilks, K2TQN, had been looking over my shoulder. I had not heard of the schooner *BOWDOIN* until I became involved in this exercise. John's article for this issue is "Where in the World is the Bowdoin?" He has some excellent photographs of *BOWDOIN* and come to find out she is on another Arctic voyage during the summer of 2008. She is now part of The Maine Maritime Academy and John describes the work of this maritime education program.

BOWDOIN at 88 feet is 39 feet shorter than *BLUENOSE II* at 127 feet, but one can tell they come from the same era. It appears as though American ships change their radio call sign often. *BOWDOIN* was assigned radio call sign WAM8653 on her run north in 2008. John, K2TQN, has more detail on the *BOWDOIN* in several issues of QST. It is all very interesting because he has access to the personal diary kept by Don Mix, 1TS. John, K2TQN, has a monthly two-page article in QST with the title "Vintage Radio". He started this in the January, 2000, issue of QST. It is a most interesting section of QST. He has many interesting old photographs each month.

THE ROYAL ORDER OF TRANS-ATLANTIC BRASS POUNDERS

Around Halifax and back in March, 1924, as a mark of distinction a society was formed called the Royal Order of Trans-Atlantic Brass Pounders. This was abbreviated R.O.T.A.B. Only those who had held two-way communications across the Atlantic Ocean could be members. There were five out of the ten amateurs in Halifax, who held that distinction in March, 1924. The five were Joe Fassett, 1AR, Art Greig, 1BQ, Art Crowell, 1DQ, Davidson, 9BL and Bill Borrett, 1DD. Davidson was a college student here in Halifax and moved to Quebec on completion of his studies. Each member was allowed to put ROTAB on his QSL cards. Note the R.O.T.A.B. in the bottom left corner of the **C**1AR card and just below the ARRL on the **C**1DQ card above.

This R.O.T.A.B. was a lot of fun and was along the lines of the Royal Order of the Wouff Hong. The ARRL editors and writers often referred to this area as Major Borrett, 1DD, and his ROTAB's. By 1925 this had spread to Montreal and Toronto where several were made members. There were 14 ROTAB's in the Maritime Division by 1925. There were 11 new members installed at the Montreal Convention in 1925 and 6 more at the third Maritime division convention at Saint John in 1926.

THE SECOND HALIFAX ANNUAL CONVENTION

The second Halifax Annual Convention took place on the weekend of March 21ˢᵗ, 1925. There were 54 delegates and guests present. Mr. A. A. Herbert, U1ES, Treasurer, ARRL, Hartford was in attendance. Arthur Herbert, known by many as Herbie, was born in Montreal but had moved to the U.S. at an early age. He became an amateur operator in 1908 and held call 2ZH before World War I. He was a member of ARRL from the beginning and had been treasurer for 23 years when he became a silent key in 1941. When the U.S. Navy contacted ARRL in World War I wanting trained wireless operators, it was he and Mr. Maxim who found 15 experienced operators for them.

Getting back to the convention, Art Greig, 1BQ, demonstrated the operation of a 5-meter transmitter. He and Bill Borrett, 1DD, had the highest score in the code contest. Art, 1BQ, won at 22-1/2 words per minute for 5 minutes and a silver medal went to the winner. Mayor Murphy presented "Old Joe", 1AR, the Murphy Cup. He was awarded the Murphy Cup as the outstanding amateur of the year 1924 and 1925. The Murphy Cup was a cup donated by Mayor Murphy of the city of Halifax for that purpose. One wonders where that cup is today. It would be a shame to learn it is in someone's garage full of rusty nails or something along that line. Joe Fassett, 1AR, and Art Greig, 1BQ, had the two best stations in this area. Cecil A. Landry held the 10AR and c10AR call signs. He is recorded in the silent key list for the March, 1952, issue of QST. They used the 10AR Broadcast license to broadcast the events of the 1925 second Halifax Annual Convention. Most of the components of this station came from the transmitter of Richard Binns, 1EB.

WEEKLY AMATEUR RADIO BROADCAST ON CHNS

Bill Borrett, 1DD, stated in June, 1926, that there would be a weekly amateur radio broadcast on radio station CHNS on 322.4 meters and that works out to 930 kilohertz in today's terminology. Bill Borrett, 1DD, was an official of CHNS and he and E. S. Campbell, 1DJ, had a weekly radio broadcast on the ARRL and amateur radio. They also offered amateur radio classes at CHNS for anyone who wanted to become an amateur radio operator. This station moved from 960 kilohertz on the AM band to 89.9 megahertz on the FM band of frequencies in 2006.

These **6 amateur stations** were the **only** amateur stations in the Dartmouth and Halifax area on the 1926-1927 lists of amateur experimental stations:

1AR	J. J. Fassett	Dartmouth
1BN	J. Rose	Halifax

1BQ	A. W. Greig	Halifax
1DD	W. C. Borrett	Dartmouth
1DJ	E. S. Campbell	Halifax
1DQ	A. M. Crowell	Dartmouth

42 STATIONS IN AREA ONE IN 1926-1927

By the time the list of amateur radio stations for the fiscal year 1926-1927 was recorded there were 42 stations in area one. The call signs went from 1AB to 1EI in alphabetical order. There were many calls that had not been assigned of course. The fiscal year was from April 1st until March 31st. In 1926 a complete list of call letters of all the Canadian hams could be purchased from the Department of Marine and Fisheries, Ottawa, for twenty-five cents.

The first Callbook I found with Pegasus on the cover, the familiar flying horse, was the June 1928 issue. It was called the Citizens Radio Amateur Callbook and contained call signs for both amateur and commercial radio stations. It stated it held calls for the stations of 83 countries. The June issue was $1.10. The one year consisted of four issues and was $3.25. Apparently the one in charge of this callbook held call sign **NU9FO**.

FIRST AMATEUR RADIO CALL BOOK

The June, 1929, issue of QST advertises the first Radio Amateur Call Book Magazine that was produced by Radio Amateur Call Book Inc., Chicago, Illinois. Both the U.S. and Canadian governments were each producing a call book that sold for 25 cents prepaid anywhere, so this new company was up against some heavy competition. Although normally anything government produced would be so far out of date by the time it was available that this fact would help the Radio Amateur Call Book Magazine. The Call Book Magazine became known the world over as "The Callbook".

The first Radio Amateur Handbook was for sale in 1926. It cost one dollar postpaid anywhere. This was changed in 1940 to one dollar postpaid anywhere in the continental United States. And in 1941 it was one dollar and fifty cents in Canada and one dollar still in the continental U.S. This Handbook contains everything one would want to know about radio.

FEMALE OPERATORS

One gets the feeling there were no female operators because everything is worded as men or boys in all the old records. There were several female operators but it is hard to believe the way everything is described or worded. On page 5 of the 1935 ARRL Handbook they state they wanted to boast a real "he-station". One wonders what that station would resemble. It was a different world back then.

The District Manager Vancouver in his notes for March, 1923, was excited in the fact they had a female operator in the division at Prince Rupert, but the District Manager for Quebec states in his notes for January, 1925, that Mrs. D. G. Sturgess, with call sign 2CN, was the first Canadian female operator. He does not record the date she was first licensed. Mrs. Sturgess was a good-looking lady from the photograph that appeared with this note.

A note in QST for May, 1923, 3CDQ (female operator) says that winding banked coils has improved her vocabulary considerably. She was an American and no doubt she is right because most of us have done a lot of cussing to a lot of equipment over the years. For some unknown reason I remembered this call sign and found out this lady's name. She was Liz Zandonini and there is a photo of her on page 18 of the January, 1979, issue of QST. In that photo Liz, now W3CDQ, is receiving a plaque for 50 years ARRL membership.

In YL News and Views (the actual name of the section), page 72 of the April, 1981, issue of QST there is a much better photograph of Liz and a full biography. She was quite a lady to say the least. She passed her commercial licence in 1917 and was an operator during World War I and went on to the bureau of standards and was involved with the time signals from station WWV. She retired from the bureau of standards in 1965. Her favourite mode in amateur radio was CW. She said she may have transmitted in phone once. Her favourite key was a straight hand key and her favourite antenna was a dipole in her attic. You do not get it more basic than that and she was out receive as many amateur radio awards for many and various radio contacts after she retired.

NOVA SCOTIA WEATHER

The dipole in the attic is the perfect antenna for a VE1 station. With the hurricanes, freezing rain, snow, high winds and weather in general for this area, every VE1 station should have one at least for a backup if nothing else. In other words, if your roof is still on your antenna should work. If the commercial electric power has failed, simply replace your power supply with a deep cycle battery.

TYPICAL HALIFAX STORM SCENES — Nova Scotia bore the brunt of a surprise storm yesterday which streaked up the Atlantic Seaboard and brought 40 to 60 mile an hour northeast winds and from four to 12 inches of wet, drifting snow. The storm knocked out power lines and communications systems and brought most transportation to a standstill. Typical scenes in Halifax, which caught a major portion of the nor'easter, are those shown above. The photo at left shows an abandoned car parked on Grafton Street near Spring Garden Road, thoroughly buried by drifting snow. The photo at right shows a trolley coach "plowing" its way along a downtown street. Trolleys yesterday maintained their service but were considerably off schedule because of stalled cars on the routes. (Staff photos by Crosby).

HARC Files

The weather is still the same and this one was on November 20th, 1955.

This is where our truck was parked for the winter of 2014-2015 and the way it looked on February 16th, 2015. That is a lot of ice around the truck from thawing and freezing. We let Mother Nature dig it out for us. This truck is the one we use to pull our 33-foot 5th wheel trailer we had Equest trailers here in Nova Scotia build us. The truck is a Duramax diesel Allison transmission unit known as a one ton dually.

This is my QSL card showing the trailer. The antenna is a Comrod AT82 whip designed and built for fishing vessels working the North Atlantic. I tune this with an SGC SG237 Smartuner. The whip is advertised as a 30 foot but measures 26 feet. It is great for long range work but rather poor for close work, like all whips. We managed to work Sam, VE1YVN, in Dartmouth, Nova Scotia from Gold Canyon, Arizona with no trouble. The yellow power cord assists in charging the batteries. The trailer has four deep cycle batteries and a solar panel on the roof to assist charging. The battery room door is the door just ahead of the wheels in the photo. The trucks electrical system also charges the batteries when traveling. It is a very nice unit that we hope will be in the family for many years.

We had Equest build this as stated and we have three Ten-Tec 50-watt HF Scout transceivers mounted in the trailer along with a Yaesu FT-1900/R/E 2-meter transceiver connected to a Larsen ¼ wave whip on the trailer roof. We also have a small AM general radio service transceiver better known by the more popular Citizens Band terminology. The CB is great for around the campgrounds.

I was going to upgrade the Scouts but decided to keep them because I am having too much fun with them.

MORE ON THE FEMALE OPERATORS

There were two female operators in the U.S. in 1910. Olive Heartberg was using call sign OHK and Mrs. M. J. Glass was using call sign FNFN according to a note on page 82 of the March, 1974, issue of QST. These two ladies were the first two American amateur radio operators. Emma Candler from St. Mary's, Ohio was licensed in 1915 and assigned call sign

8NH. Emma's husband was 8ER and principal of the local high school. She got fed up with everyone calling her OM (old man) over the air and would correct them with OW (old woman). The YL terminology had yet to be created.

QST for May, 1917, had a photograph of her station. She was known as Mrs. Candler in the language of the day and QST stated her 8NH station had been copied in thirty states. She had worked 9ZF in Denver, Colorado, and ships off Key West, Florida and Cape Race, Newfoundland. She had heard 7ZC in Lewiston, Montana, for the longest distance from her station. Her aerial was 87 feet long, 58 feet high, six wires, lead-in from the center. She was equipped for receiving long un-damped waves. The August, 1917, issue of QST listed the call signs of over 150 stations she had heard during the month of March, 1917. The Old Man mentioned in one of his rotten articles that a "wouff hong" would have eliminated the QRM that obliterated her. She was certainly well known and respected by the amateurs of that time. In another article the Old Man mentioned that the QRM was so heavy she was finding it hard to act like a lady.

It would be hard to top Miss Winifred Dow from Tacoma, Washington, who was licensed in 1917 and assigned call sign 7FG. The inspectors were very impressed with the fine station she had built, both transmitter and receiver. Winifred was 14 years old and in the 8th grade at school. After the war she was assigned call sign 7CB. One has to wonder what became of her. Leave it to me to go off on a wild hunch. Jim Spilsbury, VE7BR, of Spilsbury Tindall Radio fame left his radio manufacturing business in charge of a Winnifred while he created the accidental airline; Queen Charlotte Airlines. This Winnifred became his second wife. She was Winnifred Hope and at least twenty years too young to be 7FG/7CB.

They stated in the August, 1917, issue of QST that United States amateurs would be blessed with several hundred of the fair sex within their ranks on termination of the war. They went on to state there should be a new label for them because the girls did not like OW; for old woman. A few would tolerate OL, for old lady but the best they came up with was DG; for dear girl.

Eulalia M. Thomas, W8CNO, explains the difference between an OW, XYL and YL in an article titled XYL in the September, 1929, issue of QST. There is no mention of YL or XYL in the list of radio abbreviations created at the Washington convention of 1927 that went into effect on January 1st, 1929.

John Dilks, K2TQN, found the first YL and gives a full-page description on page 76 of the September, 2002, issue of QST. He actually reproduces the letter in colour that states the term YL was coined. She was M. Adaire Garmhausen, 3BCK, a friend of Liz Zandonini, W3CDQ. She was given this label when she submitted an article to QST titled "How to Build a Wireless Station". The letter in reply that coined the new phrase is dated May 13th, 1920. Like Liz she did not marry, was into Amateur Radio in a big way and had started as a telegraph operator. Adaire admitted she was born on May 28th, 1899 and John felt she became a silent key in the late 1990's. The W3BCK call sign is not listed in my 1974 Callbook. Liz, W3CDQ, is listed in the 1974 Callbook.

The acronyms are; OW old woman, XYL a married woman and YL a single woman.

THE IARU CHANGES ITS CONSTITUTION 1928

The International Amateur Radio Union (IARU) changed its constitution on October 30th, 1928, and the full constitution appeared in the December, 1928, issue of QST. Up until this date one joined the IARU as a member, but after that date IARU has been a union of independent national radio societies. This has acted as the medium between societies to affect the vital cooperation of the world's amateur radio operators on a world-wide scale. Individual membership was no longer possible.

Newfoundland was not part of the Maritime Division of ARRL after the summer of 1929. Newfoundland became part of the Maritime Division again when Newfoundland joined Canada in 1949. The Newfoundland Amateur Radio Association was one of the independent national radio societies that made up the International Amateur Radio Union this twenty-year period.

Everyone went a bit overboard with these international conventions shortly after the Washington Convention of 1927. There were two meetings in 1929 of these international committees. One was in Prague, Czechoslovakia, in April 1929 and the other in Madrid, Spain in September 1929. At Prague it was agreed that each individual nation would make the laws governing their amateur radio stations. They also agreed at Prague to hold another meeting at Copenhagen, Denmark, in 1931, after the one at Madrid, Spain. The meeting in 1938 was held at Cairo, Egypt.

Individual amateur radio operators around the world have been able to keep abreast of the IARU activities via the IARU News column that has been a part of the ARRL publication QST since 1925.

JOE FASSETT 1AR

HARC Files

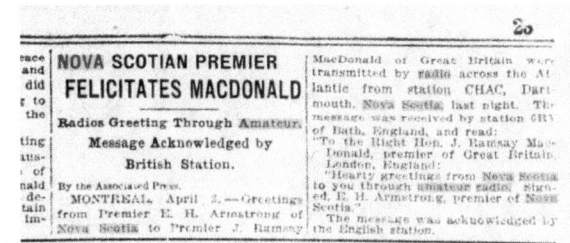

Credit Bill Elliott VE1MR

This is a newspaper clipping from the Evening Star, Washington, D.C., dated Thursday, April 3, 1924, and is self-explanatory.

Joe Fassett, 1AR, was often referred to as "Old Joe" and they claimed he was known the world over by that label. He was first referred to as "Old Joe" in February, 1925. He had one heck of a station, all hand built, that he built, and he was one fine operator. He was awarded the Murphy Cup as the outstanding amateur of the year 1924 and 1925 as stated. The cup was presented to Joe by Mr. Murphy, Mayor of Halifax at the second Halifax Annual Convention in March, 1925.

Joe Fassett, 1AR, was the first in this area to work; Belgium, Denmark, Sweden and New Zealand. He did this during a four-month period and working any one of those countries would have been a feat in itself but to work all four was unreal. He was the **first Canadian** station to work **New Zealand**. Not only that but he managed to work the west coast at the same time. He worked 5BZ on the west coast or as they were now appearing in print; c5BZ. It was felt that no Canadian station could beat his record. The big local news of 1926 was the fact "Old Joe" had relayed a message from South Africa to Hawaii. That must have been a thrill and a half to say the least.

Frankie Clarke, 1AI, New Brunswick, was awarded the Murphy Trophy for 1926. He must have been one heck of an operator in order to wrestle that from "Old Joe". I am sure one would have lost interest in the trophy if Joe won it every year. Joe Fassett became a silent key in 1956 and appears in the silent key's column of the May issue of QST.

March 9 1956

12

Pioneer Of Radio Hams Is Mourned

A man who pioneered radio in Nova Scotia and gained international recognition for his ham radio operations died at Woodside yesterday.

John Joseph Fassett, 72, former member of the county council and school board, was widely known for his work in relaying messages during the sinking of the Titanic, and for his "ham connections" around the world.

BORN IN ENGLAND

Although retired for the past year, Mr. Fassett had been employed with a sugar refinery in Woodside for 40 years as an electrical engineer.

He was born in Kent, England and came to Halifax at an early age. He moved to Woodside where employed by the sugar plant.

Mr. Fassett gained international fame for his work in relaying messages to New York following the sinking of the "Titanic." During the Second World War, he worked in conjunction with the Canadian Broadcasting Corporation in gathering information from all parts of the world for news broadcasts.

FOUNDED CREDIT UNION

He was interested in community and civic affairs. As a former member of the county council and the school board, he was instrumental in establishing the present setup of schools in his home community.

He was a member of St. Alban's church, Woodside and was one of the founders of the St. Alban's Credit Union. He was a member of the Keith Lodge, Halifax and a charter member of the Fidelity Lodge, Woodside.

He is survived by his wife, Annie; one son, Herbert, Woodside; three daughters, Mrs. Frances Sim, Vancouver; Mrs. Jean Jenkins and Mrs. Norma Taylor, both of Woodside, and nine grandchildren.

The remains are resting at Zink's funeral parlor in Dartmouth and the service will be held Saturday afternoon at 2:30 o'clock at St. Alban's church. Interment will follow in St. John cemetery, Fairview.

RADIO PIONEER DIES — Nova Scotia's first licensed radio ham operator, John Joseph Fassett, died yesterday at his Woodside home at the age of 72. Mr. Fassett received international recognition through his connections with ham operators and his work in relaying messages during the Titanic sinking. The funeral will be held Saturday afternoon.

HARC Files

FIRST MEETING OF CANADIAN DISTRICT MANAGERS

Keith Russell, 9AL, had wanted a meeting of all the Canadian District Managers of ARRL. He wanted this meeting held in the area of Winnipeg and had wanted it for some time. He managed to get one early in 1925. Bill Borrett, 1DD, Maritime District manager, J. V. Argyle, 2CG, Quebec District manager, C. H. Langford, 3XN, Ontario District manager met A. H. Keith Russell, 9AL, Canadian General Manager in Toronto; all went to Winnipeg on the same train.

While traveling to Winnipeg they managed to get all business pertaining to the east organized and ready for the main meeting. This meeting was also open to any member of ARRL. Keith advised those at this meeting that there was to be a Canadian Section in QST. The others at this meeting were A. J. Ober, 4DQ, Vancouver District manager, J. E. Brickett,

Winnipeg District manager and A. A. "Herbie" Herbert, U1ES, Field Representative at ARRL Headquarters in Hartford, Connecticut. A. J. Ober was 4DQ because he lived in Alberta.

The two main points that occurred at this meeting was that all managers, including the General Manager were to serve two years only. They were to be elected to these positions and this meant an election every two years.

The other item they wanted and had approved was that all managers across Canada were to meet on the air on a wave of 125 meters or 2400 kilohertz, every Wednesday evening at 10:30 PM Mountain Time. And all Canadian members of ARRL were advised to listen in on these meetings and participate if necessary.

This was an exclusive Canadian amateur radio frequency and it was changed in early 1926 to 52.51 meters or 5713 kilohertz in today's terminology. The reason for the change was that so many were working on 40 meters. Their antennas were cut for this band so they were not checking into the Wednesday night schedule. They used this frequency for Trans-Canada traffic. It also became the Inter-Empire frequency back then. The Empire was everything and as long as a station was part of the British Empire it could check into this net as it would be known today.

When these guys slept back then is beyond me. Most, if not all, held steady jobs and yet they did most of their amateur radio operating late at night. There was a ruling that they were not to transmit from 8 PM until 10 PM each evening and this is when they must have slept. This ruling was to prevent them from interfering with those listening to the broadcast stations that were in service. While visiting my grandfather when I was in the navy, he mentioned that we do not hear the Morse code like we did years ago. Things had improved quite a bit by 1958 compared to the 1920's and 1930's. These Wednesday night schedules worked out very well and became known among the group as the "Wednesday Night Prayer Meetings".

CANADIAN FREQUENCIES IN 1926

These were the wavelengths assigned Canadian amateur stations in 1926:

Wavelength – meters	Frequency – kilocycles
0.7496 – 0.7477	400,000 – 401,000
5.35 – 4.96	56,000 – 64,000
21.4 – 18.7	14,000 – 16,000
37.5 – 42.8	7,000 – 8,000

Wavelength – meters	Frequency – kilocycles
*52.6 – 52.51	5,700 – 5,710
85.7 – 75.0	3,500 – 4.000
197.2 – 150.0	1,520 – 2,000
I. C. W. & Telephone	
180.0 – 170.0	1,706 – 1,667

* This band was for British Empire work exclusively.

In late 1926 there was a request in large letters in the Maritime division news for all Nova Scotia Hams to meet on 52 meters from 1 to 5 PM on Sundays and from 6 to 8 PM on week days. Bill Borrett, 1DD, got fed up in early 1927 and said he was cutting an antenna for 52 meters and was going to camp there permanently to drum up some activity.

Another interesting feature of QST at this time during the 1920's and 1930's was that they left out the divisional news for each section in the newsstand issue. This involved about 16 pages and was used as an enticement to get those interested in the subject to join the ARRL and receive their own copy via subscription. Those old boys were thinking back then!

There is a good description of a 20 – 40 – 80-meter crystal-controlled transmitter in the August, 1926, issue of QST.

The March, 1926, issue of QST stated there were no licensed amateur stations in Russia but asked that the ARRL members listen for a few stations there using three letters as a call sign.

There was an experimental station at Halifax with call sign CA during the 1920's. This was a transatlantic experimental station and 1DJ was kept busy there so had little time for amateur operating.

THE FIRST NORTHERN STATION & FIRST LABRADOR STATION

The first northern station was **NC5AC** at Pond Inlet on Baffin Island. Constable M. Timbury, RCMP, a former navy radio operator was stationed there and installed this station after the fall shipping season of 1926. There was a complete description of this station including a photograph of Tim in his uniform in the February, 1927, issue of QST.

The First Labrador station was Stanley Brazil, 8AZS, at Battle Harbour in 1926. Actually, he started as BHL or VOA the Battle Harbour Coast Station call sign. Stan worked here in Halifax at Camperdown Radio. He also served as the radio operator in the Sambro Lightship. He may have been a member of HARC while working in this area like so many of the others.

NEWFOUNDLAND STATIONS WITH CANADIAN CALL SIGNS

I have never been able to fully understand the relationship between Canada and Newfoundland, one of my main interests in this history. All coast stations in Newfoundland were run by the Canadian Marconi Company in Montreal. It appears as though the ones constructed by the Newfoundland government were assigned Newfoundland call signs with the VO prefix. The others apparently were built by Canadian Marconi and received a Canadian call sign with the VC prefix.

Belle Isle was one of these stations with the VC prefix; VCM to be exact. The operators on this station were there for a year and of course there was little to do in the winter. E. Davy was one of the operators for the winter of 1926-1927 and he was assigned amateur radio call sign 9DA. This was Newfoundland real estate and I am sure anyone would agree the call should have been a Newfoundland call sign with the 8A prefix. Newfoundland was removed from the Maritime Division of ARRL in the summer of 1929 and had no connection to ARRL until joining Canada in 1949.

1927 CALL SIGNS

On February 1st, 1927 at 0000 GMT European countries were to have a two-letter prefix. The first letter was the letter **E**; Asian countries, similarly, with the letter **A**, North America **N**, South America **S**, and Africa **F**. This was according to the International Amateur Radio Union News in the American Radio Relay League's publication QST for January 1927. The stations in the Maritime Provinces of Canada became **NC1AA** to **NC1ZZ** and the stations in the Dominion of Newfoundland became **NC8AA** to **NC8ZZ**.

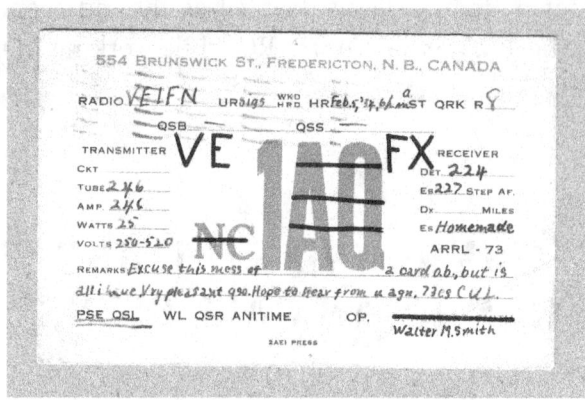

HARC Files

This will at least show what an **NC**1 QSL card looked like and is the only one I have found.

The 2AEI Press was probably in the New York area.

Stu McIntyre was the one who made the VE1FX a most memorable call.

This list of Intermediates as they were called took up a whole page of the January, 1927, issue of QST and is much too long to record in full here. Canada, Newfoundland and Labrador were assigned **NC**, **N** for North America and the **C** for the area of Canada, Newfoundland and Labrador. **NU** was the prefix for the United States and **NJ** was the prefix for Jamaica, etc. **EG** was the prefix for Great Britain, **EF** France and Monaco, etc. For four cents worth of stamps ARRL would send you a copy of these printed on pasteboard for hanging on the radio shack wall. (Radio amateurs were calling their radio stations a shack.)

The Intermediate could get lengthy for example; 1AW **NUXOA** 3AA. I'll bet you cannot break that one down so here goes. 1AW is Mr. Maxim's call in Hartford, Connecticut. The **N** is North America, the **U** is the United States and the **X** indicates it is an amateur radio station in a ship. **O** is Oceania and **A** is Australia and Tasmania. Therefore, some Aussie in a ship wants a chat with the "Old Man". Thank goodness the **NC** prefix in the call sign did not last long.

INTERNATIONAL RADIO CONFERENCES

They hold International Conferences from time to time and the first was the one in London in 1912. The American Radio Relay League (ARRL) and the International Amateur Radio Union (IARU) attend these conferences along with all of the world's various governing bodies on communications. The ARRL, the IARU and Radio Amateurs of Canada (RAC) inform the amateur radio world of any changes made that affect amateur radio. We who held commercial certificates had to report to a radio inspector and have our certificate endorsed after each of these conferences. Just before each conference terminated it was agreed on a date to hold the next, usually around five years from that date.

There was no conference after the 1912 conference until the 1927 International Radiotelegraph Conference held in Washington, D.C. The ARRL and the IARU were both present at this conference. At the 1927 Conference three key recommendations were agreed to in regard to Amateur Radio.

1. **A formal definition and recognition of Amateur Radio.**

2. **Allocation of specific harmonically related bands below 200 meters – 160, 80, 40, 20, 10, etc.**

3. **Agreement on country call sign prefixes.**

Item 1 meant that Amateur Radio would now be an official part of the radio community on an international basis. To paraphrase the writings of Mr. Maxim it was no longer in the realm of Toyland but had the dignity of an international asset.

Item 2 made it official on an international scale. The amateur community had been more or less using these bands since 1924. An article appeared in September, 1924, titled 20, 40 and 80 meters that started off by stating these waves were not new and had been used before. It went on to state there was no point in using 40 meters until one had a wavemeter to indicate where they were on 40 meters. It was indeed a different world than the one we know today. The following are the frequencies assigned to the world for the use of amateur radio.

AMATEUR FREQUENCY BANDS
assigned by The Washington Convention of 1927

Kilocycles	Width in Kilocycles	Assignment	Approx. Meters on basis factor 3	Meters on basis factor 2.998	Harmonic family for sequence of related portions		Amateur Purpose
					Kilocycles	Meters	
1715–2000	265	Amateur, Mobile, point-to-point	150 – 175	149.9 –174.8	1775	168.92	Domestic
3500–4000	500	" "	75 – 85.7	74.96 – 85.66	3550	84.46	"
7000–7300	300	Amateur Exclusively	41.1 – 42.9	41.07 – 42.83	7100	42.23	International Night
14,000–14,400	400	" "	20.83 – 21.43	20.82 – 21.42	14,200	21.11	International Day
28,000–30,000	2000	Amateur & Experimental	10.00 – 10.71	9.99 – 10.71	28,400	10.56	Experimental
56,000–60,000	4000	" "	5.00 – 5.36	4.997 – 5.354	56,800	5.28	"

QST January 1928

This was the big problem and the big worry for the amateur radio community. They knew they had to be on a frequency within a band they were assigned. One could state a band they discovered and made work. They could create an international incident if they were outside their assigned band. This is the reason each band was a harmonic of a lower band. This applied to the marine and aircraft bands as well. Not only did it make it easy to build the transmitters for these bands in that they could use the harmonics of a crystal cut for the lowest band, but should they create a harmonic over the air of the frequency in use it would be in the next higher band assigned to that service. One could often hear a station on a certain frequency that was actually transmitting on a frequency within a lower band of frequencies assigned that service years ago. I heard several over the years.

Item 3 meant the world's amateur radio stations were assigned a prefix on their call signs from the blocks of international call signs first assigned the various nations of the world at the 1912 London Radio Convention. Canada had been assigned the **VAA-VGZ** block of call signs at the 1912 convention. I have felt that Canadian amateur radio stations were given the VE prefix because the **VE** signal in radiotelegraph meant *Understood*. If nothing else, those interested in radio were able to understand it via amateur radio.

The Dominion of Newfoundland had been assigned the VOA-VOZ block of call signs at the 1912 convention. Newfoundland assigned the VO prefix in 1927. Canada and the United States assigned their prefix on October 1st, 1928, although the Convention was to be effective January 1st, 1929. The United States wanted this on October 1st so that the new call signs would be recorded correctly in the June, 1929, government call sign list.

Canada simply followed the United States and all that was known at the time was that Canada was changing to the VE prefix. The continental United States went W and the U.S. Possessions went K. The NC prefix did not last very long and the number and the suffix remained the same in each call sign. The ARRL did not know what any other nation was going to use as a prefix. ARRL listed the ITU blocks of call signs as agreed at the 1927 convention. It went from Chile CAA-CEZ to Union of South Africa ZSA-ZUZ inclusive.

In other words, Bill Bligh, **NC**1BC, simply became Bill Bligh, VE1BC. The same applied to those holding all the other call signs across Canada and around the world at this time. For example; 1AA-0ZZ in France became F1AA – F0ZZ, 1AA-0ZZ in Sweden became SM1AA – SM0ZZ, 1AA-0ZZ in Australia became VK1AA – VK0ZZ, and 8AA-8ZZ in Newfoundland became VO8AA – VO8ZZ and so on.

When Newfoundland assigned the VO prefix to their amateur radio call signs, they simply added it to their 8 amateur radio call sign they had been using such as 8AA to 8ZZ. Art Grant, VE1EP, made his first DX QSO in 1932 with Jim Moore, VO8AW. Jim was one of the first three Newfoundland stations to show up in this area in 1924. On January 1st, 1935, Newfoundland and Labrador amateur radio call signs were divided into six different areas with a digit in their call sign for each area; VO1 through VO6 inclusive. The majority, if not all of their call signs had a one letter suffix. Ernie Ash, VO8AA, became VO1A. Jim Moore, VO8AW, became VO2J.

Jim Fraser, VE1FT, and Jim Moore, VO2J, must have been good friends. They made so many contacts that one will note he did not record a date on his QSL card for contacts made while he operated St. Paul, VGS, and Sable Island, VGF, from 1937, 1938 and 1939. It cost two cents only to mail the card but as they used to say a penny saved is a penny earned or something to that effect. One wonders where the card would wind up today if they put that address on it.

The Estate of Jim Moore VO1BY, VO2J, 8AW – Carbonear, NL

The Estate of Jim Moore VO1BY, VO2J, 8AW – Carbonear, NL

The Estate of Jim Moore VO1BY,
VO2J, 8AW – Carbonear, NL

Jim Fraser, VE1FT, stated that this is a poor snap of the Ham Shack at St. Paul Island, 1939, but will give an idea.

Another point of interest is that the **V** prefix is believed to have been given the British Colonies in memory of Queen Victoria. She had been very popular and had died on January 22nd, 1901, just 11 years before the 1912 London Radio Convention.

The majority of the world's civil aircraft registration and call signs came from these same blocks of call signs, and all boats and ships received call signs that indicated the nationality of registration from these same blocks. The world's land or coast stations received a call sign from these same blocks as well and had been assigned these call signs since the 1912 convention.

All of this was to take place officially around the world on January 1st, 1929, as stated. Changing the call signs did not appear to be a big problem. If one worked a call with the prefix ZL they simply looked on the ITU list. They would realize the station was in New Zealand, and so on around the world for any other call heard.

Another thing that was changed with the 1927 Washington Radio Convention to become effective January 1st, 1929, was the Q Code and radio operating abbreviations. The main or most important Q Code change was that QSA was assigned the digits 1 through 5 inclusive. QRK remained the same without the digits: "Are you receiving me well? Are my signals good? *And* – I receive you well. Your signals are good." It took years of hard use to create these codes. They did not happen overnight. There were a number of these codes in use over the years. The navy had their Z codes; the Radio Corporation of America had their own separate Z codes, and so on.

TED MCELROY

Theodore R. McElroy first appears in QST in the January, 1922, issue. He had recently set one of his first speed records at 51-½ words per minute. One wonders how they measured the ½ word per minute with their equipment back then. One of his web sites states this was his first speed contest and that it was 56 words per minute and that makes more sense. An old cousin of mine, Merrill Roscoe, K9DQS, saw McElroy copy some code one time. They claim his best record was at Ashville, North Carolina, on July 2nd, 1939, when he copied 77 words per minute. Others record 75.2 words per minute or over 17 symbols per second. Merrill said the code stopped and Ted kept typing for another five minutes. He must have had an excellent memory of some description. Ted stated the hard part in copying code was finding a good typewriter. They claim he could type 150 words per minute and he claimed he could type three letters to his secretary's one. He manufactured his own brand of telegraph keys and most were known as a Mac-Key. He started manufacturing keys in 1934 and sold his business in 1955. His company made a lot of keys for the American military during World War II. The navy presented him with an award for the excellent keys he had made for them. He was a legend in his own time and one cannot record radiotelegraph without running into him.

Tis three O'clock

Tis three O'clock in the morning,
I've listened the whole nite thru,

The sun will soon be shining,
And I haven't heard a CQ.

QST May 1923

CD QST - MISTAKES

The one that produced the CD's of QST that I have been using for research for this exercise did not do a very good job in places as mentioned. There is no divisional report for the Maritime Division of ARRL from January, 1928, until March, 1929. There are many cases where there is no record of this report. The frustrating part is in reading something of interest. When you go to the next page it makes absolutely no sense. You look at the top of the page and realize it is page 10 of the 1927 issue, for example, and not page 10 of the issue you are reading.

In the Maritime Division report for March, 1929, we learn that Art Crowell, VE1DQ, has recently replaced Major Bill Borrett, VE1DD, as the Section Communications Manager (SCM). Art thanks the gang for the hearty election to SCM. In this same March report, we learn a couple of things. One is that the "Wednesday Night Prayer Meeting" is now held on what is 80 meters in our terminology. It is held between 3895 and 4000 kilohertz in today's language. The other is that "Old Joe" VE1AR pulled a surprise on the gang and showed up on 3500 kilohertz on phone.

There were plenty of mistakes back then. Newfoundland went to the VO prefix in the call sign in 1927 but Loyal Reid, VO8AR, was always listed as VE8AR in the list of SCM's at the front of QST. Art Crowell, VE1DQ, replaced Bill Borrett, VE1DD, as SCM for the Maritime Division in early 1929 but was listed on this list as VE3DQ. It was not until November, 1929, that they managed to get it correct as VE1DQ.

There was another change in the Maritime Division in the summer of 1929. Instead of the four SCM's; Prince Edward Island, Newfoundland, New Brunswick and Nova Scotia there is only the one; Art Crowell, VE1DQ. There is a description of this change in the July issue of QST and Art is recorded as W1DQ. The VO, VE and W prefix had recently been assigned so it was taking some time to get it accurate. There is no Newfoundland report mentioned in this description of the new Maritime Division. Someone must have realized that the Dominion of Newfoundland and the Dominion of Canada were not one and the same.

WIND AND SLEET STORM FEBRUARY 1931

The heaviest wind and sleet storm in the history of this area occurred in February 1931. It would have been the storm of the century if today's media had been involved. VE1AL did some fine work in restoring the telegraph and telephone lines. VE1BN helped, and the Canadian Pacific Railway telegraph ran a special wire to VE1AZ's house in New Glasgow to help deliveries. Gordon Arthur, VE1AX, and Don Smith, VE1CC, at the Halifax end stuck to the job and handled many important orders for Maritime Telephone and Telegraph. Gordon Arthur, ex VE1AX, became a silent key in late 1961.

MARITIME DIVISION CONVENTION JUNE 1931

Art Crowell, VE1DQ, was SCM or Section Manager of the American Radio Relay League in 1931, and he and Gordon Arthur, VE1AX, organized a Maritime Division Convention at Halifax. I do not know who recorded the record of this event but it started off rather interesting and sounds like Art Crowell, VE1DQ:

"What we lack in numbers is more than made up in enthusiasm as evinced during the Maritime Division Convention held at the Lord Nelson Hotel, Halifax, on Friday and Saturday, June 19th and 20th. With the arrival of our Canadian General Manager, Alex Reid, VE2BE, the gang for the first time had an opportunity of meeting him and received a complete report of the Board of Directors' meeting as well as Canadian affairs."

The numbers mentioned are those who are members of ARRL and of course this has nothing to do with the Maritime Amateur Radio Association (MARA). They came from all over the three Maritime Provinces. Unfortunately, this record did not record the actual number at this convention. MARA did not become an associate member of ARRL until 1934 when it was known as the Halifax Amateur Radio Club (HARC).

This record goes on to state "When old Joe Fassett and Alex Reid met it was just like the meeting of two lost brothers. That's the kind of friendship amateur radio makes." They believed that a convention was a good place to have a good time and this included a game of golf at Bedford. George Harris gave the delegates a good talk and complimented the amateurs on the little trouble they gave him and assured them they would receive fair treatment from his office. George was in charge of the local federal Department of Transport Telecommunications and Electronics Branch Empire. This included all the Coast and Aeradio stations, anything connected with radio in and around what today is Atlantic Canada. This is the George Harris who was the Warrant Officer Telegraphist in *HMCS NIOBE* during the Halifax Explosion on December 6, 1917.

Several interesting contests took place with everything leading up to the big event of the convention – the banquet. They claim it was a great feast. Major Bill Borrett, VE1DD, the former division manager, was in true form and gave a fine talk on amateur radio in the Maritimes from 1920 to 1927, and his many jokes on old Joe. Art Gregg and Art Crowell were enjoyed by all. Bill Borrett, VE1DD, then presented a beautiful cup, donated by radio station CHNS, to be presented to the Maritime amateur who worked the greatest number of stations on any band, either CW or phone that year. QSL cards confirming the QSO were to be sent to the SCM Art Crowell, VE1DQ. Ten new members were initiated into the Royal Order of Trans-Atlantic Brass Pounders, but unfortunately, they did not list these ten new members with this record. After the toasts, prize awards and initiation into the Royal Order of Trans-Atlantic Brass Pounders the convention came to a close.

Alex Reid, VE2BE, was the Section Communications Manager (SCM) for Quebec for four years. He replaced A. H. K. "Keith" Russell, VE9AL, as Canadian General Manager (CGM) at ARRL Headquarters, Hartford, Connecticut, on January 1st, 1930. Art Crowell, VE1DQ, was SCM for the Maritime Division and Alex Reid, VE2BE, was CGM throughout World War II.

PHONE ENDORSEMENT BEFORE WORLD WAR II

To go on phone before World War II one had to have a phone endorsement that required two years on CW with logs to prove it. They then had to have a Radio Inspector come in and inspect their station to make certain it was in good working order.

I have noticed in going through the records made just before and just after World War II that they do not record their activity in meter bands. They have it recorded by frequency in megacycles, yet they recorded it in meter bands in the few radio logs I found. It was mentioned in 1929 that the German's had replaced kilocycles with the term kilohertz. The Canadian Department of Transport adopted the term Hertz in 1967 but it was to remain cycles in legislation until the ITU adopted the term.

NEW MEMBERS 1933

Welcome new members VE1FO, VE1FN, VE1FK, VE1CZ and VE1FY in 1933. The only names I have for those call signs are Doug Smith, VE1FO, and John Doull, VE1FN. John became VE3ZW in 1948 and was VE1WC in 2008. George Crowell, VE1LB, was assigned the VE1FY call sign in the 1940's. He has no idea who held the VE1FY call in 1933.

If anyone else held the VE1FO call before Doug it was for a very short time. The 1926-1927 lists of call signs for area 1 goes to 1EI. There were no 1F calls assigned. This was just before the assignment of the VE prefix. Therefore, it would appear that Doug Smith was the first one assigned the VE1FO call sign.

GREETINGS GOVERNOR GENERAL 1933

The Canadian General Manager of ARRL, Alex Reid, VE2BE, congratulated all the Canadian amateurs on a fine job of transmitting greetings from the Governor General of Canada to the Lieutenant Governor of each province on Christmas Day, 1933, and hoped that this would become an annual event. MARA member Bill Horne, VE1GL, received the message here in Halifax and sent a message of greetings back for the Lieutenant Governor.

FORMATION OF THE MARITIME AMATEUR RADIO ASSOCIATION

All amateur radio operators were a member of a close-knit community and many were close friends. This was natural simply because of their mutual interest in radio. This mutual interest in radio prompted the formation of The Maritime Amateur Radio Association (MARA) in this area. P. L. Whitman, 1AI, was president of the Halifax Wireless Association in 1923. There were ten amateur operators only in Halifax at the time and apparently all ten were members of this organization. Richard Binns, 1EB, was building a receiver for the Halifax Wireless Club in 1923. One assumes the club and the association were one and the same. Richard moved to Ontario in late 1925.

Art Crowell, VE1DQ, stated in the May 1931 issue of QST that he, VE1BL (unknown), Don Smith, VE1CC, Bob Cunningham, VE1AS, and Gordon Arthur, VE1AX, have started organization plans for a local club. I have been unable to find the name for the call VE1BL. This club would become the Maritime Amateur Radio Association. The Maritime Division news for September, 1933, is the first mention of the name Maritime Amateur Radio Association (MARA). This note mentions that MARA has been reorganized and holds regular monthly meetings at the home of VE1DH and VE1DI, 21 Parker Street, Halifax. This note is from the Secretary Treasurer, VE1DH. There have been several notes on these two calls living in the same home. The October, 1933, issue states that VE1DI has joined the Royal Canadian Mounted Police. VE1DH was A. T. White according to the Minute Book for 1934 – 1937. The regular monthly meetings were still held in the home of VE1DH but he moved to Saint John, New Brunswick about the time this copy of QST was received.

The MARA meetings were held mostly in the homes of the various members. They claim most of the meetings were held in the Halifax area so it was decided to change the name of this group to the Halifax Amateur Radio Club.

CREATION OF THE HALIFAX AMATEUR RADIO CLUB

These are the minutes of the meeting on March 16th, 1934 that created the Halifax Amateur Radio Club. The minutes are signed by the Secretary Treasurer, Wes Street, VE1EK, and one will note the heading H.A.R.C. meeting of March 16th, 1934. The minutes of the meeting previous to this has the heading M.A.R.A.

HARC Files

proceed with this business we discussed the advisability of changing our name of Maritime Amateur Radio Association to one which did not appear to cover such a wide area and scope as this name suggests. Five suggestive names were voted upon with the name Halifax Amateur Radio Club proving most popular. This name polled seven votes; M.A.R.A. 34c. Dist three votes; Hfx. A.R.A. three votes; Radio Amateur Assn. Hfx Dist one vote. After adoption of our new name discussion on the constitution was continued and it was voted upon that the Secretary have this information drafted by next meeting.

Members present included: VE1DE, EK, EP, B6, FB, JN, FO, EP, 2AF and B Dowden. Newcomers desiring membership were

HARC Files

Sidney J Simon, Robt. Munro, A.D.Solomon and V.D.Baker
Dues collected amounted to $3.25 including back fee (EP) 25¢. Two did not pay their dues. (Solomon, Dowden)
Adjournment was made at 10.45 p.m. moved by B.C. seconded J.F.O.

(Sgd) A.E.W.Street (Secy)

HARC Files

In attendance:

VE1DE John MacKasey	President
VE1EK Wes Street	Secretary Treasurer
VE1FB John Roue	Assistant Secretary

In attendance:

VE1EP Art Grant

VE1BC Bill Bligh

VE1AW Cliff Short

VE1FO Doug Smith

VE1CP Trevor Burton

VE2AF	unknown	
VE1HK	Barclay Dowden Became VE1HK and then VE3TT	
-----	Sidney J. Simon unknown	
-----	Robert Munro	unknown
VE1OC	Aaron Solomon	Became VE1OC
-----	V. D. Baker	unknown

(Note that Barclay Dowden, VE1HK, had been a Radio Officer in S.S. NERISSA but was not in her when she was sunk by a U-boat in 1941)

The minutes of the meeting state five names were voted on and Halifax Amateur Radio Club was the most popular. I can count four only from the minutes but leaving the name unchanged would make five so the five choices must have been:

1. Maritime Amateur Radio Association

2. Maritime Amateur Radio Association Halifax District

3. Halifax Amateur Radio Association

4. Radio Amateur Association Halifax District

5. Halifax Amateur Radio Club

It is rather nice that they chose the name they did.

The Minute Book that records this meeting starts with the meeting of M.A.R.A. before this one on February 16th, 1934, and ends with a brief note on the meeting of H.A.R.C. on June 18th, 1937.

There were 22 members in 1934 and that was an increase from about 15 members in 1933. One will note 10 members were in attendance at this meeting that approved the HARC name and the 4 who desired membership. The following were elected during the previous meeting, the meeting of the MARA on February 16th, 1934. John MacKasey, VE1DE, was elected President replacing C. S. Taylor, VE1BV. The Secretary was renamed Secretary Treasurer and Wes Street, VE1EK, was elected replacing A. T. White, VE1DH. VE1BV and VE1DH lived outside the area of Halifax/Dartmouth and this precluded their attention to duty. VE1BV lived at Stewiacke and VE1DH moved to Saint John, New Brunswick, at this time. John Roue VE1FB was elected Assistant Secretary.

There were so few in attendance for the meeting on April 20th, 1934, that none was held and the next meeting was held on May 18th, 1934. At this meeting the quorum was changed to read "five members" and not simply "those present". The meetings were held in the homes of the various members and continued to be held on the third Friday of each month starting at 8:15-PM. A committee was formed to find a suitable meeting place rather than the various homes of the members. This committee decided to meet in the YMCA. The first meeting took place at the YMCA on December 16th, 1936. There had been a supper for the members at the YMCA before this first meeting.

HAL WARD VE1GR

Hal Ward, VE1GR, wanted the meeting day changed to Wednesday at the meeting on January 13th, 1937. This meeting turned into a clash of personalities to the point VE1GR resigned in a scene of confusion hitherto unknown at club meetings, according to the minutes. The meetings continued to be held on Friday. Hal must have got up on the wrong side of the bed and was having a rough day. He became a very highly respected and well-liked member of HARC. So much so that when he became a silent key, late 1948, the HARC created the VE1GR 75-meter DX trophy in memory of Hal Ward. It was to be awarded annually to the ham with the most confirmed foreign station contacts and is today another trophy that should still be in the possession of the club and held in the club room. Brit Fader, VE1FQ, was awarded this trophy for 1956 and again in 1957.

BILL BLIGH VE1BC

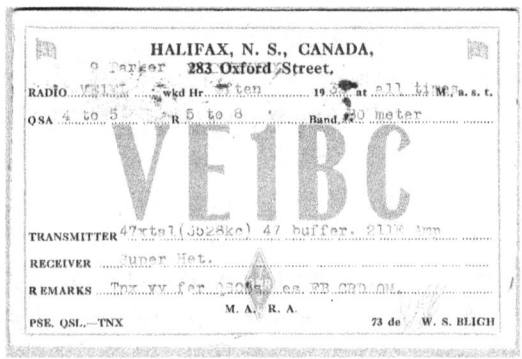

HALIFAX, N. S., CANADA,
283 Oxford Street,

S. G. "Spud" Roscoe VE1BC

This is the card of Bill Bligh mailed to Bill Lockhart, VE1EL, in Moncton at 11:30-PM August 8th, 1932. Note the M.A.R.A. at the bottom of the card.

This card cost Bill two cents to mail, but with the cost of cards and the postage it would add up and could be an expensive hobby for one back then.

Bill Bligh was working for the Canadian Marconi Company at this time.

He passed his second-class commercial Certificate of Proficiency in Radio the next year in 1933.

He moved from Berwick to Halifax in 1931 and opened his own radio repair business in Halifax in 1935.

It is stated in August, 1941, that VE1BC has the most up-to-date service shop east of Montreal.

Bill was still running this business when he became a silent key in May, 1965.

VE1EE RECORD

This is one of the best records one will find anywhere. It is on record that in March 1934 VE1EE worked a VE3 on 7 mc. He was a member of the Loyalist City Amateur Radio Club in Saint John, New Brunswick. It is a shame he was not a member of HARC because it is an outstanding record. He did this without an antenna. I believe it to be true because anything was possible and probable with that old gear in use back then.

C. S. TAYLOR VE1BV

C. S. Taylor received his license and call sign VE1BV in 1931. He was made the Nova Scotia Route Manager early in 1932. Bill Borrett, VE1DD, the Director of Radio Station CHNS presented C. S. Taylor, VE1BV, the beautiful cup donated by CHNS for the greatest number of stations worked in 1931. He had a total of 800 QSO's for the year. C. S. Taylor, VE1BV, was the last president of the Maritime Amateur Radio Association.

C. S. Taylor, VE1BV, was hardly recorded in the traffic totals for 1933 and must have slowed down some with his operating. He appears again in the December, 1933, issue of QST with a total of 74 messages. He is the only VE1 mentioned in a thank you note from the University of Michigan. He was one of the ten amateurs from Canada, Iceland, Norway, Holland and England who had relayed traffic from an inland ice expedition by this university on Greenland. Karl V. Hanse was the radio operator operating the expedition radio station NX1XL north east of Upernivik, Greenland. Karl could copy the station at the university but was unable to work it due to the low power of his station. They were also blaming it on the high terrain near and south of his location. VE1BV was under the care of a doctor in the summer of 1935 and once out of the hospital moved to Truro. Art Crowell, VE1DQ, recorded in the April, 1948, Maritime news that C. S. Taylor, VE1BV, showed up again on 7 megacycles (megahertz in today's terminology).

SID YOUNG VE1EO

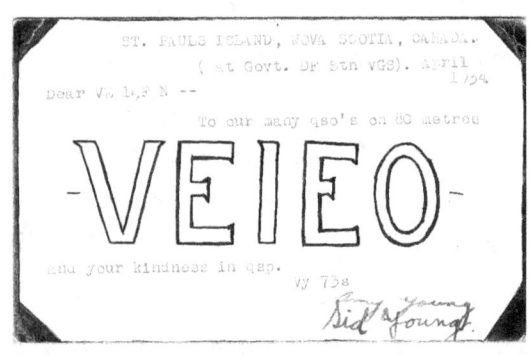

HARC Files

This is a Sid Young "homebrew" QSL card for John, VE1FN, for those skeds.

John Doull, VE1FN, had schedules set-up with Sid Young, VE1EO, on St. Paul's Island in 1934. Sid came to Camperdown Radio VCS and became a member of HARC after this tour of duty on St. Paul's. Sid was from Victoria, British Columbia, and his biography can be found from page 45 to page 49 in the book "from spark to space" by the Saskatoon Amateur Radio Club. Sid retired a radio inspector at Saskatoon with call sign VE5AJ. Art Crowell, VE1DQ, and Brit Fader, VE1FQ, both had radio contact with Sid as VE5AJ in late 1949. Sid moved to Sidney,

British Columbia and spent his last days as VE7SM. Wes Street, VE1EK, was not that happy with the input on HARC in the book "from spark to space". He had given them a lot more history than they published and he was a bit put out with the amount of coverage Sid had given himself. Sid was one who helped publish the book. We are lucky to have this record so one should not complain.

AFFILIATION WITH ARRL 1934

Affiliation with the American Radio Relay League was made on December 14th, 1934, and the affiliation certificate is shown here.

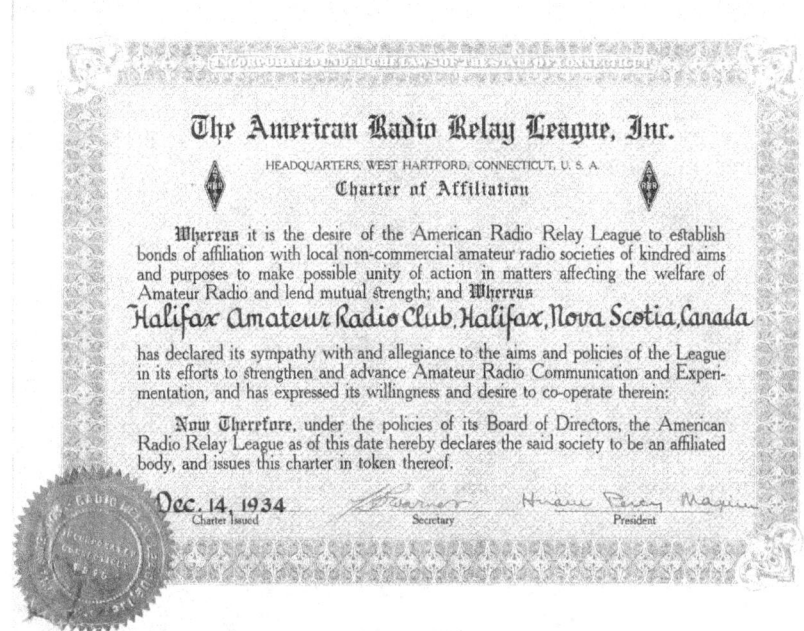

HARC Files

MARITIME OR HARC NET 1935

Bill Horne, VE1GL, was Net Control for the Maritime Net in 1935. But unfortunately, it does not state; phone, CW, the time or on what frequency. Art Crowell, VE1DQ, often recorded this net as the HARC net. It also shut down during the summer months.

TESTS 1935

RADIO TESTS
BY AMATEURS
BEING MADE

THE second half of an interesting amateur test, involving amateurs all over the United States, Canada and Europe, started at 1 o'clock in the morning of Thursday, December 19 and general two-way transmissions will be from 1.30 o'clock to 4 o'clock in the morning, Atlantic time, between Europe and the American continent. Among those taking part in these tests is H. J. Ward, well known Arundale amateur radio operator. Mr. Ward's transmitter, although only 40 watts on phone, has recorded some very good distances. Chicago in two-way communications on 1.7 meg. several times, as well as stations in Florida and Alabama.

ASKED TO REPORT

ANY listener, tuning in on the 3,500 to 4,000 kilocycle band between now and December 22 who hears any European amateur calling VE1GF during the tests will confer a favor by reporting it at once to H. J. Ward. The first part of experimental test by Canadian and American amateurs to establish contact with Europe on relatively low frequencies began on Saturday evening, at 7.45 AST, all American and Canadian amateurs operating radiophone or code stations on 3.5 to 4.0 megacycles (75 meters) for 15 minutes, after which they listened for Europe for the same length of time. Two-way transmissions took place later.

While distant reception (DX) is quite common on the high frequencies (14 and 7 megacycles) it is considered rather rare on the lower frequencies.

HARC Files

This was an 80-meter test that took place in 1935 by HARC.

1935 HAMFEST

The social gatherings known as picnics were very popular years ago. The automobile was more reliable and the roads were better in the 1930's than they had been. This meant one could travel farther and easier.

In 1935 the cooperation of other VE1 hams was requested for a Hamfest. The plans were made for the first weekend in June (King George V's birthday). It was a new venture for the group, but enthusiasm was great. The opening dinner had 89 amateurs. The guest speaker on this occasion was Major W. C. Borrett, VE1DD, and Joe Fassett, VE1AR, reminiscing on the early days. Alex Reid, VE2BE, was also present. Alex was the General Manager of the American Radio Relay League in Canada. The Sunday picnic included a game of baseball. Five-meter

MONDAY, JUNE 3, 1935

Amateur Radio
Work Described

SPEAKING on amateur radio work in Canada, Alex Reid, Montreal, of the American Radio Relay League, addressed the Maritime Division of Radio Amateurs at a banquet held in connection with a convention under the auspices of the Halifax Amateur Radio club, on Saturday night. Major W. C. Borrett was toastmaster.

Yesterday they held a field day at Sackville. One of the features of the program was the experimental work carried on with five-meter transmitters from moving cars. The organization expects to carry on further work with this type of transmitter. Mr. John Douil, the secretary, stated last evening.

HARC Files

gear was in evidence, and it performed well in demonstrations. One of the contests was to determine who could get an oscillator going first, using only wire, condenser, tube, socket and small parts, without the aid of solder or pliers. Bill, VE1BC, won this contest and Cliff, VE1AW, was second. A Morse code transmitting and receiving contest was held. Another contest was to identify the various prefixes of the world's amateur radio call signs.

MOOSE RIVER GOLD MINES DISASTER

The Moose River Gold Mines disaster in April 1936 brought international fame and through the medium of radio, the whole world heard live, on-the-scene broadcasts of the 12-day effort to rescue three men who were trapped deep underground when a mine caved in. Two of the men were rescued, and today the Moose River Provincial Park marks the site of the massive rescue operation. More of the area's mining history is on display in the nearby Moose River Gold Mines Museum.

The H.A.R.C. received high praise from the Canadian Press for their efforts in this disaster. Art Crowell, VE1DQ, and two assistants went to Moose River and passed a lot of traffic to the Canadian Press via VE1AW in Halifax. John Doull, VE1FN, was one who was at the Moose River site for two days. They did not have 12-volt equipment back then. Their equipment was powered by 45-volt dry cell batteries. Everything was CW and about all they could get out of the equipment was two to three watts. This meant they could barely work Halifax. Ralph Fraser, VE1HJ, claims Walt Wooding, VE1ET, did most of the operating at VE1AW. VE1AW was located on North Street in Halifax. There was a lot of electrical interference especially from the electric tram cars in use in the city at that time. They were planning to relay all the traffic via an intermediate station, possibly VE1AG at Musquodobit Harbour, during the hours of darkness at least, but this did not take place. These stations were on the air for 96 continuous hours handling messages from various news media reporters at the site. Bill Bligh, VE1BC, was involved as a relay station in Halifax, relaying messages to Canadian Press.

MONCTON HAMFEST 1936

The Moncton Amateur Radio Club held their first Hamfest on July 4, 5 and 6, 1936, with 135 delegates from all over Eastern Canada and the Eastern United States. Special Public Service Certificates were presented by Major W. C. Borrett, VE1DD, of Halifax to those who had assisted Canadian Press in getting news to the world of this Moose River disaster at this Moncton Hamfest. Recipients were: Art Crowell, VE1DQ, Bill Horne, VE1GL, Cliff Short, VE1AW, Walt Wooding, VE1ET, Trevor Burton, VE1CP, Bill Bligh, VE1BC, Gordon Arthur, VE1AX all of Halifax and Fred Bayer, VE1AG of Musquodobit Harbour, Nova Scotia. Special Public Service Certificates were mailed to John Doull, VE1FN, and the others who participated in this disaster and were unable to attend the Moncton Hamfest.

YACHT YANKEE RADIO OPERATOR 1936

HARC members will enjoy this. I feel confident many HARC members have heard of Irving Johnson. He had a number of large sailing yachts over the years, and he and his wife spent their lives wandering around the world with young people who wanted to learn sailing and seamanship on a "share the cost" basis. I believe all of his yachts had the same name; *YANKEE*. The last one was registered in Mystic, Connecticut.

The *YANKEE* of 1936 was a two masted schooner that had been constructed in the 1890's and served as a Dutch Pilot vessel. The old pilot service orders were to stay out until no other vessel could. By that time the pilot ship could not risk coming in close to shore, so she was literally built for all kinds of weather. She did not have an engine so was sail only.

Captain Johnson finally broke down and agreed with some personal friends that he should carry a high-frequency radio in 1936. Alan R. Eurich, W8IGQ, from Youngstown, Ohio, had a 2nd class commercial license and agreed to sail as the radio operator. This was on the "share the cost" basis of course. Captain Johnson agreed to carry the radio but was not about to spend any money on one. He had an old master oscillator power amplifier rig designed for service in 1926 on 100-meter CW installed. Alan rebuilt this old rig with an 801-driving push pull 801's to work on the high frequency marine bands. His receiver was a Sargent model 12 because Walter Evans of Westinghouse Radio Division had warned him that moisture would be a big problem in the ship. The makers advertised the model 12 as having been made primarily for shipboard use. His antenna was a 55-foot vertical Marconi.

The FCC gave Alan, W8IGQ, permission to transmit on the marine radio frequencies and listen to his amateur friends on the amateur bands. The *YANKEE* was assigned international call sign WCFT and Alan sailed around the world working his many amateur friends. They transmitted on the amateur bands and he transmitted on the marine bands or frequencies. The QRM of the high ship traffic areas, like around the Panama Canal and the English Channel gave them some trouble, but all in all it worked well. This communication was all in CW.

In 1940 Oakes Spaulding, W1FTR, was the radio operator in *YANKEE*. They spent a good part of that summer in the South Pacific visiting the islands. What a life! Oakes was transmitting with WCFT on the marine frequency of 8280 kilohertz and listening on the 40-meter amateur frequency of 7280 kilohertz in today's terminology.

The *YANKEE* had a wooden hull and my personal feeling is that Oakes would have noticed the difference in his receiver's sensitivity from that 1000 kilohertz spread. His equipment would not have been that much different from the equipment I sailed with. The transmitter and the receiver were using the same antenna and the receiver worked much better when tuned to the same frequency as the transmitter. The receiver became less sensitive the more one tuned it away from the frequency the transmitter was tuned.

Oakes had taken delivery of Andrew Young's new transmitter in Panama and they delivered it to him at Pitcairn Island. Andrew Young was VR6AY but the war was on and Andrew was British so was not allowed to transmit. ARRL warned everyone that if they worked VR6AY it was a "bootlegger". The October, 1941, issue of QST gives an excellent description of this voyage with *YANKEE* by Oakes Spaulding including a photo of Andrew Young, VR6AY, at his station.

PITCAIRN ISLAND

The sale of special issue postage stamps is the only revenue of Pitcairn Island. That and the wooden carvings the islanders carve and sell to the few ships they manage to talk into stopping off shore. The islanders now have a motor launch to go out and meet these ships. It used to be a large rowing launch. There is large surf next to the island and all the islanders, women and children included, would get in the boat and row while singing like only the South Seas Islanders can sing. At times they would disappear between each large wave until they managed to get out in the calmer water. This was a sight to remember by those fortunate enough to see it.

John Wrafter KC4ABC, EI3JD ex EL0AZ

John Wrafter KC4ABC, EI3JD ex EL0AZ

These are the Pitcairn Island boats in 1982 as photographed by John Wrafter Radio Officer in *MS STOLT INTEGRITY*. John stated they were making more money from the sale of T-shirts in 1982 than they were from postage stamps.

The only marine radio station on Pitcairn Island was a transmitter and receiver that were designed to be fitted in a ship. This station had call sign ZBP and did use CW when ships carried a Radio Officer. This station provided communication with Pitcairn Island only. The postal service was the only other means of communication to anywhere and it was via a vessel that serviced the island on occasion. I can think of a half dozen or more of the residents on the island that were good CW operators. They transmitted on 12110 and 500 kilohertz. The station was open from 0130-0200 and 1730-1800 UTC on 500 and from 1730-1830 UTC on 12 megahertz. They also had 2182 kilohertz AM with limited hours of watch keeping.

UTC stands for universal time coordinated, the modern terminology for the old Greenwich Mean Time.

Today this communication is probably VHF marine band transceiver only. The new cruise ships today do not have an HF receiver on board, so no other ship of any type would have HF equipment. These modern ships communicate via satellite and VHF only. Pitcairn Island is equipped with satellite communications but it is very expensive for them to transmit. They receive anything transmitted to them at no cost to them. Therefore, today a ship can fax the island a message giving the time they plan to stop off at the island via satellite but the island will not likely reply to this message.

This is a 1997 Pitcairn Island postage stamp showing Andrew Young, VR6AY, stating he was the first amateur radio operator on the island and started operating in 1938. Note that the stamp shows his call sign in Morse code. There were four Pitcairn Island amateur radio postage stamps in this issue and all four are shown on page 20 of the July, 1997, issue of QST. I had contact with the late Floyd McCoy, VR6AC, while sailing as VE0MO in *BOUNTY*.

HARC member Russ Latimer, VE1BPP, included his experience on a visit to Pitcairn Island in the September, 1988, HARC Bulletin.

```
                        SILENT KEY
                      RADIO OPERATOR
                   Andrew Clarence Young
                        1899 - 1988

       From faraway Pitcairn Island comes news of the recent passing of
   pioneer radio operator, Mr. Andrew Clarence Young at the age of 89.

       Pitcairn Island,  that sun-drenched  gem,  a  symbol of peace and
   beauty, located within the vastness of the world's  largest ocean. It
   was there  that Fletcher Christian, with  a small group of mutineers
   from H.M.S. BOUNTY, landed in 1790.  And  it  was  there  that Andrew
   Young was  born in  1899. It was in 1921, 131 years after the initial
   landing, that Andrew, with  the assistance  of merchant  ships' radio
   officers  that  occasionally  stopped  at  Pitcairn,  established  the
   island's  first  wireless  telegraph  station,  which  provided  a  link
   between themselves and passing ships.


       Since  Pitcairn  Island  is  located  on  a  direct  route  between
   Panama and New Zealand,  merchant  ship  traffic  is  usually active,
   although only  a few  of  the  passing ships  ever stop,  except for a
   scheduled visit with mail and supplies.

       While Andrew held  the  distinction of  being  Pitcairn's first
   "sparks", he also became the island's first radio amateur in 1938. At
   present there are several radio amateurs on Pitcairn,  all using the
   international call sign prefix VR6.

       My exposure  to these  interesting events began on 13th May 1949
   when  I  served  as  Radio  Officer  onboard  the  Canadian  freighter
   "Triberg". While  listening on  600 meters,  these Morse signals were
   heard: "VDDG DE ZBP QSA? QRK? K" - a QSO was quickly  established and
   the following day we  arrived  off  Bounty Bay and enjoyed a brief
   meeting with Andrew  and  others from  the island.  The continuity of
   that contact has endured, through correspondence with several folks
   at Pitcairn, to  the  present time.  Our visit  was somewhat shortened
   that day,  with  the  arrival,  close astern,  of the beautiful New
   Zealand Steamships Lines' liner, "Rangitiki".

       Andrew  Young  will  be  remembered  for  his  strong character,
   initiative,  and  leadership  qualities,  as  well  as  for  many
   accomplishments throughout his long life. In addition to  serving the
   island in  the field of  wireless communications, he served at various
   times as longboat coxswain, school teacher, island secretary, and SDA
   church  treasurer,  as  well  as  having  taken  a  leadership role in a
   number of courageous acts and rescues at sea.

       Indeed a pioneer in radio communications, Andrew's true hallmark
   was his  strong code  of personal  ethics and his constant companion,
   the Holy Bible.

       To mourn his passing, Andrew  leaves his  immediate  family who
   live in New Zealand, as  well  as  the  Ivan  Christian family at
   Pitcairn, to whom deepest sympathy  is  extended  by  radio operators
   around the world.
```

HARC Bulletin, September, 1988

HARC REQUESTS CLUB CALL SIGN VE1MK

Bill Bligh, VE1BC, requested that HARC obtain call sign **VE1MK** to match ARRL's **W1MK** call sign at the February, 1937, club meeting. He was advised at this meeting that Mr. Hiram Percy Maxim had recently become a silent key and the ARRL planned to replace their **W1MK** call sign with Mr. Maxim's **W1AW** call sign. **VE1AW** was not available so the club decided to go ahead and obtained the **VE1MK** call sign.

HIRAM MAXIM SILENT KEY

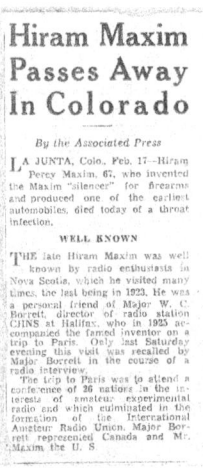

Hiram Maxim Passes Away In Colorado

By the Associated Press

LA JUNTA, Colo., Feb. 17—Hiram Percy Maxim, 67, who invented the Maxim "silencer" for firearms and produced one of the earliest automobiles, died today of a throat infection.

WELL KNOWN

THE late Hiram Maxim was well known by radio enthusiasts in Nova Scotia, which he visited many times, the last being in 1923. He was a personal friend of Major W. C. Borrett, director of radio station CHNS at Halifax, who in 1925 accompanied the famed inventor on a trip to Paris. Only last Saturday evening this visit was recalled by Major Borrett in the course of a radio interview.

The trip to Paris was to attend a conference of 26 nations in the interests of amateur experimental radio and which culminated in the formation of the International Amateur Radio Union. Major Borrett represented Canada and Mr. Maxim the U. S.

HARC Files

ARRL put their W1MK station on the air around 1923 at their office in the business district of Hartford, Connecticut, according to one item in QST. The first notice I found was in January, 1927, that described the operation of this station. 1MK remained silent during the hours of 8 to 10:30 PM so they would not interfere with the BCL's (Broadcast Listeners). Soon after the 1927 notice I found they published a monthly schedule in QST of the stations times of operation and working frequencies.

1937 CLUB STATION

The Maritime Division news section of the November 1937 issue of QST states that the VE1MK club station is a transmitter and receiver and is now in operation at the YMCA.

The traffic totals in some of the Maritime Division news sections of QST list VE1MK as having handled messages so the club station had been put to good use. The October 1938 issue of QST states that a Zepp Antenna has replaced the old single wire antenna at the VE1MK club station.

HAMFEST MARITIME DIVISION ARRL 1937

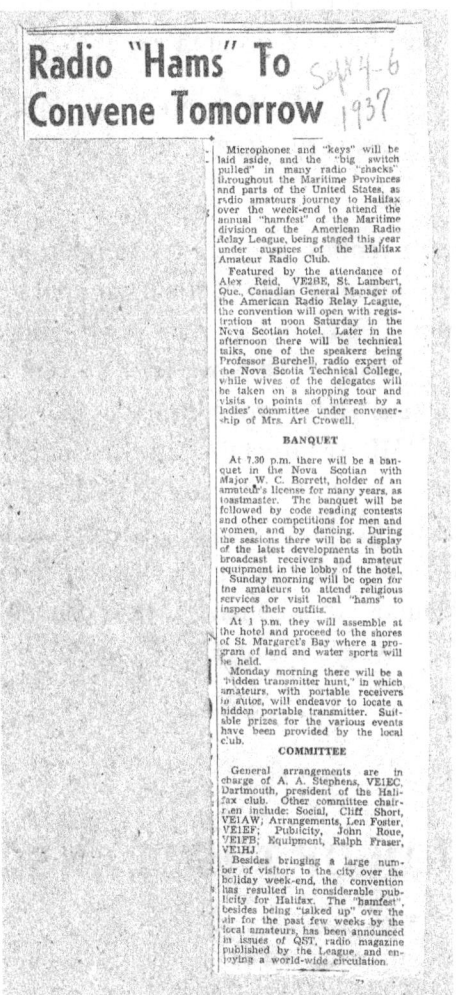

HARC Files

The Maritime Division Convention was sponsored by HARC and was held on the Labour Day weekend of September 4th, 1937. At least the ball started rolling on Saturday afternoon, September 4th, with registration at the Nova Scotia Hotel. The total number in attendance was 116 and the person who traveled the greatest distance was Mr. W. H. Lord of

Chattanooga, Tennessee. The banquet got underway at the Georgian ballroom of the Nova Scotia Hotel at 8 o'clock that evening. A. A. Stevens, VE1EC, president of HARC handed controls over to Major Bill Borrett, VE1DD, who acted as master of ceremonies. The convention was officially welcomed to Halifax by Alderman, Dr. S. H. Keshan, representing the mayor. Alex Reid, VE2BE, Canadian General Manager of ARRL spoke briefly. Joe Fassett, the grand old man of Amateur Radio in Eastern Canada, formerly VE1AR, recalled some amusing anecdotes of the early days of radio.

After the speeches were over the ballroom was cleared for the contests. A. A. "Steve" Stevens, VE1EC, won the code receiving contest and Barclay Dowden, VE1HK, won the code transmitting contest. The humourous contests brought forth many laughs and the convention adjourned after the presentation of prizes.

Sunday afternoon a picnic was held at St. Margaret's Bay with outdoor sports. During the afternoon the five-meter band was very active since many of the cars were equipped with five-metre stations.

Those attending this 1937 convention scattered around the city early Sunday evening, September 5th, and visited the radio shacks of the local operators. They reassembled for a midnight movie at the Capital Theatre and then visited the theatre projection rooms. They gathered Monday morning, in front of the Nova Scotia Hotel, after a few hours' sleep (for some) for the five-meter contests. These proved to be one of the best features of the whole convention. Ralph Fraser, VE1HJ, was in charge of the transmitter hunt and he put out a good signal, so good that the transmitter was found by VE1MA in a very short time. His second signal was more difficult to find in a large cemetery. In the afternoon different ones acted as the transmitter and were located by the fleet of five-meter equipment cruising around the streets of old Halifax. This concluded the convention and the various VE1s returned to their home QTH. Everyone felt HARC needed to be commended for a very enjoyable convention.

CREATION OF THE TRANSCEIVER

I have been messing with radio in one form or another for over fifty years and always felt the words transmitter and receiver had been combined into the word transceiver after World War II. ARRL stated that Transceiver is a trade name coined to describe a combined transmitter-receiver as manufactured by the Chicago Radio Laboratory of Chicago, Illinois, and was created about 1920. The 1938 ARRL Handbook has this to say about transceiver: "For portable work, a reduction in weight and general simplification of equipment can be made by using the same tubes for transmission as for reception. In the early days of u.h.f. work this idea was carried out very thoroughly, but with present conditions the simpler types of transceivers cannot comply with the regulations respecting stability of transmission, and are undesirable for reception because of severe radiation from the super-regenerative receiver operated at relatively high plate voltage. At the present time the "transceiver" normally uses only the audio equipment for both transmitting and receiving, the R.F. sections being entirely separate."

Modern transceivers share the same RF sections except for the RF amp. The 1956 and 1963 editions of the Handbook do not have a definition for transceiver.

HARC DUES

The monthly dues were set at twenty-five cents per meeting at the beginning of the H.A.R.C. This was changed to $3.00 per annum beginning on January 1st, 1937, and has increased to $25.00 per annum for membership in H.A.R.C. for 2008.

VE1MK AND THE QSL MANAGER

The Club operated a station through which traffic was handled three or four nights a week. One assumes this refers to the club station VE1MK after it had been set-up at the YMCA in 1937.

John Roue, VE1FB, was not only the nephew of the one who designed the sailing vessel *BLUENOSE;* he was the first QSL card manager for this area. He was the QSL card manager for the Halifax area in 1933 when ARRL created this position that is still in operation. John, VE1FB, stated he was swamped in QSL cards and asked everyone to send him an envelope for their cards. He was the QSL Manager with M.A.R.A. in 1933 and kept the job when M.A.R.A. was renamed H.A.R.C. Art Grant, VE1EP, stated he would bring your cards to you and always had them sorted with the SWL's and such on the top, with the good DX cards in the bottom of the pile. Art always turned his pile over and read from the bottom up.

ART GRANT VE1EP QSL CARD

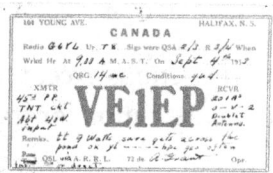

Courtesy Thomas Roscoe K8CX

This is an interesting QSL card mailed by Art Grant, VE1EP, for a contact on September 4th, 1933.

This card is to Barbara Dunn, G6YL, the first British female amateur radio operator.

This contact took place on twenty meters.

Note the power; Barbara was using 9 watts and Art was using about a 40-watt input and a doublet antenna. Barbara went on to an interesting life in radio and did not marry.

Her biography can be found on the K8CX web site.

PROVINCIAL EXHIBITIONS

MARA had a booth at the 1933 Provincial Exhibition that increased the traffic totals. VE1EX handled the traffic and he accumulated a total of 27 messages.

HARC had requested a complete amateur radio booth at the Nova Scotia Provincial Exhibition. One would assume this was the 1934 provincial exhibition but no further detail has been found.

The Halifax Amateur Radio Club had a booth at the Nova Scotia Provincial Exhibition in the fall of 1936. The Cliff Short, VE1AW, station worked very well at this exhibition and hundreds of people had their first opportunity to transmit via amateur radio. These transmissions were made on 20-meters via full duplex, and a list of many names of those showing an interest in obtaining their amateur radio license was recorded.

AUCTIONS MONTLY MEETING 1938

HARC decided to have a Radio Auction at each monthly meeting in 1938 so that members could auction off any excess radio equipment. They continued to have these auctions for years. If anyone had anything they did not want they would bring it to a monthly meeting.

A member would auction it off giving HARC a small percentage of the sale. This has been a great way to get a few dollars out of some unused piece of gear and another way to raise a few dollars for the club.

2-1/2 METERS 1939

Barclay Dowden, VE1HK, gave a very interesting and informative talk on 2-1/2 meters or 112 megacycles at a meeting early in 1939. 112 megacycles is in the Aeronautical Aircraft VHF AM phone band today. That band is from 88 to 118 megahertz. This came under an International Convention after World War II. This is the main communication band for aircraft and it is still using AM or amplitude modulation. The reason it is amplitude modulated and not frequency modulated is that the discriminator will block on the strongest signal in a frequency modulated receiver. This is called the capture effect. The detector in an AM receiver will pass all signals heard. At least in theory and the reason is if an aircraft gets in trouble, it should be able to attract attention on a frequency already in use. After World War II 2-1/2 meters became 2 meters from 144-148 megahertz.

VE1MK 1939 AND CLOSING FOR WAR

The April, 1939, issue of QST lists the VE1MK station as the Maritime Division station with the highest traffic score for the month at 74 messages handled. This list also notes that in future VE1MK will transmit O.P. (operator practice) on 3690 kilocycles each Monday, Tuesday, Friday and Saturday at 7:30-PM Atlantic standard time. The VE1MK station will undergo a bit of a rebuilding according to the Maritime Division notes in June 1939. They must have felt the station would be in use for some time and did not realize it would be shut down and dismantled in September. Ralph Fraser, VE1HJ, operated station VE1MK during the 1939 Operator Relay Station party. XTAL magazine sponsored a W/VE contest in 1939 and Ralph Fraser, VE1HJ, and Gordon Phelan, VE1KG, operated VE1MK during this contest. Therefore, the old station at the YMCA received a lot of use in 1939.

AIR SERVICES

CANADA

DEPARTMENT OF TRANSPORT

Ottawa, September 5th, 1939.

To the Licensees of all Amateur Experimental Radio Stations:

Whereas, in the opinion of the Government of Canada, an emergency has arisen in which it is expedient for the public service that His Majesty's Government in Canada shall have control over the transmission of messages by Radio Stations of all classes.

Now, in pursuance of the power conferred on me by the Defence of Canada Order and The Radio Act, 1938, and regulations issued thereunder, I hereby notify you that your Amateur Experimental Station Licence is suspended forthwith, and direct you to completely dismantle and render inoperative all equipment installed in your station.

(C. P. Edwards)
for Minister of Transport

Thanks to HARC member Terry Bigelow VE1TRB

With the outbreak of war in 1939 the amateur radio operating activities ceased. Every amateur in Canada received this letter from the Minister of Transport who was the governing body of all radio regulations at that time. One will note this letter states in part:

"I hereby notify you that your Amateur Experimental Station Licence is suspended forthwith, and direct you to completely dismantle and render inoperative all equipment installed in your station".

FINANCIAL STATEMENT 1939

```
         FINANCIAL STATEMENT HALIFAX AMATEUR RADIO CLUB

                       July 21, 1939

Receipts
     Balance from statement April 21, 1939   $66.75
     Dues                                      16.20
     10% receipts of auctions                   3.91
     Drawing June 16                            21.80
     Deposit on Club-room key by VE1BU           .25
                                                        $108.91


Expenditures
     Club-room committee (xmtr)               36.65
        "      "      "    (pair 809 tubes)    3.50
     Amateur Station VE1MK licence             2.50
     Prizes                                     1.50
     Y.M.C.A. memberships                      36.00
     Notices of meetings                        1.20
     Misc. postage for QSLs, milk,coffee,sugar 4.20
                                                       85.55
                                                       23.36
              Less 10 keys for Club-room                2.50
                   CASH ON HAND                                $20.86


Paid up members to date

VE1FB   J. E. Roue           84 Spring Garden Road
   HJ   F. R. Fraser         23 Oakland Rd.
   FO   D. A. Smith          98 Wellington St.
   KB   Harry Scott          26 York St
   HK   C. B. Dowden         49 Walnut St
   BP   Clarence Fuller      34 Pepperell St
   JH   E. S. MacLaughlin    78 Harvard St.
   KG   Gordon Phelan        30 Rosebank Ave
   FQ   Brit Fader           128 Henry St
   BC   W. S. Bligh          8 Williams St
   DQ   A. M. Crowell        69 Dublin St
   EU   Ainley Croft         3 Rose St., DARTMOUTH, N.S.
   MZ   R. S. Hart           64 Elm St
   NS   R. A. McKenzie       39 Edward St
   IB   Fritz Webb           200 Russell St

$2.00 VE1AQ  F. L.Clairmonte  72 South Park St
 1.20 VE4AWT Jim Mathews     Mess 24, R.C.N. Barracks.



            HALIFAX AMATEUR RADIO CLUB
             ROOM 50, Y. M. C. A. BUILDING
                  HALIFAX, N. S.
```

HARC Files

This is HARC on July 21st, 1939, including the 17 paid-up members.

CLUBS OPEN DURING WAR

By mutual consent the Halifax Amateur Radio Club carried on its monthly meetings. This proved to be a wise move since Halifax, an eastern Canadian port, was well known in those days, and frequently visited by amateurs from abroad, and all across Canada. HARC maintained a column in the local newspaper of interest to the amateur radio community. Thus, during this period of radio inactivity many radio friends were made by personal contact.

The chairman of the HARC transmitter committee, Ralph Fraser, VE1HJ, was instructed to dismantle the club station, VE1MK, put the entire club's apparatus in the one box, and then store it in the attic of the home of Brit Fader, VE1FQ, at the January 26th, 1940 meeting.

HARC was not the only club that did not disband during the war. W. K. "Ken" Angus, VE4VJ, of The Northern Alberta Radio Club in Edmonton had a rather lengthy but interesting letter in the March, 1940, issue of QST stating how they intended to carry on during the war as a social club. He listed 8 steps they were taking to keep this club operating.

The Wireless Association of Ontario kept going during the war. Their monthly meetings in 1943 consisted of some very interesting high level technical discussions by some very interesting speakers. The average attendance at these meetings was 83 persons interested in the science of radio. The December issue of QST for 1944 states: "The Wireless Association of Ontario, Canada's only radio club which has been carrying on continuously since 1913 is still holding regular meetings". This goes on to state their paid-up membership is 95 and lists some very interesting speakers that had participated in their monthly meetings.

Alex Reid, VE2BE, the Canadian General Manager at ARRL Headquarters had an even lengthier letter than Ken's in the April, 1940, issue stating all Canadian clubs should continue on during the war. He pointed out that the ARRL saved amateur radio at the end of World War I. He stressed the fact that everyone should keep up their membership to ARRL because they had no idea what would take place on termination of the war. It would be interesting to know how many clubs did manage to continue to the end of the war.

The ARRL and QST kept going through this war and played an important part in it. They had to cut back the size of QST 25 percent to conserve paper as did all the other publications in circulation. Not only did ARRL operate through the war but the FCC continued to examine and issue amateur radio licenses through the war.

As far as I know one could not sit for an amateur radio certificate in Canada during the war. One was not permitted to operate an amateur station so sitting for a license does not make sense. Besides so many radio operators were required during the war that if one knew enough about radio for the amateur certificate, they would have been recruited for one of the many other positions. I believe the code requirement for an amateur radio certificate in Canada was 15-WPM (words per minute) at this time. The RCAF required 12-WPM for their radio operators in their marine squadron. This squadron was the one that operated the RCAF vessels in their marine fleet. These were the vessels that serviced their flying boats, supplied high speed search and rescue, and their supply vessels that serviced their many coastal bases. There was no radio phone in the RCAF Marine Squadrons, it was strictly CW.

QST COST 1940

There was an increase in the cost of QST for Canadians with the May, 1940, issue. The cost remained 25 cents per copy for the U.S. but went up to 30 cents per copy for Canada. The cost was up again to 35 cents in Canada and remained 25 cents for the U.S. applicable to the October, 1940 issue.

AMERICANS 1940

After June 5th, 1940, it was illegal for American amateur radio stations to communicate with anyone but another American amateur station. Their DX contests were held with their own islands such as Hawaii, Samoa, Puerto Rico and even Alaska.

Conscription for the U.S. Armed Forces commenced in November, 1940, and ARRL advised its members to take their amateur radio license to the recruiting office if they wanted to enlist as radio operators. They simply felt it would be a good chance they would be recruited as radio operators but it was no guarantee.

SILENT KEYS 1941 AND THE MONTH IN CANADA

British amateur radio operators killed during the war started to appear in the Silent Key column of QST in 1941 along with the other Silent Key entries.

The ARRL made a request in the May, 1941, issue of QST for Canadian Amateurs to write in and contribute to a section titled "The Month in Canada". The first of these appeared the next month but it was Quebec and Ontario only. Apparently, no one in this area provided any detail.

ELECTRONIC KEYER

The big electronic news in the amateur world of 1940 was the electronic keyer and several made the front covers of QST. There were several plans and schematics between the covers as well. One has to ask why it took so long for these to become such a big hit with the CW crowd.

ALEX REID VE2BE ART CROWELL VE1DQ BRIT FADER VE1FQ

Alex Reid, VE2BE, remained the Canadian General Manager of ARRL throughout the war and Art Crowell, VE1DQ, remained the Section Communications Manager for this, the VE1 area.

Alex Reid, VE2BE, and the Editor of QST made a request for news from all Canadian Amateur Radio Operators for a new section in QST with the title "The Month in Canada". The first time this section appeared in QST was in the June, 1941, issue but there was no Maritime or HARC news. Apparently, none had been submitted. There was news of 300 Canadian amateur operators. This section was broken down into The Maritimes and then the rest were listed by province. The Maritimes often appeared as Nova Scotia probably because Brit Fader, VE1FQ, was the one who provided the detail.

L. J. "Brit" Fader, VE1FQ had joined the "Concert Parties Division" as a consequence of his home recording experience and traveled around a lot. He managed to collect a wealth of information on the members of the VE1 amateur radio community. Most were in one branch or another of the military and where they were stationed and what they were doing. It makes for very interesting reading if nothing else. The second time this section appeared in August, 1941, it stated there were 3,380 Canadian amateur certificates at the outbreak of war in 1939. By this time 1,700 of that number were active members of the service. Of those 1,700 more than half were commissioned officers mostly in the RCAF.

KEITH RUSSELL VE9AL

Here's a picture that should make every VE ham, and all others, too, swell with pride. Group Captain A. H. Keith Russell, ex-VE9AL, is now serving as director of technical training of the RCAF at Ottawa. Formerly a Toronto lawyer, he is one of Canada's best and oldest hams, and was the first Canadian General Manager of the League.

Page 31, September, 1943, QST

This section was in the September, 1941, issue but the next was not until the February, 1942, issue. The September, 1941, issue states that our old friend A. H. Keith Russell now VE9AL has been promoted to Wing Commander and now commands No. 4 Wireless School at Guelph, Ontario. This also lists all the former amateurs at the school with their rank and name.

In the December, 1944, issue it states Air Commodore, A. H. K. Russell has retired and resumed the practice of law and that he had been on active service with the RCAF since September, 1939. The RCAF claimed they had sufficient members to handle the war in Europe and in Japan so they discontinued recruiting further members. This meant that Air Commodore Russell's request for retirement was possible and accepted.

He returned to his law practice in Toronto after five years' service in the RCAF and a job well done.

AMERICAN ARMY AIRWAYS COMMUNICATIONS SYSTEM 1944

An excellent article on the American Army Airways Communications System with the title "The Great Spiderweb" by Pfc. H. D. Colson, 78th AAF Base Unit and S/Sgt. R. C. Fleischman, Technical Advisor, appeared in the October, 1944, issue of Radio News. This paragraph is from page 92 in that article. This article was mainly describing the construction of site Crystal 2 that became Frobisher Bay, Baffin Island, and in 2009 is known as Iqaluit, in the territory of Nunavut.

America was indeed fortunate to have a tremendous reserve of radio hobbyists or hams—men with radio know how requiring no extended technical training—for without these men, those "kids around the corner," and the innumerable radio repairmen, this gigantic enterprise probably would have failed at the outset. So pressing was the demand for hams during those days that Uncle Sam waived basic training for many of these men. Hams were hurried overseas to all parts of the world in a ceaseless stream of task forces; there was a desperate rush to get stations on the air. How they got the stations on the air is a story of monumental courage, patience and intestinal fortitude; during that period equipment was scarce, secondhand, make-shift and difficult to transport. The manpower shortage was discouragingly acute.

SENDING A LETTER TO A PRISONER OF WAR

The September, 1941, issue of QST describes in detail how to send a letter to a former Toronto amateur now a prisoner of war in Germany. QST listed several amateur radio prisoners of war from Britain, the British Commonwealth countries as well as those from other nations including the American amateur radio operators during the war.

LITTLE NEWS FEBRUARY 1942

The only news in February, 1942, was a small article from VE2 another small article from VE3 but a fairly good one from VE4. One of the hams in VE4 felt the Canadian government should permit Canadian hams to operate civil defense on 112 megacycles and ultra-high frequency like the American hams. It was also noted that VE4AHY was now a Sergeant in Ottawa teaching the girls in the Canadian Army. That would be a hard way to fight a war!

Alex Reid, VE2BE, stated he would be glad to hear from more of the chaps in various training schools so must have felt this section was going to continue. Unfortunately, there were no "This Month in Canada" sections from time to time during the war. They were a wealth of information and terminated with the December, 1945, issue of QST. The January, 1946, issue went back to the prewar system of recording with a section in Station Activities titled Canada and broken down to Maritimes, Quebec, Ontario, Vanalta and Prairie.

They claimed that seven of eight radio service men in England were now serving in the armed forces in 1942.

AMERICAN AMATEUR RADIO OPERATION SUSPENDED

The January, 1942, issue of QST states that all American amateur radio operation has been suspended as of December, 8th, 1941. The last page of that issue states the Army Signal Corps needs radio operators and lots of them. If anyone submits an amateur radio or commercial radio license to the recruiting office, they will be guaranteed a job in radio.

W1AW kept transmitting bulletins and some 2,000 hams were soon reactivated for various civil defense councils operating on 112 megacycles and UHF. The Halicrafters Company was advertising a UHF receiver with the highest frequency at 165 megacycles.

There was a complete description including illustrations so one could build a 112 megacycle Pack Set ready to go and this is what the majority were doing in 1942. QST listed the frequencies for the War Emergency Radio Service (WERS) as:

112,000 – 116,000 KC

224,000 – 230,000 KC

400,000 – 401,000 KC

I have no idea why they listed them in kilocycles and not MC – megacycles.

QST was urging hams to enlist in the armed forces and one city in the United States gave a subscription to QST to each ham that enlisted. ARRL also advertised in 1942 the fact that if anyone knew a competent amateur operator in the service and not in radio, they would have them transferred to a radio job. The ARRL must have had plenty of clout.

There was also a shortage of equipment and the armed forces would buy factory made transmitters from any ham who would sell. QST ran blank forms one could use to list the equipment for sale and the price wanted. ARRL apparently bought the equipment for the government.

The ARRL stated that the Halicrafters Company alone improved and increased production in 1942 to the point it would have taken seven years in peace time. The Halicrafters Company advertised as "improved in war for better peace time reception".

One Sergeant in the U.S. Army requested a frequency for an exercise one day and was given his favourite prewar 80-meter frequency purely by chance. The amateur radio operators in uniform would often get together for a Hamfest. They often contacted one another when in a group. One would send CQ on an oscillator if one was available or simply whistle CQ. QST printed several photographs of these that had taken place anywhere and everywhere.

The ARRL was making a plea for any metal stating amateur operators were probably the biggest group that collected this material. They went on to state that they should donate the metal now. They would be able to buy it for a few cents once they won the war. If they did not win the war, they would not need it.

The editor of QST stated that any government that was ready for war was not a democracy and that to me makes as much sense as anything I have heard or read about any war. ARRL felt there were around 12,000 American amateur radio operators in the service by early 1942 or about six months after Pearl Harbor on December 7th, 1941. There were a lot of photographs and detail on the various radio schools in the American Services in the pages of QST. There were lists of those amateurs missing in action; those killed were listed in Silent Keys and several pages of those serving, their unit and location. It is amazing some of it was not considered classified information at the time.

CANADIAN HAMS LISTED IN QST

The March, 1943, issue of QST lists two- and one-half pages of known Canadian Amateurs in the service. This lists them by call sign, name, rank if known, and location and has them broken down to Army, Navy, Air Force, Royal Canadian Corps of Signals and Royal Air Force. I found it rather amazing they were allowed to print such a list and this continued whenever anyone sent the detail to Alex Reid, VE2BE, who kept requesting this information.

THE CIVIL AIR PATROL

The Civil Air Patrol (CAP) was another important part of the American Coastal Defense during the war. A good many amateur radio operators provided the communications for this organization. The August 1942, issue of QST gives the complete detail on making a radio station for the aircraft used for the various patrols. The pilots, mechanics, radio operators and any others were all civilian volunteers. No doubt a number of these stations were built by the amateur radio operators. I found this fascinating. The radio station consisted of a receiver that would tune the band from 200 to 400 kilocycles. This is where one found the old Radio Range Stations that first entered service in 1927. The aircraft transmitter was fix-tuned, AM of course but on 3105 kilocycles. The aircraft transmitted on that frequency only and listened to the appropriate Radio Range on their receiver. They claimed this aircraft radio station had a range of 40 miles. The international frequency for this service had moved down to 3023.5 kilohertz when I operated one of the last Radio Range Stations. Radio transmitters could often be heard on the second harmonic when transmitting. In this case the second harmonic of 3105 was 6210. The marine bands for ship to shore radio had been expanded and 6210 became a working frequency for passenger ships and this was the reason for the switch to 3023.5. Think of the many aircraft that had to have their radio transmitter moved and retuned to this new frequency? Experience tells me that many did not get moved and no doubt the stations involved monitored both frequencies for a time until this settled down and became well known. There is an excellent detailed description of this old Radio Range Navigation System in the February 1944, issue of QST. This was the first Aeronautical Navigational Aid.

CAP was assigned two frequencies; 3105 kilocycles and its second harmonic 6210 kilocycles. There are detailed instructions for building a small transmitter for these frequencies to be used in a small aircraft in the May 1945, issue of QST. It states that this should make a good 80- and 40-meter transmitter with slight adjustment on termination of the war.

3105 and 6210 kilocycles were the only transmit frequencies in Amelia Earhart's Lockheed 10E aircraft when she disappeared in 1937. It is truly amazing that girl managed to get as far as she did on the little, she knew about what she was doing. We often said it was amazing how forgiving the aircraft could be when we operated aeradio.

The Civil Air Patrol is still in service and is a part of the United States Air Force Auxiliary. There were 64,000 members in 2008. They participate in various programs such as search and rescue. It would be a great experience for any person interested in aircraft.

THE FIRST USE OF SILENT KEY

The first use of the term Silent Key appeared in March 1927. Up until then these monthly entries were listed as obituaries. There were less than a dozen a month recorded in this section up into the 1940's. During the war the amateur radio operators in the service were listed with their rank, name and call sign. There were quite a few from Britain, the British Commonwealth countries and a few others from various nations along with the American service men.

PRISONERS OF WAR MISSING IN ACTION LISTED IN QST

During World War II QST had three lists of Amateur Radio Operators and all three were usually on the same page. These three listed the Silent Keys, Prisoners of War and those missing in action. QST had another section called Gold Stars. This section described those amateur radio operators complete with a photograph in uniform that had been killed in the armed forces. One or two Canadian servicemen were included during the war. The Silent Key column contained around a dozen names up to and including the 1960's. I know of several American amateur radio operators that did not make the silent key lists, so the lists do not give an accurate indication of those who have become a silent key.

DOUG SMITH VE1FO

This is what is recorded in the book "from spark to space" by the Saskatoon Amateur Radio Club and was written by Wes Street, VE1EK:

RCAF.COM

This is a Beaufighter Aircraft, the type Doug Smith flew in the RAF.

One of the faithful and enthusiastic members of the H.A.R.C. before the war was Doug Smith, VE1FO. He joined the R.A.F. in 1938 and during the war was cited for bravery and credited with knocking out important enemy trains in North Africa on several occasions. On his last raid anti-aircraft batteries felled him. His grave is located in Misura, Libya. In commemoration and remembrance, the Halifax Amateur Radio Club has claimed his call, VE1FO, as their station call.

Actually, Doug joined the RAF in August, 1939, just previous to the outbreak of the war in September. Misura is the same as Misratah, Libya or located within Misratah.

This was recorded in the April, 1942, issue of QST by Brit Fader, VE1FQ, but unfortunately the newspaper clippings were not included:

"I enclose a few newspaper clippings about the exploits of one of the local gangs. Doug Smith, VE1FO. VE1FO was one of the main instigators and backbone members of the local club. He was one of the pioneers of five-meter work in the Maritimes, and was also the local emergency coordinator for this part of the Maritimes. We regret very much the fate that has befallen him, and hope that there is a remote chance that he is safe and a prisoner of war, rather than a casualty."

This appeared the next month and of course we all know the outcome today:

"Last month VE1FQ mentioned the experiences of Doug Smith, VE1FO, in Africa. We didn't have room for the complete story then, which was probably just as well, because now we have a happy ending to add. Probably the story is best told by quoting the following CP Cable from Malta:

"A young flying officer from Halifax, N.S., is not fighting the Battle of Libya single- handed, but it looks as if he was trying to win it with the aid of his observer. . .

"The Haligonian opened his 'offensive' Dec. 1 on the Tripoli-Bengasi Road, along which he made six diving attacks on enemy trucks, and then shot up four gasoline tankers, two of which caught fire. During the operation he was attacked by two Italian fighter planes, but evaded them.

"On the following day he returned to the fray and had a set-to with an Italian plane near Sirta. After half a dozen bursts the enemy aircraft landed on a beach. The hustling Canadian calmly photographed the machine, then shot it up until it caught fire, after which he photographed it again!

"Later he attacked some 20 motor trucks. Most of them had machine guns, 'but they were unmanned as the gunners seemed to go along with the drivers who hopped out and ran about 50 yards, throwing themselves in the sand.'

"Next the Canadian pilot swooped down on five gasoline tankers and trailers, setting them all on fire. He also took photographs of two of these trailers in flames. 'Machine guns were getting too accurate for further photographs,' he reported apologetically to his commanding officer..."

That story was about Flying Officer D. A. Smith, VE1FO.

Hardly had the story of VE1FO's heroism in the battle of Libya reached these shores when word came that he was "missing" on active service. For some time, he was given   up for lost, and not only the ham fraternity but a host of friends throughout Nova Scotia   paid tribute to his gallantry, initiative and courage.

Then came the happy ending mentioned in the beginning – happy, that is, in comparison with the fate originally feared. Early this year another report came through, saying that Flying Officer Smith had been located. He is now a prisoner of war "somewhere in Italy."

Brit, VE1FQ, included this in his August, 1942, report:

> "I am sorry to report that, since sending in my last report, word has reached us to the effect that Doug Smith, VE1FO, who was a prisoner of war, has passed away. He died from wounds received when his plane was shot down in Africa, and will be sadly missed by the VE1 gang, especially the Halifax hams, who knew him so well."

Since Doug is buried in Libya, he must have been held a prisoner of war with the Italians there and not in Italy. These reports could be confusing with all that was taking place. We trust he did not suffer and it would be nice to know the name and the fate of his observer. It is not shown in the above photograph but there was a glass dome on top of the fuselage behind the pilot of the Beaufighter where the observer rode.

Doug Smith Drive in the city of Halifax is named for Doug. It is at the site of the old Halifax airport that was in operation during the 1930's just off Chebucto Road.

HARC MEMBERS SERVING IN THE ARMED FORCES

S. R. Kenny, VE1CS, of Halifax was lost in a destroyer believed to be *HMCS OTTAWA* on September, 13th, 1942. He was one of 113 members of the Royal Canadian Navy and 6 members of the Royal Navy lost when this ship sank from enemy action. This was Canada's first *HMCS OTTAWA* and the most popular name for a Canadian warship. *HMCS OTTAWA* in 2008 is the fourth vessel of that name.

C. D. Kenny, VE1JS, a telegraphist in the Royal Canadian Navy was the brother of S. R. Kenny, VE1CS. VE1JS was a member of HARC. The Telegraphist was renamed Communicator Radio in 1950, Radioman in 1960 and Radio Sea after unification in 1968.

Leading Telegraphist W. E. Robinson, VE1NP, from Halifax lost some Ham Gear he had been working on for use after the war in the loss of a corvette believed to be *HMCS LOUISBURG*. This Corvette was lost from enemy action in the Mediterranean Sea on February 6th, 1943. 37 Royal Canadian Navy and 5 Royal Navy crewmembers lost their lives in this action.

A Leading Telegraphist was normally the highest-ranking radio rating in a Corvette and therefore Robinson was probably in charge of the radio room. He would have two or three Telegraphists under him to handle the operating. There would also be a coder or two to decipher the incoming and code the outgoing message traffic. There would also be a couple of Signalmen to handle the signal flags and Morse code with the signal lamps. A senior signalman called a Yeoman would be in charge of the signalmen. The Telegraphists were also taught "Wig Wag" (transmitting Morse code with two flags). The communications department in any ship was a fairly large department when you add them all up. I have often wondered what became of them on termination of the war. A few became amateur radio enthusiasts, a few became commercial radio operators, and the rest simply had enough and had no further interest.

Len Foster, VE1EF, was president of HARC for a few months in 1934. He was a member of the Signal Corps of the Canadian Reserve Army during World War II with the rank of Sergeant. He was one of the older members of HARC and had served in the Signal Corps of the U.S. Army during World War I. At one time he was stationed at the large radio school the U.S. Army operated at Fort Monmouth, New Jersey.

BEER CAN ANTENNA

A good antenna I found in use during World War II was a vertical constructed from 55 empty beer cans at the Army Training Center at Fort Clark in Texas. It was well guyed and had withstood many wind storms and also worked well. This was constructed by a few hams consisting of a few young always hungry boys serving in the army.

The beer can antenna continued on after the war to the point one police station in the U.S. was using one on the roof of their police headquarters. There is all the detail one would ever need to make a beer can antenna in the November, 1955, issue of QST on pages 26 and 27. The ARRL felt the beer can antenna had run its course in the June, 1956, issue of QST. They stated they had the minister of a church write in and wanted to know where he could get 82 beer cans and a kid wrote in wanting to know if orange juice cans would work. They did not start making these cans out of aluminum until after 1956.

TEACHING MORSE DURING WAR

The single most activity that created the most work during World War II had to be teaching Morse code. There were classes anywhere, everywhere, to anyone and everyone. It is amazing the world did not have everyone with a ham ticket on termination.

CLOSING SUPPER MEETING 1945

Eddie MacLaughlin, VE1JH, stated HARC held their closing supper meeting for the summer months of 1945 with hams from all over the Dominion in attendance. All branches of the services were represented. One of Halifax's leading manufacturers donated their cafeteria and recreation hall to the club for the occasion. Highlights were the splendid meal, motion pictures, sing-song and the piano playing by Ken Warren of the Royal Canadian Navy. There were also others there who would be joining the ham fraternity when the bands are open.

Page 59 September 1945 QST

We would like to thank Maty Weinberg, KB1EIB, for scanning this photo in QST

The HARC supper meeting was attended by:

Left to right, Front row:

A.J. Neilson, VE4OE; Don Bain, VE1LZ;

Second Row:

D. Coppe, VE1NO; C. Underwood, VE4AFG; W. Robinson, VE1NP; Ken Warren; W. Street, VE1EK; E. MacLaughlin, VE1JH; N. MacKeigan, VE1AG; M. Purvis, VE5AJU; M. Koz; Cpl. Ames, VE3HS; M. Fitzgerald, VE1HP; R. Hart, VE1MZ;

Third Row:

Stewart, VE4UH; R. O'Connell, VE4ABU; N. Looker; G. Brown, VE1EV; M. Prior, VE4UB; F. Higgins, VE1KY; G. Cooke, VE1NW; E. Harrington, VE1NQ; Sgt. Patterson; Sgt. Duvar; M. Armstrong.

Fourth Row:

S. Mendlesohn; O. Sandoz; W. Bligh, VE1BC; E. Schaffer; H. Yeadon; Cpl. North; Cpl. O'Donnell; R. Morrison; C. Wigle, VE3WL; J. Whitely, VE1OK; Sgt. Cann; M. Pearce.

Fifth Row:

F. Totten, VE1JK; C. Kenny, VE1JS; J. Burke, VE1NE; Len Foster, ex VE1EF; F. Webb, VE1DB; W. MacLean, VE1EY; H. Bishop, VE1OB. Members of the club missing from the picture are A. Baxter, VE5AJV; Les Peppin; W. Wooding, VE1ET.

ARRL REQUESTS RECORD OF SERVICE DURING WAR

The ARRL kept requesting every Canadian and American amateur radio operator to record their record of service during World War II via the forms they mailed in each issue of QST. One wonders how many or how much success they had.

CLUB CALL SIGN VE1FO

The HARC VE1MK club call sign was not retained after the war. A motion was made by Wes Street, VE1EK, at the monthly meeting of HARC on Friday, January 18th, 1946, that HARC apply for call letters VE1FO for the club station as a memorial to the late Flying Officer Doug Smith who gave his life in the 1939-45 conflict. The call letters VE1FO was the call assigned to the late Doug Smith when he was the most active worker and member of the H.A.R.C. This motion was seconded by Ralph Fraser, VE1HJ.

78 Harvard Street,
Halifax,N.S.
Jan. 2I st. I946

Mr. G.F. Harris,
Dept.Of Transport,
Halifax,N.S.

Dear Sir.

The Halifax Amateur Radio Club hereby make application to
have the station call changed from VE1MK to VE1FO the call
letters of the late Douglas Smith.

As you no doubt know Doug. was one of our most active members
and he gave his life in the service of his country.

We wish to obtain his call letters to xxxxxxx serve as a
Memorial to his memory.

Yours faithfully,

E.S. MacLaughlin   Sect'y
The Halifax Amateur Radio Club.

HARC Files

HISTORY VE1MK CALL SIGN

The VE1MK call sign is but one of many call signs that I have a fond memory from years of experience. The VE1MK call sign was assigned to Marshall S. Killen, North Sydney after the war. Marshall worked for Western Union for many years. He visited many countries over the years where he operated with a variety of call signs. He had been VE3CDK and was VE3KK in 1980 when he was awarded the ARRL certificate of merit for outstanding contributions to amateur radio.

My memory of the VE1MK call sign is mainly when Maurice Spicer was assigned the call. Maurice had worked on the highway as a young man and lost his eyesight from a dynamite blast. He married Mabel, his nurse, and they ran a store at Harbourville. They made some of the best homemade ice-cream. The village will never forget the time Smokey Wagstaff had Maurice drive his pickup down through the village while he hid so he could guide him. He even had Maurice wave to those he met on this trip. It created quite a sensation. Mabel and Maurice ran a nursing home when they moved to Berwick and two of my grandparents died in their nursing home.

OPENING AMATEUR RADIO AFTER WORLD WAR II

The ARRL stated in the Editorial of the November, 1945, issue of QST that W1AW had received limited authorization to go back on the air. The editor also warned all amateurs to not put any money into gear for 112 megacycles because it was going to be moved to 144 – 148 megacycles and that 5-meters would become 6-meters from 50 – 54 megacycles. As this issue went to press, they just had time to squeeze in these last two pages:

MORE BANDS!

U.S.A. and Canada on November 15th Open 10 and 5-Meter Bands and Four Microwave Bands; 2½ Shifted; International DX Restored

JUST as this issue of QST is ready for the bindery, and with barely time enough for us to slip in this extra sheet, the Federal Communications Commission for the United States and the Department of Transport for Canada on November 9th have simultaneously announced important actions restoring amateur radio on frequencies above 28 Mc. The actions are effective at 3 A.M., E.S.T., on November 15th.

The FCC action is covered by its Order 130 and replaces the temporary authorization of last August under which we operated until Nov. 15th. While it is expected that by early December FCC will be able to set up the machinery to issue new station licenses (and begin the renewal and modification of old ones), such facilities are not yet available. The only action possible at the moment is therefore to continue a temporary authorization to those of us already licensed. Station licenses that were valid at any time between Dec. 7, 1941, and Sept. 15, 1942, are validated for another six months — until 3 A.M., E.S.T., May 15th. (During that time there will be FCC instructions on how to apply for renewals.) Such stations are then authorized to operate on a newly-stated group of frequency bands. The action applies to all areas under FCC jurisdiction except the central, southern and western Pacific areas. Unfortunately, at the time of releasing the order military clearance had not been completed for Hawaii and the U.S. island possessions in the Pacific, and they are excluded. (The prohibition is but temporary and it is possible that it will be lifted even before Nov. 15th. K6 amateurs should keep themselves informed by listening to W1AW's broadcasts. Here are our new frequency bands after Nov. 15th:

TEN METERS

The postwar band 28-29.7 Mc. is opened in its entirety to c.w. The portion 28.1 to 29.5 is available for a.m. 'phone (A-3), while f.m. 'phone may use from 28.95 to 29.7 Mc. The 'phone figures are reportedly derived from some FCC postwar planning and do not represent ARRL suggestions. It is needless to say that 'phone stations should observe them carefully.

FIVE METERS

We open up temporarily on our old band: 56-60 Mc. is available for c.w., i.c.w., a.m. 'phone and facsimile, and 58.5 to 60 Mc. for f.m. 'phone, precisely as before the war — until March 1st. At that time, subject to further FCC order, television is to vacate our new band 50-54 Mc. and it will be assigned to us in lieu of 56-60.

TWO AND A HALF SHIFTED

The new band 144-148 Mc. now takes the place of 112-115.5 and no operation on the latter is now permitted. The new band is available for c.w., i.c.w., 'phone and fax, and also for f.m. 'phone and f.m. telegraphy. But in some areas part of the band is still in use for military control circuits, with the result that amateurs within 50 miles airline of Washington, D. C., and Seattle, Wash., are denied the use of 146.5 to 148 Mc. For them the band is temporarily only 144-146.5 Mc. (It is probable that, when K6 is reactivated, the same thing will be true within 50 miles of Honolulu.) Let all hands take careful note of these figures and, where indicated, keep clear of the military portion. Full bandwidth is to be expected in these areas in a few months.

MICROWAVES

We do not yet get our assignments at 220-225, 420-450 and 1145-1245 Mc. They are temporarily held up because of some conflicts but further news is to be expected in a few weeks. We do get the remaining four microwave bands,

2,300 to 2,450 Mc.
5,250 " 5,650 "
10,000 " 10,500 "
21,000 " 22,000 "

and they are open to all imaginable types of transmission except pulse, i.e., c.w., i.c.w., a.m. 'phone, fax, television, f.m. 'phone and f.m. telegraphy.

WAR RESTRICTIONS REPEALED

The Commission then canceled a handful of its temporary wartime restrictions of unhappy memory. Gone now are Order 72 and its amendments, which prohibited communication with foreign countries, and Order 73 and its amendments, which forbade portable and mobile operation below 56 Mc. Also off the books are Orders 87 and 87-A, which closed us down and took away our frequencies, and Order 87-B which instructed that no further station licenses be issued or modified. The way is thoroughly cleared. We may work foreign DX if we can find it on ten, and we may work v.h.f., u.h.f. and s.h.f. with Canada and Mexico. Huzzah and hooray!

CANADIAN ANNOUNCEMENT

The restoration of Canadian amateurs was accomplished by a press statement by the Honorable C. D. Howe, Minister of Reconstruction, who announced that the seven bands of frequencies above enumerated would be placed at the disposal of Canadian amateur radio effective Nov. 15th. Only the over-all band figures were mentioned, with no stipulation of subdivision by types of emission. The band 56-60 was only allocated temporarily, he said, and would be replaced in approximately six months' time by 50-54 Mc.

"At the outbreak of war, an order was issued suspending the operation of all Canadian amateur radio stations," the Minister said. "This order has now been rescinded and, effective November 15, 1945, all amateur experimental station licenses which were in force immediately prior to the war are reinstated and will be effective until March 31st next. It is essential, however, that all who hold a 1939 Amateur Experimental Station License must first obtain permission from the nearest Government Radio Inspector before going on the air. Radio Inspectors have likewise been instructed to furnish full information to prospective amateurs."

The Minister stated that every effort was being made by Canada and the United States to clear other frequency bands for radio operations, particularly the 3.5-4, the 7-7.3, and the 14-14.4 Mc. bands. Announcement would be made at the earliest possible date as to final postwar frequency allocations to amateur stations but in the meantime it was essential that amateur radio operators confine their activities to the frequencies now released.

There were approximately 4,000 Canadian amateur radio operators at the outbreak of the war and the Honorable Mr. Howe paid tribute to the manner in which they had foregone their interesting hobby, in conforming with the governmental order, so as to enable these frequency bands being utilized by the armed forces or other essential war services. "Canadian amateur radio operators have contributed materially to this country's war effort," he said. "Most of our amateurs were young men and they responded enthusiastically to the call of their country, especially during the early stages of the war when the Armed Services urgently needed large numbers of radio operators.

"They served at sea with the Royal Canadian Navy and the Canadian Merchant Navy. They served with the Army, the Royal Canadian Air Force and the Government radio services. They also provided the essential instructor personnel for the training of Canadian and Empire airmen under the British Commonwealth Air Training Plan. Many have paid the supreme sacrifice. Several have won high honors for gallantry on the field and a few have risen to the higher executive brackets in the Armed Services. We have just cause to be proud of our amateur radio operators."

And so, fellows, we take another major step toward restoration. There will be further developments at short intervals. Make it a habit to listen for W1AW, which will always give you the newest news on these matters — Mondays through Fridays at 8, 9 and 10 P.M., E.S.T., on 3555, 7145 and 14280 kc., by special authority of FCC and the special cooperation of the armed forces. Meanwhile, may you find juicy DX on 10 meters, nice bending on 5, and lots of fun with your microwave ideas.

— K. B. W.

K.B.W. – Kenneth B. Warner, W1EH (Managing Secretary, ARRL) Editor QST

The last two pages of the November, 1945, issue of QST.

D. F. Taylor, VE4QV, and Douglas E. Kerr found this item in Canadian newspapers —

"Rationing officials in some parts of the Maritimes would like very much to meet that fellow, whoever he is, who first gave radio amateurs the nickname of 'hams.'

"As you will recall, the newspapers a few weeks ago published stories with headlines to the effect that restrictions on radio 'hams' had been relaxed.

"Some dear old ladies apparently read only the headings, didn't stop to consider what the 'radio' reference could mean, and immediately besieged their butchers with a demand that he wrap up a ham for them without benefit of coupons and no more fooling about it.

"The Prices Board asserts, just in case anyone doubts the butcher's word, that authentic ham of every type is still rationed."

QST March 1946 page 33

There were various letters given to various days: VE (Victory Europe) meaning the day the allies gained victory over the enemy in Europe. VJ (Victory Japan) meaning the same for Japan. One amateur radio operator claimed November 15th, 1945, should be given the title VA day. He probably would have received a lot of votes on that. The Dominion of Newfoundland had their VA day one month earlier on October 15th, 1945.

HARC 1945 – 1946

The Maritime VE1 news portion of "The Month in Canada" section of QST was recorded by Eddie McLaughlin, VE1JH, when it appeared in QST the last part of World War II. Eddie started this item in the December, 1945, issue as "The opening meeting of the Halifax Amateur Radio Club, for the 1945 – 1946 season was held October 19th with twenty-nine members in attendance."

CONSTITUTION AND BY-LAWS OF THE HALIFAX AMATEUR RADIO CLUB

CONSTITUTION   Article 1

The name of this organization shall be the HALIFAX AMATEUR RADIO CLUB. Its purpose is to serve as a means of social contact among its members, to promote good-will, and to extend knowledge of thr radio art.

Article 2

Members shall include any person expressing a desire to become a member and stating their interest in Amateur Radio. Applications shal be submitted for approval to the Club for election of a majority vote at the next meeting.

Article 3

The officers of the organization shall consist of the following; President, Vice-President, Secretary and Treasurer. These officers together with the Chairman of any Standing Committee shall act as the Executive Committee.
Officers shall be elected at the regular meeting in the month of January in each year, for the term of one year. They shall take offi at the meeting in the month of February following their election and shall continue in office until their successors are elected and installed.

Article 4

The regular meeting of this organization shall be held on the third Friday of each month, except for the month of December.
A special meeting may be called at any time by the Executive Committee.

Article 5

This Constitution may be amended by a two-thirds vote of the licensed amateur members present at any meeting, the proposed amendment having been previously advertised by sending a copy to the last address of record of each paid-up member at least two weeks previous to the date of the meeting.

BY-LAWS   Article 1

BY-LAWS   Article 1

Section 1   The PRESIDENT shall be a licensed amateur and a member of the ARRL. He shall preside at all meetings of the Club and of the Executive Committee. He shall perform all other duties pertaining to his office.

Section 2   The VICE-PRESIDENT shall perform the duties of the President during the latters absence. The Vice-President shall also be responsible for making all arrangements for meetings places approved by the Executive Committee.

Section 3   The SECRETARY SHALL keep the minutes of all meetings, prepare and mail notices of meetings, maintain a mailing list, receive applications for membership, receive all monies due the Club and perform such other duties as shall be assigned to him by the Executive Committee.

Section 4   The TREASURER shall receive all monies from the Secretary and deposit them in a bank approved by the Executive Committee. He shall make all disbursements after they have been approved by the Executive Committee. He shall submit a report of all receipts and disbursements at each meeting.

Section 5   The EXECUTIVE COMMITTEE shall be responsible for all matters of policy of the Club. It shall review all reports of conduct unbecoming a member of the Club, and if such reports are, in its opinion, sustained, shall submit the case to the body of the Club at a regular meeting

Page 2

CONSTITUTION AND BY-LAWS OF THE HALIFAX AMATEUR RADIO CLUB

BY-LAWS   Article 2

Section 1   The name of a candidate shall be proposed by a licensed amateur paid-up member and submitted for approval. The name shall be brought before the Club by the Secretary at the next regular meeting. A vote of the Club shall be taken and with a majority the candidate shall be declared elected to the Club

Section 2   Licensed amateur members shall have the privilege of voting providing that their dues are not in arrears more than three months

Section 3   A member charged with conduct unbecoming a member of this Club may be expelled by a two-thirds vote of the licensed amateur members present at a meeting if the Executive Committee has found the charges are sustained by its investigation

Article 3

Dues of this Club shall be Three Dollars ($3.00) per year, in advance, payable at the January meeting

Article 4

Section 1   Five licensed amateur members shall be considered a quorum at any meeting, any special meeting having been previously advertised by mail or telephone

Section 2   A majority of the Executive Committee shall constitute a quorum for an Executive meeting

Article 5

Section 1   Roberts Rules of Order shall be used when questions of parliamentary procedure arise

Section 2   Special Committees may be appointed from time to time as the need for them arise. They shall be appointed by the President and shall meet the approval of the majority of the Executive Committee

Section 3   The Club shall not attempt to govern the action of its members in the operation of their amateur radio stations but shall co-operate by offering assistance in any reported case of violation of operating regulations

Article 6

It shall be a permanent policy of the Club to support the American Radio Relay League. This shall be expressed by affiliation therewith and by participation in League activities

Article 5

Section 1   Roberts Rules of Order shall be used when questions of
            parliamentary procedure arise

Section 2   Special Committees may be appointed from time to time as
            the need for them arise. They shall be appointed by the
            President and shall meet the approval of the majority of
            the Executive Committee

Section 3   The Club shall not attempt to govern the action of its
            members in the operation of their amateur radio stations
            but shall co-operate by offering assistance in any
            reported case of violation of operating regulations

Article 6
            It shall be a permanent policy of the Club to support the
            American Radio Relay League. This shall be expressed by
            affiliation therewith and by participation in League
            activities

Article 7

Section 1   The By-laws may be amended at any regular meeting of the
            Club by a two-thirds vote of the licensed amateur members
            present, the proposed amendment having been previously
            notified to members by mail at least two weeks previous
            to the meeting

Section 2   These By-laws may be temporarily suspended or altered in
            case of emergency by a majority vote of the Executive
            Committee subject to the approval of two-thirds vote of
            the licensed amateur members at the next regular meeting

HALIFAX, N S
    March 15, 1946   re-drafted from original dated May 18, 1934
                                        amended Sep 28, 1934

HARC Files

This was recorded on two pages of legal-size paper and too large for my scanner so I had to reproduce it in four sections. The top and bottom of page one and the top and bottom of page two. You will note some duplicates but better than missing something.

CANADA MATCHED U.S.A. ON FREQUENCY BANDS

The Canadian Department of Transport matched the United States Federal Communication Commission in the matter of frequency band openings. Canadian amateurs were permitted on 80-meters (from 3500 to 3700 kilocycles) on April 1st, 1946, with a maximum power output of 50-watts, and subject to the condition that there was no interference to U. S. military radio services. It took some time to get the United States military to release all the amateur frequencies but Canada went along with the United States.

CANADIAN AMATEUR OPERATORS CERTIFICATE

The Canadian Amateur Radio Operator License was but one grade, the Amateur Certificate of Proficiency in Radio. It remained in force indefinitely or until a new examination was required by a change in law or treaty. All Amateur <u>Station</u> Licenses, containing the call sign, expired on March 31ˢᵗ each year and had to be renewed for the next 12 months.

U.S AMATEUR CALL SIGN AFTER WORLD WAR II

The ARRL was trying hard to create a new amateur call sign for the United States after World War II. They did not like the idea of a six-character call such as WA1ABC. They felt five, such as W1ABC was plenty long enough. They knew they had to make a change from the pre-World War II assignments. We know they had to go to the six-character call in 1958.

CANADIAN AMATEUR CALL SIGNS 1946

The ARRL announced the Canadian split in the Western Provinces in the December, 1945, issue of QST. On April 1ˢᵗ, 1946, the Canadian amateur call sign prefix was changed as follows:

VE1 New Brunswick, Nova Scotia and Prince Edward Island
VE2 Quebec
VE3 Ontario
VE4 Manitoba
VE5 Saskatchewan
VE6 Alberta
VE7 British Columbia

As one can see this is a big change from the 1919 list. Quite a few in the western provinces and territories were able to keep their two-letter suffix. For example, Jim Spilsbury was VE5BR before the war and became VE7BR from this division of the call signs. Jim started his radio manufacturing business in 1941 and retained the VE7BR call for the rest of his life.

The only exception to that statement is the RI suffix in the call sign. Anyone holding that suffix before the war had to get a new call sign. The RI suffix was held for Radio Inspectors. VE1RI is not recorded in the 1949 VE1 Callbook and the RI suffix is not recorded in any of the Canadian calls in the 1963 callbook.

VE8A-L   Yukon
VE8M-Z   North West Territories

I do not know what to make of this VE8 split in the call signs. When I passed through Edmonton on my way to the Yukon in June, 1963, I exchanged my VE1AGN call sign for a Yukon call. I cannot remember the VE8 call I requested but was told it was not a Yukon call. I settled on and received VE8RM; the "Red Monkey". The 1963 Callbook does not agree with any split in the VE8 call sign because there are Yukon and North West Territories stations throughout the complete VE8 block. I have to agree there was a split of some description, at least in the Radio Inspectors office in Edmonton.

VE9 Experimental

The VE9 call sign no longer communicated with the other amateur stations. This call was in use by various organizations for their own experimental purposes. One of the local electronic companies here in the Halifax area had call sign VE9AEC if I remember correctly, but I have no idea what use they made of it. I heard it on the marine frequencies so assume they were experimenting with some marine equipment.

VE0 amateur station in a Canadian ship

The VE0 prefix was first assigned in 1954. The first vessel assigned this prefix was *HMCS IROQUOIS* with call sign VE0NA. This was our first *IROQUOIS*, the tribal class destroyer built during World War II. A lot of my amateur radio operating has been with the VE0 prefix. To be truthful about it the call makes little sense to me and Canada is the only nation to my knowledge that has a special prefix for this use.

MARITIME MOBILE AMATEURS

This meant the Royal Canadian Navy was ahead of the United States Navy by two years. According to ARRL in the October, 1956, issue of QST the first U.S. Naval ship to receive permission to operate an amateur radio station was *USS ELDORADO* in 1956. There was no special prefix for these calls. The operator had to use his own call and stroke it with the oblique stroke and add MM for Maritime Mobile. The first was W0TWT/MM but they did not state his name. One would find that in the Callbook.

In 1957 the United States Navy Department gave permission for eight naval and coast guard ships to operate amateur radio stations. All eight were headed for the Arctic area in connection with the Distant Early Warning Line (DEW line). The operators in all eight used their own call sign with the MM after the oblique stroke. Once stationed in the Arctic MM was replaced with the Alaska amateur radio prefix KL7. One was the *USS ELDORADO* only with W4CMP/MM this trip, according to QST for August.

The February, 1959, issue of QST has a few photographs of the first American navy amateur operators. It was in 1959 that the United States permitted the rest of their vessels to operate maritime mobile on various bands in various areas of the world. The operator used his own personal amateur radio call sign with the oblique stroke and the letters MM, or Maritime Mobile spoken after their call sign in phone.

In 1962 I received a proper amateur radio license for a Canadian ship, VE0MO, but I was only allowed to transmit from 14,000 to 14,250 kilohertz in the 20-meter band and all of the 15- and 10-meter bands while on the high seas or in international waters. My rig was a Collins KWM-1 that operated on CW and USB (upper sideband) only and on the 20-, 15- and 10-meter bands only. 14,000 to 14,100 was CW only. The U.S. operators were not allowed below 14,200 on phone. There was a gentleman's agreement among the amateur community that sideband would stay out of the 50 kHz section from 14,200 to 14,250 kHz.

"Fred understands what they're saying since he converted to single sideband"

QST, May, 1955

This is the top of a Walter Ashe Radio Co. advertisement on page 127. The Walter Ashe Radio Co. joined the Dodo Bird and a lot of other things many years ago.

AM was still the main phone operation back in 1962 and I can still remember some of the old timers telling some of those with the "Donald Duck" equipment to take a hike. They refused to communicate with them. The editor of QST describes this in detail in his editorial of the October, 1960, issue. In the end I simply hung out around 14,130 kHz while on phone and any U.S. stations that wanted a contact worked me in CW. This refusal to participate in progress, that I assume is the proper terminology, was not new in amateur radio. When CW first appeared and was all the rage with those who were moving on, a lot of the old-time hams used to put "The Spark Forever" on their QSL cards. I believe this is called "Bucking the System" and no doubt one could find a little of that in everything imaginable.

The Department of Transport was the organization in Canada in charge of these things in 1965. They made a change with the amateur stations in ships on the high seas so that they could operate from 14.0 to 14.35, 21.0 to 21.45 and from 28.0 to 29.7 megacycles. That would have made life much easier for me signing VE0MO three years earlier in 1962.

C. E. Young, Lunenburg

VE0MO

The black line from the top of the mizzen over to the top of the main can be seen in this photograph.

That is the top part of the Marconi inverted L main antenna.

When the VE0 prefix was first assigned it was split up so that VE0NA to VE0NZ were warships. VE0MA to VE0MZ were the ships that were not warships. It was not long before this became a mess. Somehow a ship on the East Coast was assigned the same call sign as a ship on the West Coast. Therefore, the naval vessels on the East Coast were assigned VE0NEA to VE0NEZ and the West Coast naval ships were assigned VE0NWA to VE0NWZ. I find it hard to believe anyone knew what took place with the VE0M prefix at this time. When I went aboard *CCGS TUPPER* as Radio Officer in 1973 she had already been assigned amateur call sign VE0MBC and I simply renewed it. I held amateur call sign VE1AMK at the time and received the VE1BC call in January, 1975. If nothing else VE1BC worked well with VE0MBC.

VE0MO is the call letter of H. M. S. *Bounty*, the three masted sailing vessel built expressly for the film, *Mutiny on the Bounty*, as the ship sails on her present voyage to a number of the leading ports of the world. The *Bounty* began her tour in Vancouver, British Columbia, then went to Victoria, then to Seattle, then to San Francisco. She left the latter city on July 9th and is sailing to the east coast via the Panama Canal. After stops in New Orleans, Miami and possibly Washington, D. C. she will proceed across the Atlantic Ocean, stopping in London and paying visits to several English Channel ports. She will return to New York in November just prior to opening of the new film.

Of special interest to radio amateurs, is the ship's radio operator, "Spud" Roscoe, who is operating the KWM-1 on board. QSL cards will be issued to all amateurs making contact with the *Bounty*. Roscoe will be on the air at various times, on 20, 15 and 10 meters, sideband only.

Those making contact are to send their cards to Metro-Goldwyn-Mayer Studios Amateur Radio Club, Culver City, California, from where they will be forwarded directly to H. M. S. *Bounty* in whatever part of the world she happens to be.

Page 67 CQ November, 1962

I am very proud of this VE1BC call sign. When Bill Bligh became a silent key in May, 1965, his XYL, Evelyn, traded her VE1OW call sign for his VE1BC call sign and held it for ten years. Bill had done a lot for those of us sailing in *HMS BOUNTY*. Bill had a chart in his store window in Halifax where he kept a record of our daily position. He would have our wives over so they could talk to their spouse on the ship via amateur radio.

I remember it well!

HMS *Bounty*, the three-masted sailing vessel built expressly for the film *Mutiny on the Bounty*, has left Vancouver, British Columbia, and is proceeding to the east coast of the United States via the Panama Canal, with stops along the way. After a trip across the Atlantic to England, she will return to the States in November. VE8MO is operating aboard, on 20-, 15-, and 10-meter sideband and c.w. If you work the *Bounty*, send your QSL to the Metro-Goldwyn-Mayer Studios Amateur Radio Club, Culver City, California. "Spud" Roscoe is the operator.

Page 61 QST September, 1962

When we "paid off" *BOUNTY* all the crewmembers chipped in and bought a nice engraved silver tray for Bill and Evelyn. The ship's purser, the late Brian Backman was instrumental in the wording of the engraving. I no longer remember the exact wording but I mentioned this tray to Evelyn a few years before she became a silent key. At the time she was living in an apartment and stated the tray was stored with the rest of their possessions and she had no idea where it was. I was hinting around to see if she would donate it to HARC because it is another item that I feel should be held in the club room. There are many pieces of silverware out there that mean nothing today to anyone but the members of HARC. A nice glass cabinet holding these items would mean a lot to HARC. The HARC members that received these items were the ones who made this club what it is and has been.

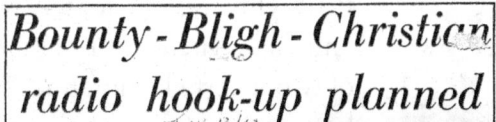

Bounty - Bligh - Christian radio hook-up planned

—Across 7,500 miles

By LYNDON WATKINS

An attempt will be made early today to arrange amateur radio hook-up spanning more than 7,500 miles from Tantallon, Nova Scotia to the tiny South Pacific island of Pitcairn.

If it is successful another chapter in the saga of the famous Mutiny on the Bounty will be completed. For the first time since the decks of the three-masted brigantine echoed to the sounds of rebellion, a Christian will speak to a Bligh. And their conversation will take place via the Bounty, the Lunenburg built replica of the famous British armed merchantman, now Panamabound, 250 miles off the west coast of Mexico.

NO CONTACT

The meeting by radio became a possibility yesterday when William Bligh, a descendant of the celebrated captain, who is a leading Halifax radio and television engineer, talked to the 23-year-old radio operator on the Bounty, Spurgeon George Roscoe of Kentville.

During their conversation Mr. Roscoe said he had been trying to raise a Fletcher Christian — direct successor

Mr. Bligh at his radio VE1BC

of the mutineer — living on Pitcairn. He operates an amateur radio station there, but as yet no contact has been made with the ship.

An attempt to bridge the distance between Mr. Bligh's home and the 2-square-mile Commonwealth island 3,270 miles east-south-east of Tahiti, was to be made early today when short wave radio reception in the area is at its best. Mr. Bligh said he hopes to arrange a three-way conversation between himself, Mr. Christian and the Bounty.

"It will certainly be quite

an historic moment," he said. This will be, Mr. Bligh thinks, the first time that a member of the two families have spoken to each other in modern times.

BURNED

Fewer than 250 people live on Pitcairn. Most of them are directly descended from the original seamen who followed Christian in his bid to set up an ideal "share - and - share-alike" society.

The mutineers ran the Brit-

—Please Turn To Page 6, Col. 3—

Bounty-Bligh-Christian hook-up planned

(Continued From Page One)

ish warship ashore and burned her before casting Captain Bligh and 17 members of his crew adrift in an open boat. Forty - one days later they reached Timor, an island 3,618 miles away in the Dutch East Indies.

Accounts of what happened to the islanders in the years immediately following the mutiny differ but today there are a considerable number of families on the island with the name Christian.

FAMILY LINK

Halifax businessman William Bligh — named after the famous captain — claims a family link with his eminent ancestor who following his return to England became an admiral and later governor of New South Wales.

Mr. Bligh has traced the

first Nova Scotian connection with the family to a Thomas Bligh who came to the province from Philadelphia in 1729 or 1800. He settled at Lakeville near Kentville, and married Margaret Foot in 1806. There was a connection between Thomas and the Blighs of Cornwall, England, from whom Captain Bligh originated. "I think he was probably a cousin," Mr. Bligh said last night.

The Tantallon man was the first "ham" to speak with Mr. Roscoe aboard the Bounty from his native Nova Scotia.

LOW POWER

"I picked him up on the 20-metre band on 14,130 kilocycles. They are using fairly low power but I got him on a single side band, which is a fairly new development in shortwave broadcasting and a big improvement on the previous AM reception.

"I could only talk with "Spud" — Mr. Roscoe — for a short time as he is being called by amateurs from all over the world. A Canadian licensed station aboard a sailing ship is something of a novelty. Having to be on watch on his normal duties and also operating the amateur set must keep him pretty busy," he said.

A native of King's County Mr. Roscoe spent five years in the RCN before spending a year at the Radio College of Canada in Toronto. He obtained the six-month assignment as radio operator aboard the Bounty shortly after completing this training.

Married a year ago, his wife is a nurse at the Kentville Sanatorium. "It was wonderful to hear of him. The last news I had was when the ship was in San Francisco," Mrs. Roscoe said yesterday.

Contacts Bounty But Fails To Get Pitcairn

An attempt to arrange an amateur radio hook-up between Tantallon, Nova Scotia, and Pitcairn Island—a distance of more than 7500 miles—was unsuccessful last night, reported William Bligh, descendant of the celebrated captain of the Bounty.

Mr. Bligh, a well-known Halifax radio and television engineer, has been in contact with the Lunenburg-built replica of the Bounty which is now off the Mexican coast enroute to the Panama Canal.

He talked with the radio of-

ficer aboard the Bounty again last night. "He couldn't raise Pitcairn either," said Mr. Bligh, "I guess they must take life pretty easy on the island."

Spurgeon Roscoe, Nova Scotian radio officer of the Bounty said the ship was somewhere in the Pacific off Mexico and expected to be in the Panama Canal in twelve days time. All aboard were in good condition.

Mr. Bligh hopes to raise Pitcairn Island this evening. By strange coincidence, the radio operator on that island is Fletcher Christian, a direct successor of the mutineer who led the revolt against Captain Bligh. The present Mr. Bligh hopes that radio links will be possible this evening on a three-way basis—between Tantallon, the Bounty, and Pitcairn Island.

HARC Files

The Fletcher Christian in that article was Tom Christian, VR6TC, and he was CW only and his transmitter was crystal controlled on 20-meters. He worked "the pileup" 10 kilocycles the other side of his fixed frequency. I had no way of contacting him because the KWM-1 would transmit on the frequency it received only. I managed to work Floyd McCoy, VR6AC, with a lot of help from Laurie Parkhurst, VE7IT, on 14130 kilocycles. Bill Bligh, VE1BC, said he could only hear enough of Floyd's transmission to know someone was on frequency. He was unable to work Floyd and Floyd could not hear him. Both were using a beam and Floyd's had an "Armstrong rotor". Floyd went out and adjusted it in the rain but it did not help. There was a McCoy in the original mutineers but I forget his position on the original *BOUNTY*. The mutiny took place in 1789 and when the mutineers were found in 1808 Adams was the only one left. The rest had killed each other off fighting over the few women they had taken from Tahiti. They changed the prefix of the Pitcairn Island call sign from VR6 to VP6 for some unknown reason.

Joan Roscoe

It was rather cramped in the radio room in *BOUNTY*. This is me operating the Collins KWM-1. The speaker is above the desk to the right. The RF went from the rig to a tuner mounted above the desk to the right and on to the main antenna switch. I had a choice of two inverted L antenna. They both were from the mizzen mast over to the main mast. The main antenna crossed at the top of the masts and can be seen in the photograph above. The top of the main mast was 118 feet above the surface of the water and I fed the vertical portion of this antenna on the starboard or right side of the ship. The other antenna went up on the port side and crossed at the main yards. This one crossed at the top of the lower square sail on the bottom of the main mast. The main antenna is the one I used the most. The other was known as the emergency antenna and was used mainly for the small low power crystal tuned 2-megacycle

AM radiotelephone. The range of this radiotelephone was more than a megaphone by a few miles only.

The metal cabinet to my left is the power supply to the main radio station that is right behind the camera. The closet above this is my clothes closet, the one where the bottle of perfume blew up that I was bringing home to my wife Joan from France. The engine room was just below this and it could get a bit warm. With lots of heat from the main movers (the main engines), the duty generator, plus the heat from the power supply the perfume simply "blew its cool". I was very fortunate and simply had the glass from the bottle to clean up. I had the best smelling radio room anywhere for a while. But it did not stain the three nice British wool skirts and the three dolls I thought it might. The dolls were of native dress. One was British, a French one and my favourite the Spanish flamingo dancer I purchased in the Canary Islands. One had to pacify Joannie so she would tolerate this foolishness of going to sea.

The key to the right of the filing cabinet in the lower right corner of the photo looks like the key to the radio room seven-day clock. We wound them every Sunday in those days; there were no battery-operated clocks. As a point of interest one of those old mechanical clocks I sailed with lost a few seconds north of about 32 north and gained a few seconds south of the same line or vice versa each day. I still say the magnetic field of the earth had something to do with that. We found it rather interesting and it gave us something to talk about.

Joan Roscoe

These are some of the QSL cards received while sailing in *BOUNTY*. The captain enjoyed showing these cards to various visitors. This is just above the radio officer's bunk and the VE0MO station is just to the right in this photograph. The main radio station is just behind the camera.

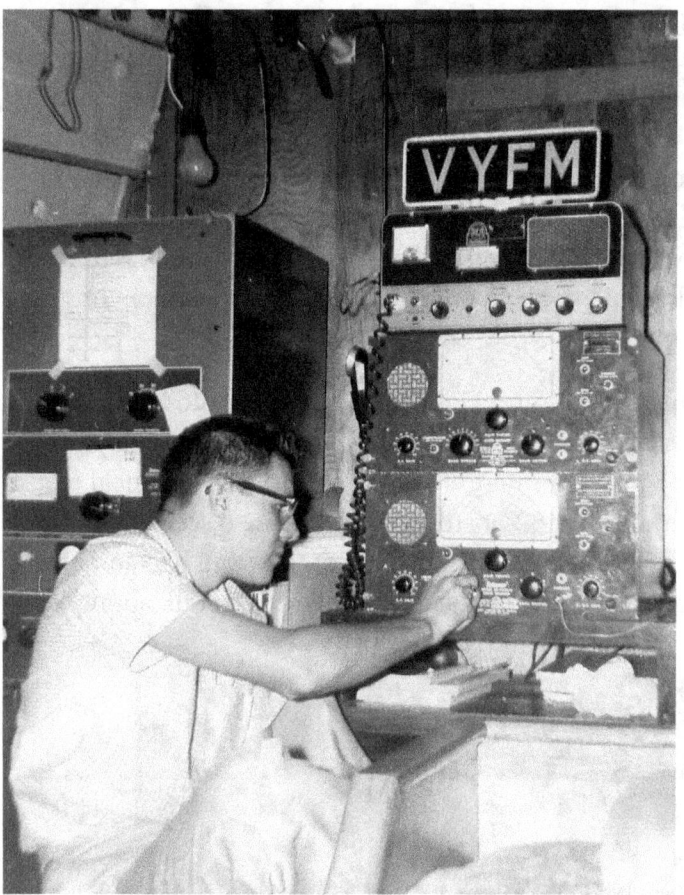

Joan Roscoe

It is too bad that I am in this photograph but I am lucky to have it. My wife Joan took this from the radio officer's bunk with her back to the QSL cards. This is the main station with the MF transmitter on the left. Two identical marine receivers one mounted on top of the other with the low power 2-MHZ radiotelephone on top of the two receivers. The HF transmitter is on the right just out of the photograph.

With the exception of the AM radiotelephone this equipment is from the RCA 5U Marine console of the 1950's. The MF Transmitter operated from 350 to 515 kilohertz, 250 watts of CW or Modulated CW (MCW). The HF transmitter operated from 2 to 24 megahertz, 300 watts of CW only. The two Superheterodyne receivers operated from 85 to 560 kilohertz and 1.9 to 25 megahertz. The two-megahertz radiotelephone had eight crystal-controlled channels of amplitude modulation (AM). This involved sixteen crystals. One for each receiver frequency and one for each transmit frequency. With this type of radiotelephone, we did not seem to have the proper channel for whatever it was we wanted to work. I would often receive the station we wanted to work on one of the main receivers and let that station choose which of the transmit frequencies that suited them. This was a typical "rock bound" or crystal-controlled station of that era including the AM broadcast band missing from the main receivers.

This is the MF transmitter and the main operating position of the main station. The Collins KWM-1 is to the left just out of the photograph. That is the speaker for the KWM-1 just above the typewriter. I had the light bulb above this transmitter for test purposes. If one held the key down on the MF transmitter, using that light bulb as part of the transmission line, it was so bright you could hardly see when *BOUNTY* was on an even keel. When she rolled over to one side, I saw her go far enough that there was no light in that bulb. This was the effect of the wooden hull with the sheet of copper on her bottom as the electrical ground. The wire dangling is hanging from the main antenna knife switch mounted on the deckhead (ceiling). That wire may be from the "homebrew" antenna tuner for the KWM-1.

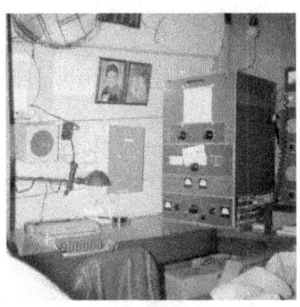

Spud Roscoe

In 2009 one can request any suffix for a VE0 call sign prefix and this has been the case for years. When the VE0 prefix was first assigned it was to be assigned to ships that transited international waters. One can see from the present assignments that this is no longer valid.

VO Newfoundland and Labrador

When Newfoundland joined the Dominion of Canada in 1949 their VOA-VOZ block of international call signs became part of the Canadian international block of call signs. The Newfoundland amateurs continued to use VO, one digit and a one letter suffix as late as 1957.

THE VO CALL SIGN

On April 1st, 1957, Newfoundland Amateur Radio Stations were assigned a two-letter suffix with the following prefix:

VO1 Newfoundland

VO2 Labrador

Jack Willis, VO2NA, reported in the Maritime News for July, 1957, that they were recovering from the confusion of the reassignment of the call signs. Ernie Ash became VO1AA with this change. Ernie is recorded in the silent key column for February, 1974.

AMATEUR OPERATING 1946

"Ham" operator ranges from Persia to Hawaii

Radio Experimental Station VE 1 ID, operated by D. Boyd Burgess, Berwick; is being heard in many countries, both east and west. Since January 1, when he went on the air after having re-assembled his equipment which had been in disuse under war regulations, Mr. Burgess has contacted 29 different countries and carried on conversations with operators of various nationalities. Most recently, he has been in contact with operators in Persia, Hawaii, Greece, Italy, England and Scotland; and listened to signals originating from New Zealand, Australia, Guam, Phillipine Islands, Okanawa and also American marine boats in the Mediterranean and Pacific oceans. Mr. Burgess, a member of the American Radio Relay League, says currently only one radio band is available to Canadian operators, but more will be released as post-war conditions normalize.

The Berwick Register April 1946

Boyd Burgess, VE1ID, worked these stations on 10-meters.

Ham activities returned to normal again in 1946, and a Hamfest was held on the Labour Day weekend of that year.

Thanks to Don Watters VE1BN

This is a photograph of the first post war Hamfest taken at Palmer's Lodge, Bedford, Nova Scotia, in 1946. The gentleman with very little hair in the front centre is Ed Tilton, W1HDQ, from the ARRL.

Ed Tilton, W1HDQ, conducted the monthly QST article "The World above 50 Mc".

A number of his antenna arrays and pieces of equipment were featured on the front covers of QST.

Sam Harris, W1FZJ, replaced Ed on this article in QST in October, 1960.

Ed is recorded in the Silent Keys Column for June, 1997.

The young lad 2nd from the left in the very back row is Don Watters, VE1QG, later VE1BN and now a silent key.

The following is recorded on page 116 of the November, 1946, issue of QST.

QST November 1946

VE1FG was A. S. Watters from Bridgetown, Nova Scotia. Art Grant, VE1EP, won the DX trophy and Clarence Roach, VE1EA, from Windsor, Nova Scotia was runner-up. Eddie MacLaughlin, VE1JH, won the high-speed copying contest.

RCAF AMATEUR RADIO SYSTEM

The May, 1946, issue of QST describes the Canadian Air Force Amateur Radio System (AFRS). This was a system designed by A/C Keith Russell RCAF (retired), VE3AL. It involved a system of amateur radio stations across Canada to teach and keep radio operators up to date on Air Force procedures, in order to have trained radio operators in case of an emergency. One did not have to be a member of the reserve Air Force. All they needed was an amateur radio license and able to work both phone and CW. Most of the members holding supervisory positions in the system were former members of the RCAF. A. H. Keith Russell, VE3AL, became a silent key in 1960 and is recorded in the Silent Key column of the May, 1960, issue of QST.

ALTERNATE SCM AND DIRECTORS' CANADIAN DIVISION

Throughout World War II most states started listing an alternate SCM (section communication manager). Leonard Mitchell, VE3AZ, was listed as the alternate CGM (Canadian general manager). At the same time the Western Canadian Divisions were listed as Vanalta and Prairie but each province had their own SCM. Art Crowell, VE1DQ, was the one and only SCM for Maritime Division and there was no mention of the three provinces of Nova Scotia, New Brunswick and Prince Edward Island. After the war William H. Butchart, VE6LQ, replaced Leonard Mitchell, VE3AZ. This terminology was replaced with the title Director Canadian Division and Vice Director Canadian Division and VE2BE and VE6LQ continued to hold these positions. Reg Town, VE7AC, replaced Bill Butchart, VE6LQ, and Bill Savage, VE6EO, replaced Reg, VE7AC while Alex, VE2BE, was Director Canadian Division. Alex, VE2BE, was still Director Canadian Division in 1959 when Noel Eaton, VE3CJ, replaced Bill Savage, VE6EO, as Vice Director Canadian Division.

Alex Reid, VE2BE, moved up to Vice President of ARRL and this was first recorded in the July, 1960, issue of QST. Noel Eaton, VE3CJ, moved up as Director of the Canadian Division and this left the position of Vice Director vacant. It was vacant until late 1961 when Colin Dumbrille, VE2BK, became Vice Director.

Noel Eaton, VE3CJ, was a member of the Eaton family that owned and operated the chain of T. Eaton Company stores across Canada. He paid a visit to the clubs in the VE1 call area including HARC in January, 1964. He attended a regular meeting held by each club. He did a repeat of this visit in VE1 and VO1 Land in 1968.

Noel was first licensed as an amateur radio operator in 1937. He was the 6th president of IARU. He was president of IARU and also vice president for international affairs with ARRL. Noel retired in 1982 and this is recorded on page 54 of the May, 1982, issue of QST. Dick Baldwin, W1RU, from Maine replaced Noel as president of IARU and as vice president for international affairs with ARRL. Noel is recorded in the Silent Keys Column of the January, 1997, issue of QST.

There is a biography of Noel in the June, 1968, issue of QST. He had retired by then as President and General Manager of the Eaton Knitting Company Limited. He had lived and held call signs; VP5BP Cayman Islands and Jamaica, ZF1BP Cayman Islands, 6Y5BP Jamaica and G3SDA. He had held a string of official offices in both ARRL and the IARU by 1968.

Noel Eaton, VE3CJ, remained elected as Canadian Director ARRL for the 1970 to 1971 election year. George Spencer, VE2MS, replaced Colin Dumbrille, VE2BK, as Vice Director from the same election. Noel Eaton, VE3CJ, became a vice president of ARRL in 1974 and George Spencer, VE2MS, replaced him as Canadian Director. Howard Cowling, VE3WT, replaced George Spencer, VE2MS, as vice director of the Canadian Division. Howard, VE3WT, had a change of positions at his place of employment and felt his new position could present a conflict of interest so resigned. Ron Hesler, VE1SH, was appointed to replace Howard and finish out his term to January 1st, 1976.

Noel Eaton, VE3CJ, was made Director Emeritus of the ARRL in 1984 according to Canadian Newsfronts in June, 1984. Noel died on September 28th, 1996, at the age of 86. The April, 1998, issue of QST has a lengthy description of Noel and a good description of the dedication of the VE3CJ Memorial Station in the Legion Hall in Burlington, Ontario.

Ron Hesler, VE1SH, was elected as the Director of the Canadian Division and took over that position on January 1st, 1976, without a vice director. There had been a mail strike and the ballots for vice director were delayed in the strike. Bill Loucks, VE3AR, was elected after the strike when the ballots were able to be counted. Bill moved up to Canadian Division Director when Ron, VE1SH, resigned on July 18th, 1976. ARRL stated Ron resigned due to personal reasons but Holland Sheppherd, VE3DV, stated in his final Ontario News in July, 1976, that Bill, VE3AR, faced the same problems as Ron and one and all should get behind the league. I remember when this took place but I did not know the full detail on what was going on. I remember Ron, VE1SH, had been a member of the army during the war. It was Ron who created the Canadian Radio Relay League in 1976 that became the Canadian Radio Relay League Incorporated in 1979 and left ARRL in 1988.

Anything Canadian is pretty small in the greater scheme of things and we should have remained a division of ARRL. With that organization we were part of the biggest voice in amateur radio. The only thing I had against it was the fact they were trying to change the Canadian amateur radio licenses to resemble the American licenses. Why we have to have so many licenses in amateur radio is beyond me. No one wants to be part of something with more licenses and exams hanging over their head. We in the commercial radio world had two certificates; the second and first class. No one elevated themselves to the first class. There were three of us only who wasted the time, effort and money to operate radio in Canada with a first-class certificate in 1978. Actually, this record was so poor that after 1966 very few Canadian commercial operators had a license of any description. That is another story. I was one third of the first-class performance in 1978. To me a certificate of proficiency in Amateur Radio as it was in the beginning is more than sufficient. The part that scares one is the fact most members of Radio Amateurs of Canada (RAC) are members simply to use the service of their QSL Bureau.

In the September, 1976, issue of QST the ARRL stated Ron Hesler, VE1SH, had withdrawn his resignation and that he was reinstated as the Canadian Director and Bill, VE3AR, the vice director to finish their terms on January 1st, 1978. Ron's XYL Donna was VE1YX. Donna is recorded in Silent Keys for August, 1981. So, one assumes Ron had some personal problems at home as well. Ron became a silent key on November 12th, 1987 while CRRL Atlantic Region Director. He had remarried and his second wife was Ellen.

Art Crowell, VE1DQ, was SCM for the Maritime Division until replaced by the election that voted Doug Johnson, VE1OM, the position of SCM on February 15th, 1954. Art's last Maritime Division report was the one printed in the April, 1954, issue of QST. This was a record of over 25 years continuous service and an all-time record at the time in ARRL. HARC presented Art and his XYL with an engraved silver tray in a gathering to celebrate this achievement. Shortly after Doug, VE1OM, was elected SCM Fritz Webb, VE1DB, was made assistant SCM.

A photograph of Art Crowell, VE1DQ, at his station and a biography on him can be found on page 10 of the February, 1955, issue of QST. Art became a silent key in 1962. This is listed in the January, 1963, issue of QST.

They had a surprise testimonial dinner for Alex Reid, VE2BE, in Montreal at the Ritz Carlton Hotel on March 14th, 1950. There were 65 amateurs in attendance and Alex recounted some of the highlights of his twenty years as Canadian General Manager at ARRL Headquarters. ARRL stated that Alex was the "dean" of the ARRL board of directors at that time. Alex, VE2BE, and his XYL paid the local gang a quick visit while on a motor tour through VE1 land in 1953.

The first listing I found for Alex was as 2BE in March 1923. He was listed then as an old ARRL man and a government radio inspector and when not too busy took a hand at a key. He was made manager of the Quebec division and first appears in that capacity in May 1926. He held three amateur radio call signs in 1926; 2BE, 3BE and u2CHK. Alex was still recorded as CGM in the November 1953 QST. Alex Reid, VE2BE, was made an honorary Vice President of ARRL in the July, 1966, issue of QST and had served as a Vice President for many years.

Alex Reid, VE2BE, was removed from the honorary Vice President position on the masthead to the Silent Keys column of the March, 1967, issue of QST. Alex had passed away on January 27th, 1967. QST ran a lengthy obituary complete with photograph on page 72 of the April, 1967, issue. Sympathy was relayed via the Maritime News section of the Canadian Division news on page 134 of the same April, 1967, issue.

NAVY AMATEUR RADIO CLUBS

There was a note in QST in the Maritime News of April, 1946, for all naval personnel interested in amateur radio to contact Chief Petty Officer Holland Shepherd, at the Communication School, *HMC Dockyard*. "Shep" Shepherd, VE1RR, was the first president of amateur radio station VE1HO when it started in the basement of the Communication School in 1946. He was also on the executive of HARC and president in 1954. Shep had quite a career and was very involved in Amateur Radio wherever he traveled. He was stationed at

Washington, D.C., and signing VE1RR/W3 in 1958. He was transferred to Ottawa in 1964 and assigned VE3EZY and this he changed for a two-letter call, VE3DV as soon as possible. Shep replaced Roy White, VE3BUX, as SCM Ontario in August, 1970. Shep was president of the Emergency Communications Advisory Committee (ECAC) and was replaced by W. H. Parker, VE5CU. The ECAC was a part of ARRL and an appointment by them. Shep became a silent key in 2007 while still holding the VE3DV call sign.

VE1HO started with a small homemade transmitter on 10-meters only and one of the famous RCA AR-88 receivers. VE1HO moved from the dockyard in March 1948, when the Communication School moved to *HMCS STADACONA*. There were two amateur stations, VE1RN and VE1HO in the same building; the Communication and Electrical building. VE1RN was the call sign of the Electrical School. When war surplus equipment became available the members of these clubs' started stations in their homes and the membership of the clubs declined to the point the VE1RN station closed.

VE1HO was retained and the transmitter was upgraded to a 500-watt Marconi PV500. The station was given its own room, Room 229 at this school. These were naval schools and the membership was constantly changing, but one of the mainstays was Sea Cadet Tom Clahane, VE1SP, an HARC member and president in 1960. Tom went on to graduate from the Nova Scotia Technical School in 1953. Chief Petty Officer Harold Jacques taught anyone interested in obtaining an amateur radio certificate. I have been unable to locate his call sign. VE1HO had their new rig going on 14 mc phone with a 4 – 125 final in the fall of 1950. One of the operators was Bill Murray, VE7YY.

When the Communication School moved from *HMCS STADACONA* to *HMCS CORNWALLIS* in September 1952 the VE1HO station moved with the school. The VE1HO station was there early 1952 before the official date of the opening of the school at *CORNWALLIS*. The VE1HO station received $600.00 from the ship's fund for materials and tools to build a transmitter of about 350 watts output. The receiver was a Canadian Marconi CSR-5. The station was "on the air" from 2030 to 2230 Mondays, and from 1930 to 2230 on Thursdays around 7040 kilocycles CW with an output of 150 watts CW only. Chief Petty Officer Jack Mooney was not only the club's 1st Vice President but the club's instructor. He taught both basic theory and radio principles to anyone not familiar with amateur radio. I have not found Jack's call sign.

Cornwallis Museum

This is station VE1HO at *CORNWALLIS* sometime in the late 1950's or early 1960's. I have been unable to learn the date of the photograph or identify those within. All three in this photograph are wearing the badge that was created around 1955 for the CR or Communicator Radio trade. This trade was renamed Radioman in 1960. The lad on the left is reading a copy of The Callbook. This book listed every amateur radio station by call sign and gave the name and address who was assigned the call sign. The advertisement on the back cover of the book is that of the RCA organization advertising their latest tubes. RCA placed these advertisements on the back of these books for several years. The receiver in front of the one wearing the headphones is a Canadian Marconi CSR-5. The speaker above the receiver is for that receiver. One can note the knobs to more equipment in front of the Wren – the female member. These knobs may belong to a Canadian Marconi PV-500 transmitter.

The three QSL cards just above the speaker with the VE1HO call break down to: W9WKU, Dewitt Jones, who became W4BAA at Captiva Island, Florida in the early 1970's. W6AWT is Bartholomew Molinari, San Francisco. And G2AND is R.H. Broadbent, Huddersfield, Yorks, England if I have read the call correctly because it is hard to read.

The photograph described above was taken at the time I served in the navy. Can you identify those within the photograph? They would be within a year or two of my age and the one with the headset was likely a licensed amateur radio operator. I was in the Communicator Supplementary branch that was renamed Radioman Special in 1960. Our badge was much the same except the solid knob where the wings of mercury join was an illustration to signify a medium frequency shipboard direction finder loop. We did not serve at the communications school in *HMCS CORNWALLIS*. We had our own school and headquarters in a cow pasture outside of Ottawa known as *HMCS GLOUCESTER*. Our amateur radio club station had call sign VE3GLO. This station was equipped with the Canadian Marconi PV500 transmitter and a Hammarlund SP600 receiver.

Those old receivers did not have a built-in speaker and one had to use a speaker as shown in this photograph.

HMCS GLOUCESTER closed in 1972 and the site has more or less returned to that of a cow pasture.

In the mid 1960's the members of the *CORNWALLIS* station VE1HO were very active with the Annapolis Basin Amateur Radio Club with call sign VE1VT. Chief Petty Officer "Moe" Lake, VE1PX, Petty Officer Terry Sullivan, VE1AOP, and civil service member Hal Surette, VE1ALL, were the leaders of this club. This club created many amateur radio operators and much interest in amateur radio in Digby County, Nova Scotia.

Another amateur radio station was installed in the Electrical School in 1952 at *HMCS STADACONA* with call sign VE1NN that operated on 14-mc phone mostly. The majority involved in these naval amateur radio stations at *STADACONA* were also members of HARC. In 1965 this VE1NN station was on the air from noon until 1 PM during the week only.

This Communication School for Radiomen moved to Esquimalt, British Columbia, in 1966 two years before the official date of unification of the Canadian Armed Forces. This unification must have taken over a decade. The first attempt was a tri-service guard formed in Ottawa for General Norris, USAF, and head of NORAD in March 1960, according to the

Naval Historian. I was the right marker for that guard, the biggest laugh most of us had during our military career.

NEWSLETTER

HARC has had a newsletter for many years. Binx Fisher, VE1AFN, was editor in 1948. Doug Johnson, VE1OM, was the editor from 1949 until 1952 and Brit Fader, VE1FQ, was the editor from January 1953, until December, 1986. This newsletter was known as "The Bulletin". The name of this Bulletin was changed to "The Reflector" in 1989 and the first edition was the September 1989 edition. The Reflector is published every month there is a monthly meeting. There is no monthly meeting in December, July and August. Lynn Bowser, VE1ENT, is the editor in 2009 and has been for several years. The issue for January 2008 was Volume 69 Number 1. Lynn VE1ENT is still the editor in 2021.

HOBBY SHOW 1948

The HARC booth at the YMCA Hobby Show was a huge success in early 1948. Brit Fader, VE1FQ, was busy on 14-Mc phone and Don Bain, VE1LZ, was busy 3.5-Mc CW. They passed a total of 54 messages using the VE1FO/1 call sign and some of these messages went as far west as Vancouver and Victoria, British Columbia.

CANADIAN REGULATIONS 1948

These are the Canadian Regulations created in 1948. These are the regulations that created the two classes of Amateur Radio Certificate; the Amateur Radio Certificate and the Advanced Amateur Radio Certificate. One could get a partial phone endorsement on the Amateur Certificate after six months as stated. These are the classes so many of us lived with until the present system. There was a big movement in Canada and the United States back then to keep any type of phone out of any amateur band below 25 mc. In other words, keep the entire amateur bands on HF below 25 mc as CW only.

CANADIAN REGULATIONS

The Canadian amateur regulations for the new license year beginning April 1st carry a 50-kc. expansion of both the 3.5- and 14-Mc. a.m. 'phone assignments and some similar expansions of n.f.m. assignments. The frequency assignments of Canadian amateurs, from 3.5 to 22,000 Mc., are precisely those of the United States and, as in this country, the full widths of all bands are assigned to A1. A0 emission is permitted only above 144 Mc. and not in the 11-meter band as in U.S. A2 may be used on 11 meters and above 50 Mc., and every amateur license carries rights to use A3 'phone and f.m. 'phone and telegraphy above 50 Mc. A4 and A5 are not permitted anywhere. Portable stations are permitted only above 28 Mc.

Except on special authority, Canadian amateurs may not use 'phone below 50 Mc. Upon special application, licensees who have been active for at least six months on frequencies below 29.7 Mc. may be authorized to use A3 and n.f.m. 'phone in the 11-meter band and in the 10-meter band above 28.2 Mc. Licensees whose stations have been in active operation at least a year and who pass an advanced 'phone examination and higher-speed code test can get special authority now to operate A3 in the frequencies 3750-4000 kc. and 14,150-14,350 kc., and n.f.m. 'phone in the ranges 3800-4000 and 14,150-14,250 kc. In n.f.m. the deviation must not exceed 3000 cycles. All 'phone stations operating below 50 Mc. must possess a visual means of indicating overmodulation.

QST Page 39 May, 1948

These are the rules for 1948. One has to agree it was a different world back then than the one we enjoy today from a lot of hard work by the ARRL.

U. S. HAMS CAN'T OPERATE IN CANADA

U. S. hams planning vacations in Canada have been asking us lately if they can get permission to operate their ham rigs in Canada on the basis of their U. S. amateur licenses. Sorry, OMs — no can do. Under the laws of Canada only British citizens can be licensed (just as under U. S. laws only U. S. citizens can get ham tickets). As a matter of fact, the Canadian Department of Transport is anxious that we stress this point to avoid disappointment to W/K amateurs visiting Canada. They also point out that all transmitting equipment in cars entering Canada is sealed by the customs authorities at the point of entry, and have suggested that particular care be exercised to see that the seal remains intact, since otherwise the vehicle is subject to confiscation.

While on this subject we might add that ARRL had some hopes of promoting reciprocal licensing arrangements between the two countries and to that end made formal representations to the Department of State last year. But the idea of noncitizen licensing was turned down at high level as not being in the national interest at this time, and that has killed the whole thing for the indefinite future.

Page 30 QST August, 1948

Canadian amateur stations were allowed to go mobile in late 1948 but one had to keep a radio log while mobile. Small log books for mobile use were soon made available by the various organizations. They also had to notify the radio inspector at the station's home address and the radio inspector at the location where they would be mobile or portable. They were allowed to go portable for one month per year only.

**PORTABLE/MOBILE IN
CANADA**

Just as we go to press, an agreement in the form of a treaty has been concluded between Canada and the United States clearing the way for eventual authorization for U. S. amateurs to operate while in Canada — and vice versa. After the agreement is ratified, and after the Department of Transport on behalf of Canada and FCC on behalf of the United States have promulgated any necessary regulations, licensed radio operators of three categories will be permitted to use their equipment in either country: amateurs, civilian pilots, and land mobile (highway radiotelephone) stations. Don't get your hopes too high for mobile in Canada this vacation season, however; ratification still has to be accomplished by the two countries — a process which will probably take several months, at least, in this country and possibly six months or more in Canada, according to CGM Reid.

Page 38, QST, March, 1951

This is 1951 so we are slowly gaining and as one can see there was a lot of red tape in order to get anything done back then.

HARC HAMFEST 1949

In 1949 Halifax celebrated its Bi-Centenary (1749-1949) and the Halifax Amateur Radio Club went all out on a celebration. Again, a Hamfest was arranged for the Labour Day weekend. This proved to be most enjoyable and successful, but the planning was on a rather elaborate scale that necessitated caution in the planning of future conventions. 250 attended this Hamfest.

1749 - HALIFAX BICENTENARY 1949

John Brown VE1DD

This is the program for the 1949 Hamfest and our thanks to HARC member John Brown, VE1DD, for providing this copy. This copy is now held in the HARC Library.

* * * PROGRAMME * * *

SATURDAY, SEPTEMBER 3

2.00 p.m. - 6.00 p.m. -
Registration. Convention headquarters, mezzanine floor.
Personal QSO'S, introductions, and display of technical
equipment in the Saloons.

6.30 p.m. -
Afternoon Tea for XYL's - YL's.
Dinner in the Ball Room with Holland H. Shepherd, VE1RR,
as Chairman.
Dinner music by Richard Fry, CBC artist, at the organ.
Roving Mic with Ron Hart VE1MZ.
Greetings - Doug Johnson, VE1PQ, President, The Halifax
Amateur Radio Club.
Greetings - Mrs. Sid Johnson, VE1WJ, President Halifax
Ladies Dit and Dah Club.
Tribute to "Silent Keys".

Civic Welcome - His Worship, Mayor Gordon S. Kinley, of
Halifax.
Address - Mr. George F. Harris, Dist. Supt. of Radio, Dept.
of Transport.
Address - Mr. Art Crowell, VE1DQ, SCM for the Maritimes.
Address - Mr. Alex Reid, VE2BE, Canadian Gen. Manager
of the American Radio Relay League.
Address - Mr. Byron Goodman, W1DX, Assistant Technical
Editor of QST, representing ARRL Headquarters.

Presentation Brown Holder DX Trophy by Alex Reid.
Presentation of the VE1GR Memorial Trophy by Alex Reid.
Drawing of Door Prizes.
15 - Minute Recess.

SUNDAY, SEPTEMBER 4

9.45 a.m. Code Contest.
10.30 a.m. Address by Byron Goodman, entitled "Single Side
Band Telephony".

11.30 a.m. Meeting of Maritime Phone Net in Salon, with
Harley Richardson, VE1IE, Chairman.

1.00 p.m. Gather outside Nova Scotian Hotel and proceed to
picnic grounds at head of St. Margaret's Bay. Races,
sports, contests, prizes.

MONDAY, SEPTEMBER 5

10.30. a.m. Talk by Officer in Charge RCAF (Sea-Air-Rescue
Centre)

12.00 Noon Break up of Convention.

There is a wealth of history here, especially around George Harris. It was he who alerted
the world to the Halifax Explosion in 1917 as the Warrant Officer Telegraphist in *HMCS
NIOBE*. Mr. Harris was made an Honorary Member of HARC at the December 16th, 1936,
club meeting. He was in charge of all radio activity in 1949 in what today are the Atlantic
provinces of Canada. All the old timers were glad to see "Old Joe" Fassett ex **C**1AR who
managed to attend this convention. Eddie MacLaughlin, VE1JH, won the door prize at this
convention, a Hammond 10-meter Beam.

Note the 10.30 a.m. Address by Byron Goodman, W1DX, entitled "Single Side Band
Telephony". There is a nice photo of Byron taken before this Hamfest on the front cover of
the January, 2008, issue of QST. The January, 1948, issue of QST was filled with Single Side

Band Telephony. The Editor stated in his editorial that it was the most significant development in amateur radio. There were not only several articles on SSB in this issue, but there were detailed instructions on several pieces of SSB equipment one could build for the Ham Bands. It was known as Single Sideband Suppressed Carrier (SSSC) for the first years it entered service. The first record of an HARC member using SSSC that I found was Tommy Baker, VE1SF. He showed up transmitting this mode on 80-meters in early 1951. Byron Goodman, W1DX, and Don Mix, W1TS, of the schooner *BOWDOIN* fame were Assistant Technical Editors at ARRL for QST.

HARC Files

TEENS OPERATING 1949

One of the news items of 1949 was a three-way contact planned between W3OVV, W9FZE and VE2TA. W3OVV had worked W9FZE and they were hoping to make contact with VE2TA soon. W3OVV was ten-year-old Jane Bieberman. Forty years later she was an attorney. Her husband was Frank, W6SWM, and son David, KB6JHT. She was still listed with the W3OVV call sign in 2008.

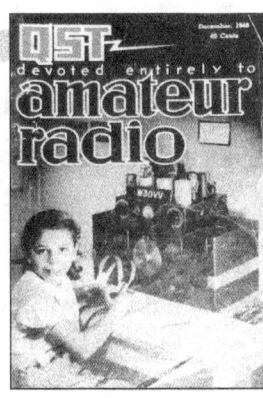

Then...and now. In December 1948, Jane DeNuzzo, W3OVV, appeared on the cover of QST (left) when she was a 12-year old amateur. In 1998 (above) Jane is still an active ham. You'll hear her during DX contests or, if you find yourself in the Gainesville, Florida area, on the local repeater.

Page 20, December, 1998, QST

W9FZE was nine-year-old Kent Lattig and for some reason they did not give Frank's last name, but he was eleven years old and held call sign VE2TA. All three could handle CW in excess of 15 words per minute. This is a record that should be recorded every time it happens with hopes of getting more young people interested in this fascinating hobby. The reason it was recorded was because the executive was starting to worry about the age of the average amateur radio operator. Both W3OVV and W9FZE had their photographs on the front cover of QST. The same story we hear today that most amateur operators are too old and there needs to be something done to encourage more young people into the ranks. It is hard to believe, but this took place in 1949. It may have started then and has simply continued on to the present time.

FIRST FP8 STATION

The first amateur radio station operated on the French islands of St. Pierre et Miquelon was W3BXE operating FP8AA in the summer of 1949.

WRCN 1951

The female members of the Royal Canadian Navy were known by the acronym WRCN and were called "Wrens" the same as their Royal Navy sisters. The first members were recruited during World War II and discharged at the end of the war. The Royal Canadian Navy started hiring female members again in 1951. The daughter of Ralph Pattison, VE1RP, was Kay Pattison who held regimental number WR8145 and was in the first class of WRENs at *HMCS CORNWALLIS* in 1951. She was the first post-World War II WRCN hired by the navy.

This first class was known as W1 and the next was known as W2. Both classes became members of my old trade in the navy, that of communicator supplementary. R. R. Pattison, VE1RP, was an active executive member of HARC and served as the club president in 1952. I managed to contact Kay's niece, a nephew and daughter but was unable to contact her direct. She was a "couple of stand ease's" ahead of me but I know some who served with her. I was hoping to get a photograph to include with this project. One of she and her father in her father's ham shack, especially with her in her sailor suit, but it was not to be.

The navy called it training but I am convinced it was a form of brain washing. We monitored the complete radio spectrum and had to identify everything we heard. There are those who still refuse to talk about it because we were brain washed into keeping our mouths shut. Those who were not in our trade had no idea what we were doing. Some of the guesstimates were worth a good giggle to say the least. They claim we were part of the National Research Council and simply camouflaged in the navy. It was one of the best jobs I had within the many years I have spent with radio.

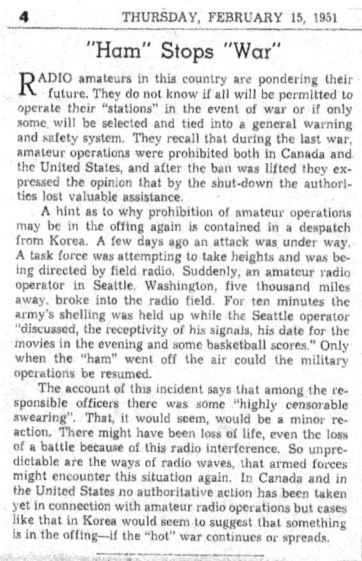

HARC Files

The Korean War no doubt had something to do with rehiring the WRCNS and this news-paper article makes that clear.

THE ULTRA HIGH FREQUENCIES

The Ultra High Frequencies and operating on these frequencies was big news back in 1940, but of course Canadian amateur stations were shut down for the war at that time. I find it rather amazing the amateur radio community was so much farther ahead of the com-mercial community. It appears to have taken several years for the commercial world to take over what had been created by the amateur community.

AMATEUR RADIO NETS

The term net was created by the Army and Navy for their system of communication back in the 1920's. The editor of the June, 1929, issue of QST states that the amateur radio opera-tors should look at this system of communication for their use. He felt it an excellent idea because it involved the one frequency only.

QST for March, 1940, listed a total of 3 pages of ARRL nets. They were listed as:

Name of Net,
Frequency,
Hours of Operation
Name and call sign of each member

At the end of this listing was another full page of ARRL Trunk Lines that were still in opera-tion. These were listed as:

Trunk Line A, B, etc.
Frequency
Stations and Routing
The operating hours of each line

In 1948 there were two local nets; the Maritime Net on Mondays at 7-PM on 3835 kilo-cycles and the Fundy Net on Tuesdays at 7-PM on 3803 kilocycles.

The Maritime Phone Net is listed as 3830 kilocycles in 1951 and in 1953 it was recorded as meeting nightly at 7 PM local time as a traffic net. One assumes it was still meeting on 3830 kilocycles. It was announced in March, 1954, that the Maritime Phone Net met daily on

3750 kilocycles at 7-PM local time. This has been the time and frequency it has met since with the exception of a few temporary changes.

OUTINGS AND PICNICS

There are a number of outings and picnics found in the records. A most successful and happy outing was held recently by the HARC under the direction of Brit Fader, VE1FQ, is recorded in the December, 1948, QST. One was held in 1953 at Bayswater Beach that was a big success. Don Bain, VE1LZ and A. E. Candy, VE1BF, and their parties arrived by motor launch. This was in addition to a social and lobster supper that included the wives and YL's at the summer QTH of Don Bain, VE1LZ, that same year. It is very hard to keep these outings and picnics sorted out because they are recorded so many times in so many ways.

CANADA WIDENS 75 METER BAND

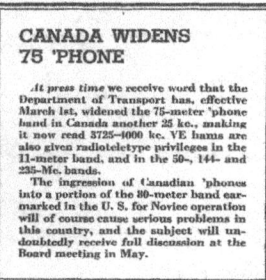

CANADA WIDENS 75 'PHONE

At press time we receive word that the Department of Transport has, effective March 1st, widened the 75-meter 'phone band in Canada another 25 kc., making it now read 3725-4000 kc. VE hams are also given radioteletype privileges in the 11-meter band, and in the 50-, 144- and 235-Mc. bands.

The ingression of Canadian 'phones into a portion of the 80-meter band earmarked in the U. S. for Novice operation will of course cause serious problems in this country, and the subject will undoubtedly receive full discussion at the Board meeting in May.

Page 21, April, 1951, QST

We would often hear CW novice stations working each other around the Maritime Net frequency of 3750 kilocycles at 7 PM the local net time. I do not remember it causing any real problem. The United States created the Novice and Technical class Amateur License in 1951.

RECIPROCITY AMATEUR RADIO LICENCE

CANADIAN RECIPROCITY

For quite some time now there has been hanging fire in Washington a treaty between the governments of Canada and the United States which, among other things, would permit the operation by amateurs of one country in the territory of the other. In mid-May, the U. S. Senate finally having ratified, diplomatic representatives of the two governments met to bring the treaty formally into force. This brings a step nearer the time when we in the U. S. shall be able to operate fixed, portable or mobile in Canada, and when our VE associates will be able to do likewise in our country. It appears now that only minimum additional regulation is required before the privilege becomes a reality — perhaps simply taking the form of specific notification blanks filed with appropriate officials. We'll have all the dope for you when available.

Page 32, QST, July 1952

FAX MAIL-STAR **19**
Friday, February 9, 1951

Remove Red Tape On Radio Use

OTTAWA, Feb. 8—(CP)—Canada and the United States today agreed to remove legal red tape, blocking the free operation of mobile radio transmitters and equipment in each other's territory.

The agreement, benefitting civilian aircraft, amateur radio operators and persons at border points with transmitters in their vehicles, was signed here by Transport Minister Chevrier and Ambassador Stanley Woodward of the United States.

HARC Files

This reciprocal agreement with the United States came into effect on May 15th, 1952, and for years we had to fill out FCC card 410 and wait for their approval to operate in the United States. I had to do it in 1962 while signing VE0MO from a Canadian ship in order to operate while in United States territorial waters as any other Canadian amateur station.

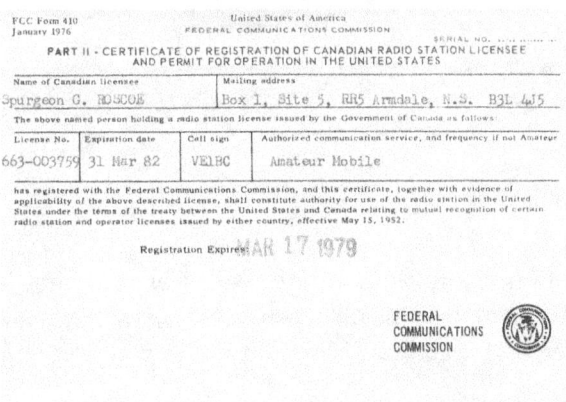

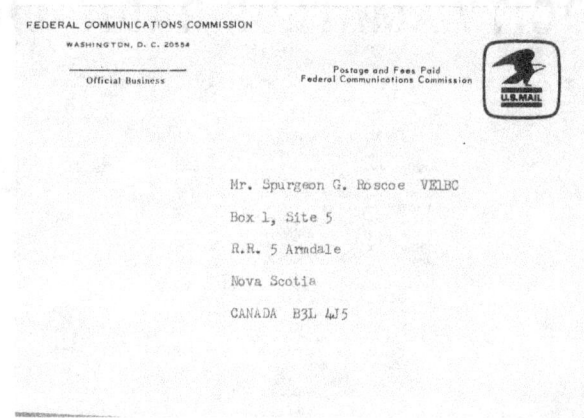

This is my copy of my FCC 410 in order to operate in the United States in 1979.

The November, 1978, issue of QST stated that the United States and Canadian governments had agreed to allow their CB/GRS operators and their amateur radio operators to operate in either country without permission. This did not become legal until January 21st, 1980, according to QST on page 8 of the February, 1980, issue. If you go south today make certain you have your license with you.

21-MC AMATEUR BAND

CHANGES IN CANADIAN REGS

In mid-July the Canadian government assigned to its amateurs the privilege of operating 'phone on the frequencies 21.2–21.45 Mc. As you will see elsewhere in this month's "Happenings," this is 50 kc. more than the request which ARRL has made to FCC for U. S. amateurs, thus following the usual pattern of Canadian 'phone suballocations extending a few kilocycles lower than U. S. assignments on each 'phone band. This suballocation is available to Canadians with "unrestricted radiotelephone privileges," a license similar to our Advanced Class.

Canada has also removed some of its mobile restrictions, dropping the requirement that mobile operation for more than one month's duration be reported to the district radio officer and doing away with the restriction that Canadian amateurs could not operate mobile in more than four months during any year.

Page 34, QST, September, 1952

Doug Johnson, VE1OM, and Ray Wilson, VE1WL, were the first two stations to operate on the 21-mc band in the VE1 call area and both were long serving members of the HARC executive.

40-METER PHONE

CANADIANS GET 7-MC. 'PHONE

The Department of Transport at Ottawa, we learn through Director Reid, in early January made 7200–7300 kc. available to Canadian holders of 'phone permits as of the 12th. This action has two significant aspects. One is that in effect the 40-meter voice band in Canada is made "Class A" — that is, requires the special 'phone license, just as 75 and 20 still do there, which of course now differs completely from FCC policy. The second is that, for the first time in our low-frequency bands, Canadian 'phones have an assignment identical with that of Ws, instead of the customary additional segment.

Page 34, QST, March, 1953

THE FIRST TRANSISTORS

Now where is that audio amplifier?

Page 115, QST, August, 1955

"I JUST FOUND MY MULTI-PURPOSE TEST INSTRUMENT! IT WAS HIDDEN BETWEEN A COUPLE OF PAGES IN THE INSTRUCTION BOOK!"

Page 19 QST, April, 1969

On February 13th, 1953, W2JEP and W2YTH made the first transistor to transistor contact. This was accomplished on 40-meters with a power input of 60 miliwatts. On April 1st, 1953, the Morris Radio Club in W2 land made radio contact with mobile units of this club within a ½ mile radius of this station. The station consisted of a two transistor AM crystal controlled transmitter and a two transistor super-regenerative receiver on 10-meter phone. The receiver was built by W2ZKE.

HURRICANE EDNA

Clyde Robbins, VE1DW, Yarmouth, and Brit Fader, VE1FQ, Halifax, were congratulated for a job well done in December, 1954. They had passed many messages on weather and damage reports from hurricane Edna via emergency power.

When I think of this hurricane the first thing I think of is the old apple tree trying to uproot itself next door to our home in the Annapolis Valley. I did not know an apple tree could bend that far without breaking off or coming out of the ground. The old tree survived the hurricane.

This is a Bob Chambers cartoon of the amateur activity in this area during hurricane Edna sent to me by Mike VE3GFN. This cartoon appeared on several HARC call books.

There were three hurricanes that more or less affected this area at that time. Hurricanes Carol, Edna and Hazel. Carol came up through Quebec. Hazel came up through Ontario and Edna came up through New Brunswick, the only one of the three to do damage in this area. There is excellent detail on the amateur radio emergency operations on all three starting with the January, 1955, issue of QST.

There was a good description of hurricane Camille that came ashore between New Orleans and Mobile, Alabama on Sunday August 17th, 1969, starting on page 56 of the January, 1970, issue of QST. They stated that hurricane was the most severe that had come ashore in the United States to date. Camille created a lot of destruction and left hundreds dead or injured from 190-MPH winds.

AMATEUR LICENCE 1950'S

The amateur radio license of the 1950's was of legal size and printed on both sides. It would not fit on my scanner and I have tried to copy it in four parts as follows:

DEPARTMENT OF TRANSPORT

Amateur Experimental
Station
1953 - 54

Call Sign V E 1 F O

CANADA

N⁰ 2488

"Licence to use Radio"

Issued in accordance with the provisions of The Radio Act, 1938, and the Regulations made thereunder.

```
T.V. Burton (For Halifax Amateur
Radio Club,
P. O. Box 663,
Halifax, N.S.
```

is hereby authorized to establish and operate an Amateur Experimental Station at

```
11 Crescent Ave.,
Armdale, N. S.
```

from the date hereof until the thirty-first day of March, 1954, subject to the provisions of The Radio Act, 1938, and to the Regulations heretofore or hereafter made thereunder, and to the conditions, if any, stated on the back of this licence.

VE1FO License front top

Frequency Bands		Types of Emission
1.800 —	1.825 Mc/s	A1
1.875 —	1.900 Mc/s	A1
1.900 —	1.925 Mc/s	A1
1.975 —	2.000 Mc/s	A1
3.500 —	3.725 Mc/s	A1
3.725 —	4.000 Mc/s	A1
7.000 —	7.200 Mc/s	A1
7.200 —	7.300 Mc/s	A1
14.000 —	14.150 Mc/s	A1
14.150 —	14.350 Mc/s	A1
21.000 —	21.200 Mc/s	A1
21.200 —	21.450 Mc/s	A1
26.958 —	27.282 Mc/s	A1 A2
28.000 —	28.200 Mc/s	A1
28.200 —	29.700 Mc/s	A1
50.000 —	54.000 Mc/s	A1 A2 A3 F1 F2 F3
144.000 —	148.000 Mc/s	A1 A2 A3 F1 F2 F3
220.000 —	225.000 Mc/s	A1 A2 A3 F1 F2 F3
420.000 —	450.000 Mc/s	A1 A2 A3 F1 F2 F3
1215.000 —	1295.000 Mc/s	A1 A2 A3 F1 F2 F3
2300.000 —	2450.000 Mc/s	A1 A2 A3 F1 F2 F3
3300.000 —	3500.000 Mc/s	A1 A2 A3 F1 F2 F3
5650.000 —	5925.000 Mc/s	A1 A2 A3 F1 F2 F3
10000.000 —	10500.000 Mc/s	A1 A2 A3 F1 F2 F3
21000.000 —	22000.000 Mc/s	A1 A2 A3 F1 F2 F3

Authorized for radiotelephone operation in accordance with radiotelephone conditions (d) and (e) on back hereof.

C.M. Williams Radio Inspector

No transfer of this Licence or of any rights hereunder shall be made by the Licensee.

Date April 1st, 1953.

for Minister of Transport

VE1FO License front bottom note signed by George F. Harris
The radiotelephone endorsement is by Charlie Williams

CONDITIONS TO BE OBSERVED BY THE LICENSED STATION

THIS LICENCE MUST BE POSTED IN A CONSPICUOUS PLACE IN THE STATION

1. This licence authorizes transmission and reception, only on the frequencies authorized herein.

2. All transmissions must be in plain language.

3. Amateur Experimental Stations must be so operated as not to interfere with the working of any government or commercial coast, land, ship or aircraft station, or with the reception of broadcasting.

4. The licensee may permit any person to take part in radiotelephone transmissions provided that he, the licensee, is present and retains physical control of the station.

5. The input power to the antenna shall not exceed 500 watts. For purpose of this requirement, the final amplifier shall be considered to be operating with an efficiency of 70 per cent.

6. If the herein licensed station uses equipment capable of operating with a power input to the final amplifier in excess of 400 watts, such equipment must have meters of a standard manufacture and accuracy permanently installed therein for measuring the plate voltage and current to the tube or tubes supplying power to the antenna.

7. The use of Frequency shift Telegraphy (radio-teletype) is authorized in the frequency bands 7.15-7.2, 26.958-27.282, 50.0-54.0, 144.0-148.0, and 220.0-225.0 Mc/s, subject to the conditions that the frequency shift does not exceed 850 cycles and that the station identify itself by the manual transmission of its call sign at least every five minutes.

8. The use of frequencies in the bands 1.800-1.825 Mc/s and 1.875-1.900 Mc/s is limited to stations located in Ontario, Quebec, New Brunswick, Nova Scotia, Prince Edward Island, Newfoundland, Labrador and Districts of Keewatin and Franklin. The input power to the antenna on these frequencies must not exceed 250 watts between sunrise and sunset and 100 watts between sunset and sunrise.

9. The use of frequencies in the bands 1.900-1.925 Mc/s and 1.975-2.000 Mc/s is limited to stations located in Manitoba, Saskatchewan, Alberta, British Columbia and District of Mackenzie. The input power to the antenna on these frequencies must not exceed 250 watts between sunrise and sunset and 100 watts between sunset and sunrise.

10. The carrier from a transmitter operating on frequencies below 144 Mc/s must be suppressed during periods of reception. Except for brief tests and adjustments which must be identified by the station call sign, the emission of an unmodulated carrier is not permitted on frequencies below 144 Mc/s.

11. The direct modulation of an oscillator with a frequency stability less than that obtainable with crystal control is prohibited on frequencies below 144 Mc/s.

12. The operation of radio equipment for the remote control of model aircraft, boats, etc., is permitted under this licence on all assigned frequencies above 53.0 Mc/s.

13. Power input to the antenna in the 420-450 Mc/s band must not exceed 50 watts.

VE1FO License back top

RADIOTELEPHONE TRANSMISSION (A3 F3)

Radiotelephone transmission (A3 F3) may be used in the following frequency bands subject to conditions (a), (b), (c), (d), (e) and (f) indicated:

1.800—1.825 Mc/s (a), (b), (c), (d), (f)	7.200— 7.300 Mc/s (a), (b), (c), (d), (f)
1.875—1.900 Mc/s (a), (b), (c), (d), (f)	14.150—14.350 Mc/s (a), (b), (c), (d), (f)
1.900—1.925 Mc/s (a), (b), (c), (d), (f)	21.200—21.450 Mc/s (a), (b), (c), (d), (f)
1.975—2.000 Mc/s (a), (b), (c), (d), (f)	26.958—27.282 Mc/s (a), (b), (c), (d), (f)
3.725—4.000 Mc/s (a), (b), (c), (d), (f)	28.200—29.700 Mc/s (a), (b), (c), (d), (f)

(a) The station shall at all times be equipped with a reliable frequency measuring device and a visual means of indicating overmodulation.

(b) The transmitter shall be of a type which is crystal controlled or which has a stability and constancy comparable to that of crystal control.

(c) The modulation system shall be so designed and operated as to ensure intelligible speech, must not in any case exceed 100 per cent and must not disturb the frequency stability of the transmitter. The deviation of the carrier frequency shall not exceed plus or minus 3000 cycles in transmitters employing frequency modulation.

(d) The licensee shall have been the holder of an Amateur Experimental Station Licence for at least one year during which period his station shall have been in active operation and provided he passes an examination in advanced radiotelephone theory and operation and a code test of not less than 15 words per minute.

(e) The licensee shall have been the holder of an Amateur Experimental Station licence for at least six months during which period his station shall have been in active operation on frequencies below 29.7 Mc/s.

(f) The retransmission of signals from a station with limited telephone privileges, by a station with full telephone privileges, on a restricted band, is prohibited.

NOTE—Licensees desiring authority to use radiotelephony in the above frequency bands must submit application to the nearest District Radio Office.

PORTABLE OR MOBILE OPERATION

This licence authorizes the operation of one portable or mobile installation in a passenger automobile owned by the licensee, or at a temporary location, for communication with any licensed Amateur Experimental Station, including the home station, provided the following requirements are complied with:—

(a) The equipment at the home station and at the portable or mobile station shall be operated by persons holding Certificates of Proficiency in Radio of at least Amateur grade, unless otherwise authorized by the Minister of Transport.

(b) Under no circumstances may equipment be operated from any aircraft or registered vessel.

(c) Portable or mobile operation must be identified in radiotelephony by the call sign followed by the word "Portable" or "Mobile" and a figure indicating the call sign area in which the operation is taking place, i.e. "VE3XYZ PORTABLE 4." Radiotelegraph transmission must be identified by the call sign followed by the oblique stroke and number of call sign area of operation, i.e. VE3XYZ/4.

(d) Whenever portable operation is to extend a period of forty-eight hours, a written notice containing full particulars must be forwarded to the local Radio Inspector in the licensee's home District and if operation is in another call sign area, the District Radio Office in that area must also be advised.

(e) Portable operation must not extend beyond one month in any period without obtaining authority for continued operation and in no case shall it exceed a total of four months in any fiscal year.

(f) The portable or mobile equipment shall be available for inspection at the licensed address of the station whenever required by Departmental Radio Inspectors.

Whenever a change of address is made, this licence must be forwarded for endorsement to the District Radio Office with particulars of the change.

DISTRICT RADIO OFFICES

St. John's, Nfld., Marshall Building	Winnipeg, Man., Room 539, Public Building
Halifax, N.S., 7th Floor, Dominion Public Building	Edmonton, Alta., 10138-100 A Street
Montreal, P.Q., Room 302, 901 Bleury Street	Vancouver, B.C., 209 Winch Building, 739 West Hastings St.
Toronto, Ont., Room 366, Dominion Public Building	Ottawa, Ont., Room 2215, No. 3 Temporary Building

HARC Files

VE1FO License bottom back

Note that Trevor Burton, VE1CP, was the custodian of this license as the secretary of HARC.

The following is a brief summary of changes that will be noted on our new 1954-55 licenses: -

1. Mobile operation will be permitted hereafter in any type of motor vehicle and, please note, in pleasure yachts.

2. Authority may be obtained to carry out television experiments on frequencies above 420 megacycles.

3. Frequency shift teletype will be permitted on all amateur frequencies other than those on which phone is permitted.

CLARENCE ROACH VE1EA

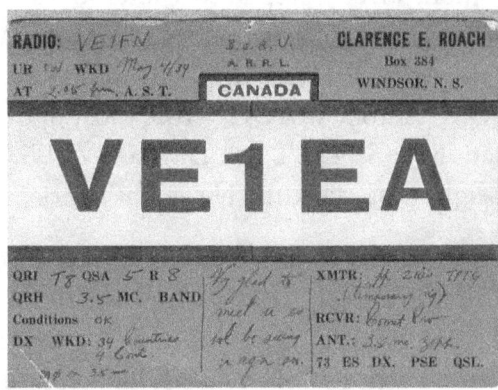

HARC Files

Clary had been a member of HARC and was recorded as living in Halifax when he became a silent key. Doug Johnson, VE1OM, stated in his Maritime News for May, 1955, that Clary was a 160-meter trans-Atlantic pioneer and had made a number of DX records on that and other bands.

SCM SERVICE WITH ARRL

Art Crowell, VE1DQ, was SCM until Doug Johnson, VE1OM, was elected on February 15th, 1954. This was a record of over 25 years continuous service and an all-time record at the time in ARRL. HARC presented Art and his XYL with an engraved silver tray in a gathering to celebrate this achievement. Shortly after Doug, VE1OM, was elected SCM Fritz Webb, VE1DB, was made assistant SCM. Shortly after Fritz became an assistant SCM Arron Solomon, VE1OC, was also made an assistant SCM. The June, 1956, issue of QST started to list D. E. Weeks, VE1WB, from St. Stephen, New Brunswick, as the VE1 SCM on the masthead but Doug Johnson, VE1OM, was listed as SCM in the Maritime News Section. D. E. Weeks, VE1WB, was listed as SCM in the Maritime News Section of August, 1956, with Fritz, VE1DB, and Arron, VE1OC, as assistant SCM's. This was changed to Don Weeks, VE1WB, in the October, 1956, issue. Harley Grimmer, VE1MX, Fairview was listed as SCM on the masthead but Don VE1WB recorded his final Maritime News in the December, 1966, issue of QST. Harley, VE1MX, thanked Don for his ten years of recording Maritime News in his first Maritime News of January, 1967. Bill Gillis, VE1NR, replaced Harley in the April, 1968, issue of QST. Harley had been transferred by his employer to Pointe Claire, Quebec. Walter Jones, VE1AMR, replaced Bill, VE1NR, with the December, 1971, issue of QST. There had been no Maritime News for 7 months in 1971 so we have little detail on 1971.

HARC member Aaron Solomon, VE1OC, replaced Walter Jones, VE1AMR, as SCM and his first Maritime News was in the January, 1976, issue of QST. Don Welling, VE1WF, replaced Aaron on the masthead of the April, 1980, issue of QST and Don's first Maritime News was in the June, 1980, issue of QST. Aaron, VE1OC, wrote most of the Maritime News articles of 1986 as ASM (acting section manager). Leigh Hawkes, VE1GA, replaced Don, VE1WF, on January 1st, 1987, and Leigh's first Maritime News was the one for April, 1987.

Aaron Solomon, VE1OC, announced at the HARC monthly meeting held on September 19th, 1979, that neither he nor Ron Hesler, VE1SH, would be running to hold their positions in the upcoming ARRL elections. QST League Lines stated that Ron Hesler, VE1SH, had resigned in July, 1980, and that Mitch Powell, VE3OT, had replaced Ron as CRRL Director. Ron resigned at the May 9th, 1980, meeting of CRRL.

Ron Hesler, VE1SH, was back again in 1986 when he was elected as the Regional Director of CRRL. Ron was first licensed in 1937 as VE1KS and he held the call until 1957 when he moved to Montreal and became VE2QF. He returned to Sackville, N.B., in 1966 and became VE1SH.

The SCM was replaced with the Section Manager (SM) in 1983.

SOLAR POWER 1955

The first solar powered radio station appears on the front cover of the September, 1955, issue of QST. It consisted of a small transistorized receiver and transmitter much the same size as a modern code practice oscillator. It operated on the 160-meter band at 1800 kilo-cycles. There was a complete description with all the detail one would want to build their own copy on page 11.

YOUNGEST HARC MEMBER 1956

Wes Street, VE1EK, was giving code practice to the HARC members at their monthly meetings in 1956. At 15 years of age Mike Goldstein, VE1ADH, was the youngest member of HARC in 1956 and had operated VE1FO/1 in the W/VE contest on September 29th and 30th. One had to be 15 years of age in order to get a ham license back then. VE1FO was at the bottom of the Maritimes listing on this contest with 3,591 points, but I feel confident Mike enjoyed this experience. Mike is the nephew of Ron Hart; W9IVP ex VE1MZ. Mike is VE3GFN in 2008. HARC member Dave McClafferty holds the VE1ADH call sign in 2008. Mike claims Dave is just holding the call for him.

RAFT L'EGARE II

The raft L'Egare II departed Halifax and drifted across the Atlantic Ocean in 1956. The raft was 30 feet by 17 feet and had one square sail. The Maritime Net held a daily sked with it until it was out of range. It then communicated with Gus Roblot, FP8AP, on the French Islands of St. Pierre et Miquelon. Actually, Gus stood by the Maritime Net until after their sked with the raft. This made it convenient for the local amateur operators who wanted an FP8 contact.

http://www.purr-n-fur.org.uk/featuring/adv20.html

The radio on the raft L'Egare II

The raft did not have a radio call sign. There was nothing legal about this whole operation and they simply used the name L'Egare over the radio. I was under the impression they used phone but in a CBC documentary on the experience one will see radio operator Marc Modena using a hand key strapped to his leg. They had purchased the war surplus radio in Montreal for $15.00. It used a hand-cranked generator and one of the crew had to crank it while Marc operated. The phone portion of the radio was so poor Marc had to use CW. They were all so malnourished and far out of shape when they reached England that it took two of them to crank this generator. Marc could send CW but was unable to receive CW. He had served in the French Navy and this indicates to me that he was a signalman. The signalmen were the ones that sent and received Morse via a signal light. I have seen signalmen send CW but they had to have someone receive CW for them. The top speed of a signalman on the light was around 13 words per minute.

Fred, VE1FA describes the radio: "the radio is a BC-654 transmitter-receiver, part of the SCR-284 radio equipment. They were first used in North Africa in 1942, then throughout the war by army, navy, and most of the Allied forces. It covered 3800-5800 KC, and put out about 17W of AM or 24W of CW. Pretty basic, but effective! They were run on batteries, generator, or hand-crank generator. I read somewhere that about 150,000 were made, and after the war you could buy a complete unit for $15-20!" In 1956 the 80-meter band was from 3500 kilocycles to 4000 kilocycles and as one can see the only amateur radio band this radio could operate on. One could operate CW on the full band but AM radiophone only from 3800 to 4000 in the United States and down to about 3740 to 4000 in Canada.

Lea.hamradio.si

BC-654

Lea.hamradio.si

BC-654

I find it hard to believe the Maritime Net would work the raft without a call sign. The regulations were followed closely back then. A gathering of amateur radio operators each evening at 7 PM local time on frequency 3830 kilocycles in 1951 was known as the Maritime Net. There is a record that this changed to 3750 kilocycles in 1954 and this is the frequency of this net today. Marc was able to tune in this net each evening so may have transmitted just above 3800 and listened on 3750.

I wish there had been more detail left by the club on the actual operation with this raft. One thing is for certain; Marc was restricted to the AM portion of 80-meters the only amateur band available with a BC-654. Two years prior to this the Canadian government created the maritime mobile amateur radio license with the VE0 call sign prefix. This was created for Canadian vessels traveling in international waters and one had to stay above 14,000 kilocycles or the bottom of the 20-meter band. This radio would have been of no use since it was capable of going as high as 5800 kilocycles only. This radio would survive the environment of this raft and a proper amateur radio station would be lucky to have survived to the Grand Banks. The BC654 broke down several times on the voyage but they managed to get it going via trial and error.

Mike, VE3GFN, told me that he remembers this but all he can remember is someone on the Maritime Net calling the raft in French. He does not remember the frequency the raft was using in reply.

A Mr. Patterson, at the Dartmouth Slips helped Henri Beaudout build the raft in Dartmouth out of cedar wood electric light poles one sees around the country. The poles were lashed together with rope. Mr. Patterson introduced Henri to Aaron Solomon, VE1OC. HARC member Aaron Solomon, VE1OC, was congratulated in QST for a fine job of public relations work with this L'Egare II incident. QST did not describe what he had done and this is the only written record I have found on this raft excursion.

There were originally four crewmembers and two kittens. The kittens were named Puce and Guiton. The SPCA had learned of the kittens and they were told they were not allowed to go on the raft. They simply hid the kittens and when no one was around hiding them on the raft. The crew on the raft was suffering from what today is known as Post Traumatic Stress Disorder from their service in the French military during World War II. They simply stated do not be a sissy, grin and bear it back then. The kittens they found were a big help in helping them cope with this disorder and the reason for making this trip with the raft.

The four crewmembers were Gaston Vanackere, Marc Modena, Jose Martinez and Henri Beaudout. Jose became ill and had to be removed by the fisheries vessel INVESTIGATOR II when 80 to 100 miles off the Newfoundland Coast on the Grand Banks. The INVESTIGATOR II had no radio contact with the raft. The Department of Fisheries had been alerted of the raft in the area with a sick crewmember and had alerted INVESTIGATOR II. They managed to contact the raft via flares the raft fired.

I would be willing to bet the only radio in the INVESTIGATOR II was a Canadian Marconi CN86 Seaway. This was a "rock bound" crystal tuned 2-mhz AM radiotelephone. It was either that or one of Jim Spillsbury's 2-mhz AM radiotelephones like the Spillsbury-Tindal MRT 600 model. This was a receiver/transmitter more or less identical to the CN86. Jim started his radio business when he was VE5BR and became VE7BR when British Columbia changed to the VE7 prefix. Actually, Jim named his floating workshop, his boat, FIVE B.R. after his ham radio call sign. He was quite a character and used this boat workshop to run up and down the British Columbia coast installing and repairing radio receivers and transmitters.

The standard 2-wire inverted L antenna so common to vessels of that time or day. This ran from the top of the main mast to the top of the mizzen mast. INVESTIGAOR II had no small mizzen mast so likely simply had a wire antenna from "Monkey Island" the top of the cabin to the top of the only mast. INVESTIGAOR II was part of the Department of Fisheries

Research Fleet. Captain Cain, who was the Bosun in INVESTIGATOR II during this incident, spent his working career as a member of that fleet. INVESTIGATOR II definitely had no radio that could work the BC-654 in the raft.

The antenna on the raft was a 60-foot wire from the top of the cabin to the top of the mast.

The L'Egare II and crew on arrival at Falmouth, England

The kittens and the other three made it all the way across the Atlantic. Gaston had a 16 MM movie camera and a rubber dingy. He was able to take movie film of this adventure from the rubber dingy they carried on the roof of the raft's cabin. The dingy was kept connected to the raft via a rope tether while Gaston filmed the raft. One can see this dingy on the cabin roof in the photo below.

The publication The Raftsmen

This is Cyril towing the raft L'Egare II out Halifax Harbour

A local fisherman, Cyril Henneberry from Sambro had towed the raft out with his fishing vessel PROMISE. Cyril happened to be in Halifax with his fishing vessel and the L'Egare II crew asked him to help them out the Harbour. Cyril was reprimanded for this because the raft had not cleared the country the proper legal way.

Henri wrote a book on this experience. I tried to find the book on the internet. The title was L'Egare II and the book was withdrawn soon after publication because of libel action according to the internet. This is wrong. The English version of the book was "The Lost One" and there were technical errors made in the translation from French and Henri had the book withdrawn. I managed to purchase a former library copy of the book "The Lost One" via Abe Books on the Internet.

Henri complains of not being able to find a time signal for navigation on the radio. They did not have a chronometer and three wrist watches only in order to keep time. All three were indicating different times. The BC654 could receive the WWV time signal on 5000 kilocycles because this transmitter was in the area of Washington, D.C., in 1956. The Bureau of Standards station WWV spent the first forty years in Washington, D.C., and moved to Fort Collins, Colorado ten years later in 1966. If the raft could not receive this signal during the day, they should have been able to receive it at night. They would not be able to receive radio station CHU Ottawa that transmits a continuous time signal because it does not transmit on a frequency within the receiving range of the BC654. Those on the raft definitely wanted nothing much to do with the real world because they did not carry a small battery-operated AM broadcast receiver that were common in 1956.

Those on the raft transmitted several times blind but it would have been more or less a waste of time in the AM portion of the 80-meter amateur radio band. The receivers tuned to that portion of the radio spectrum would not have their BFO on and would not hear the 24-watt output of the BC654 on CW.

The BC654 was an illegal radio station and not likely anyone would have paid any attention to their signal if they had heard it. It is a shame Marc had not up-graded to an amateur radio station with a proper operator license and station license. The thousands of amateur radio stations in 1956 would have tried to make contact with the raft. Not only that but any number of them would have given them all kinds of help including equipment for the voyage.

Another item of interest is that it is very hard to make electrical contact from a radio to the sea on a wooden hull; in other words, an electrical ground. I doubt a wire from that radio dangling in the water would have made any difference in the quality or quantity of the signal put out by the BC654. A wooden hull vessel has a sheet of copper attached to its hull down close, as close as possible, to the keel. Attaching this copper sheet to the radio equipment can be a real challenge. It has to be sufficient so that the equipment can actually know it is connected. In other words, if the strap of copper from the equipment to this sheet is small the equipment will not see it or feel it.

Henri mentions in the book "The Lost One" communicating with VIFG Sable Island. VI is the prefix of a station in Australia but there were radio operators on Sable Island in 1956. They were operating their station VGF mainly passing messages to radio station VBQ in CW. The majority of these messages were weather observations. Radio station VBQ was in the Post Office building on George Street in downtown Halifax. Aaron Solomon, VE1OC, must have made this communication with Sable Island possible. Aaron is not mentioned in the book "The Lost One". Aaron and his VE1OC station are described in a movie of this voyage made by the Canadian Broadcast Corporation.

One assumes Aaron could speak French and it was he who communicated with the raft via the Maritime Net and the one on Sable Island was also able to speak French. If only a written record of this communication had survived.

On October 5th, 2013 I met Henri Beaudout who was visiting Cyril Henneberry and his family in Sambro. I asked Henri why he made the trip and he stated he had been an 18-year-old kid in the French army and was sent to Germany at the end of the war. He said he saw and did things no human should have seen or done and he just wanted to see if he could get his head screwed on right. He claimed this experience helped a lot. He also remembered Aaron Solomon and asked several questions about him that I could answer because I knew Aaron.

This was the second attempt Henri had made to cross the Atlantic. The first was in 1955 when he had a raft, L'Egare, built in Quebec and drifted down the St. Lawrence River. He fetched up on Newfoundland and then went over to the Magdalene Islands. From there he drifted down to Saint Pierre et Miquelon and was washed up and wrecked on some rocks.

This second attempt was successful when he had the raft built in Dartmouth. He is now hoping to have a miniature replica of the raft built and placed in the Maritime Museum in Quebec (Musee Maritime du Quebec at L'Islet-sur-mer, Quebec).

Henri is the only one still living of those who attempted this crossing on the raft. He also hopes to have some English versions of a book available soon. There are many articles on this raft crossing on the internet. Simply bring up RAFT L'EGARE II in Google. Members of HARC were involved and this incident is part of the club history. It was truly an amazing feat.

Henri had the raft shipped back to Montreal but had no proper storage for it and it simply rotted to the point there are only a few small pieces left today. The Duke and Duchess of Bedford adopted the two kittens and they terminated their life in the lap of luxury.

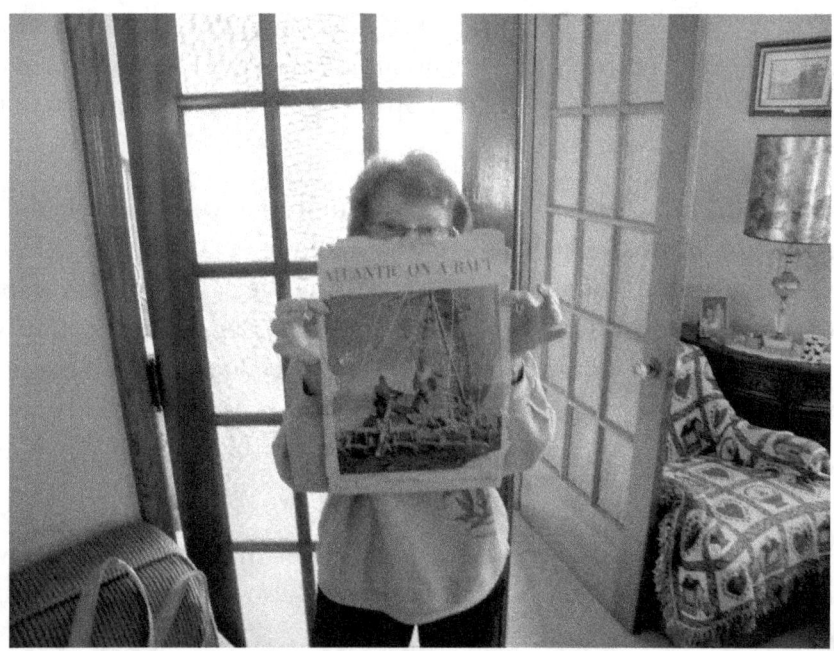

This is Pearl Henneberry, Mrs. Cyril Henneberry holding up a copy of one of many publications that printed stories on the raft in 1956.

This photograph of Pearl was taken on October 5th, 2013.

This is Captain Cyril Henneberry of PROMISE, on the left, VE1BC in the middle and Henri Beaudout on October 5th, 2013.

One can imagine what these three would look like if they removed 57 years off their age.

Cyril towed the raft out Halifax Harbour on May 24th, 1956, and the raft was towed into Falmouth, England, on August 21st, 1956, 88-days later. Henri said they had no radio contact with anyone after they lost contact with Gus, FP8AP. They did transmit a few messages blind but did not learn of anyone receiving these messages.

Page 92, QST, November, 1970

Gus Roblot, FP8AP, owned the above boat and made 145 crossings with it between the French colony of St. Pierre and Newfoundland. He did this from 1966 to 1970. Gus is just behind his son-in-law and grandchild in the photo. The boat was wrecked in 1971. Gus then moved the cabin ashore at St. Pierre to be used by visiting amateur radio operators.

Sauli Arosankari, VE1AIH ex OH5SG, was president of HARC in 1974. At the January 16th, 1974, monthly meeting he described a trip to FP8 and that he had operated in a contest from Gus's boat cabin. He said he had a good time and made 600 or so contacts with a score of 163,000 plus, but he failed to name the contest.

The caption of this photo reads left to right: Walt, W1KNU/ FP8BI, and Gus Roblot, FP8AP, famous and endeared for

HARC Files

135

his unstinting assistance to DXpeditioneers, on board Gus's 26-foot launch. Unfortunately, I could find no date for this.

In 2016 it was decided to mount a plaque in memory of this outstanding feat at Kings Wharf in Dartmouth where L'Egare II had been constructed. On Thursday November 3rd, 2016 a group gathered with Henri Beaudout at this site to learn of this monument.

This is the group gathered at Kings Wharf, Dartmouth on November 3rd, 2016. One can see Rose Marie Comeau Mahar and Henri Beaudout up at the front of the room answering questions from the grade 4 and 5 class of school children gathered to learn of this voyage. Henri does not speak English and Rose Marie is translating just like she did when they were building the raft in 1956.

Rose Marie Comeau Mahar

This is Captain Cain of the INVESTIGATOR II bent over chatting with Henri Beaudout. Captain Cain is from Newfoundland and came to this gathering to meet Henri and the first time they had met in person.

VE1BC showering the praises of HARC to the group. The only other amateur I know was in the group was VY2WP.

The school children who gathered with this group had built this model of the raft that is behind Henri and presented it to him.

This hardcover book by Ryan Barnett was published in 2017 by Firefly Books Limited 50 Staples Avenue, Unit 1, Richmond Hill, Ontario L4B 0A7. ISBN 978-1-77085-978-4 It has a lot of photographs if one does not like reading. It shows a newspaper clipping of Aaron, VE1OC, on page 83 and a drawing of him operating his station on page 91.

THE RUSSIAN SPUTNIK 1957

The Russians sent up their first satellite, the Sputnik, in October, 1957. Our branch in the navy knew something was taking place a couple of years before this. Two Russian Air Force radio operators were chatting too much. I was hoping to learn which HARC member was the first to hear Sputnik but all Don Weeks, VE1WB, recorded in Maritime News was the fact it had created much interest in the VE1 area, and many were able to receive its signals. QST ran several interesting articles on this subject in the November and December, 1957, issues. Sputnik created a lot of favourable publicity for the world of amateur radio. The amateur radio operators were the main source of information for the news media on this and the early satellites.

The Russian government wanted to publicize this as much as possible and asked all amateur radio operators around the world to send reports on Sputnik to Radio Magazine in Moscow. Aaron Solomon, VE1OC, stated he did not know of any local hams that were participating. Most of them stated that if the Russians wanted to know anything all they had to do was monitor the amateur radio transmissions they could hear.

LOST MESSAGE

Message By Short Wave Disappears

By The Canadian Press

TORONTO, Dec. 30 —Lost or stray-ed, somewhere in the 14 feet separating radio stations VE3ABW and VE3RY of Toronto, is an urgent and highly perishable short-wave message. When last heard from it was heading rapidly west. Finder please return it to Toronto immediately.

And that is what is worrying Geoffrey Light, owner and operator of Station VE3ABW, as he sits before as neat and compact a set as has ever "raised" Czechoslovakia or attempted to send New Year's greetings via the airlines to VE3RY, his next door neighbor.

"I might just as well have gone down into the basement and yelled it through the hole in the wall. It's what I usually do when I want to talk to him," he explained.

VIA RELAY

"BUT instead of that some one sold me the idea of sending him the seasons greetings via a Vancouver-Los Angeles-Miami-Halifax relay. And here I am, one day before New Year's, with that doggone message probably hung up someplace down there in Texas. That desert increases the distance between relay stations and cuts down reception," he said bitterly.

But even if he can't reach VE3RY by radio in time for New Year's, "Jef" still has an imposing list of official direct contact confirmations from scores of foreign operators. These multi-colored little placards from nearly every country the average student learns about in geography, occupy many square feet of wall space near his set.

POINTS TO CARDS

"I COULD have put those up, too," he added, as if called on to defend his "ham" status, and pointed to about 40 more cards stacked neatly on top of his set. "But the luck rip the dickens out of the wall."

The parting impression "Jeff" gave as he peered dejectedly into the bowels of his $150 brain child, listening intently to the staccato wails for some word of the "lost" message, was that of a competent craftsman betrayed into risking his proverbial all, for a cause that didn't really matter.

HARC Files

This is cute but unfortunately there was no date on it.

2 METER EXPERIMENTS

Bill Bligh, VE1BC, did a lot of experimenting on 144 megacycles in 1947. He managed to work Mount Uniacke from Halifax. He, Oscar Sandoz, VE1QZ, and Tommy Baker, VE1SF, did a number of tests they called the 26-mile test according to QST. One assumes this meant they worked each other with 26 miles of separation and tried various adjustments in one form or another. VE1QZ, VE1SF and VE1BC were all living in the city of Halifax.

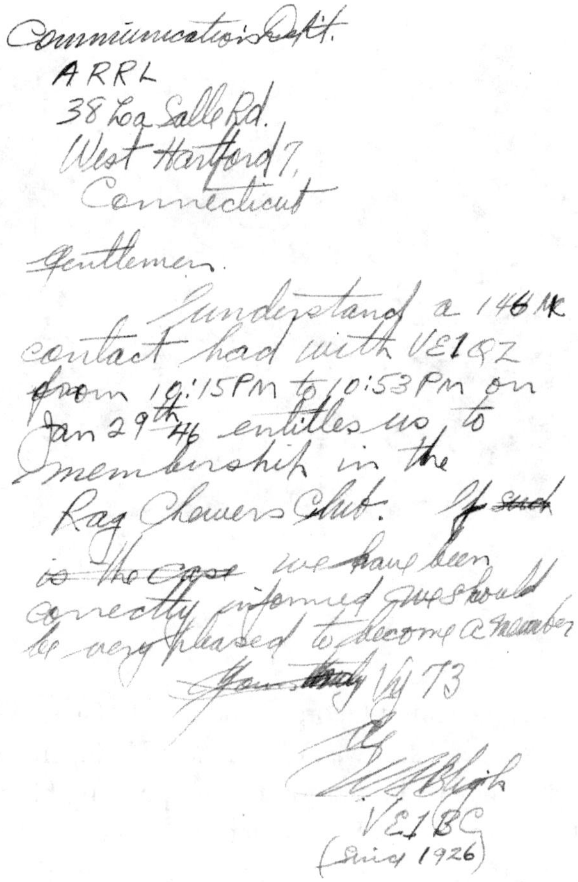

Communication Dept.
ARRL
38 La Salle Rd.
West Hartford 7,
Connecticut

Gentlemen.
I understand a 146 Mc contact had with VE1QZ from 10:15 PM to 10:53 PM on Jan 29th 46 entitles us to membership in the Rag Chewers Club. If such is the case we have been correctly informed we should be very pleased to become a member

Yours truly 73

VE1BC
(since 1926)

Thanks to Jim Guilford VE1JG this was found in an old piece of radio equipment

Bill, VE1BC, was still conducting these tests in 1951 and had managed to work Harry Bath, VE1MA, in Middleton and Boyd Burgess, VE1ID, in Berwick according to Brit Fader, VE1FQ, in the Bulletin for March, 1980.

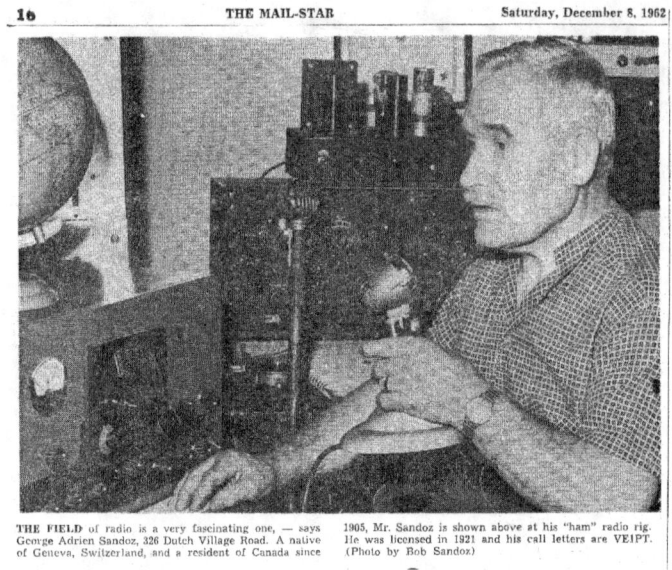

THE FIELD of radio is a very fascinating one, — says George Adrien Sandoz, 326 Dutch Village Road. A native of Geneva, Switzerland, and a resident of Canada since 1905, Mr. Sandoz is shown above at his "ham" radio rig. He was licensed in 1921 and his call letters are VE1PT. (Photo by Bob Sandoz)

HARC Files

Oscar Sandoz, VE1QZ, was the son of George Sandoz, VE1PT ex 1AH. When the article above appeared in the newspaper Oscar was VE3CRA and living in Ottawa. Oscar was very big with 2-meters in 1947. They had a photo of him operating his station in the August, 1947, QST. He was listed as the first VE1 contact on 50 megacycles. He had 63 stations worked and G. B. Grant, VE1QY, in Yarmouth had 53 stations worked. By November, 1947, Oscar was listed with 88 different stations worked and had worked 19 states on 50.1 mega-cycles. He had also made many good W contacts on 144.30 megacycles. In addition to this he had worked 51 countries on 20-meters including Nyasaland, ZD6, Palestine, ZC6, and Liechtenstein HE. Nyasaland is Malawi with call sign 7Q in 2008. Palestine or most of it is Israel with call sign 4X or 4Z and Liechtenstein has call sign HB0 in 2008. In January, 1948, they stated that they felt Oscar was the first VE1 to work a G station, one in England, on 50 megacycles. When Oscar Sandoz, VE1QZ, was really bored one could hear him working 14-megacycle phone brushing up on his Spanish.

Bill Bligh, VE1BC, was also quite active on HF while doing these 2-meter tests. He managed to work XZ2DN in Burma on 28 megacycles at this time. Burma is also known as Myanmar in 2008 but still has the XZ prefix along with XY.

An interesting article in 1951 describes a 2-meter 4-pound station for use wherever man can go on foot. This describes one of the first walkie talkie radios that is larger than a current HF transceiver and has a telephone style handset that appears to be for both transmitting and receiving.

TRANSCEIVER

The first time I found the word transceiver was in the October, 1926, issue of QST. On page 36 of that issue there is a description complete with schematic and all the information one needs to build a transceiver. This one was built in a plywood box covered in leather and measured 18-3/4 x 16 x 10-1/2 inches. With a set of 7 coils the entire amateur range above 12 meters could be covered. It did not state what it weighed unfortunately, but one assumes the one in 1951 was an improvement.

ARRL stated that Transceiver is a trade name coined to describe a combined transmitter-receiver as manufactured by the Chicago Radio Laboratory of Chicago, Illinois, and was created about 1920.

BCI AND TVI

THE HALIFAX AMATEUR RADIO CLUB

Sec. H. Shepherd.
First Ave.
Fairview.
Hfx. County
Sept 19th, 1947

Major W. C. Borrett,

Station Director,

Broadcasting House.

Dear Major:-

    I have been directed by the Executive Committee of the Halifax Amateur Radio Club to get an explanation of the incident which occurred during the broadcast of the Track and Field Meet held at the Wanderer's Grounds, Saturday September 6th, 1947.

    During this programme interference was experienced on CHNS's portable equipment by a signal calling "Test 14 can you hear me? Over". Your announcer identified this signal to be of amateur origin and made, what we consider, derrogatory remarks. It has been found in the course of investigation, that this signal originated from an authorised Naval transmitter in H.M.C. Dockyard.

    We feel that your station coverage is such that we cannot overlook the slight to the Amateur fraternity as a whole and our club would like a statement on this unfortunate lack of control over your on-the-spot-broadcast.

Copy:- R.I. Harris   Sincerely,
       Federal Bldng
       Hfx.

BROADCAST BAND
C H N S
960 K.C. 5000 WATTS

SHORT WAVE BAND
C H N X
6130 K.C. 500 WATTS

MOBILE UNIT
V D 2 O
2080 K.C. 100 WATTS

CHNS
OPERATED BY
The MARITIME BROADCASTING Company. Ltd
WILLIAM C. BORRETT.
MANAGING DIRECTOR

REPRESENTATIVES :
MONTREAL
ALL CANADA RADIO
DOM. SQUARE BLG.

TORONTO
ALL CANADA RADIO
VICTORY BUILDING

NEW YORK
JOS. WEED & CO.
350 MADISON AVE.

GREAT BRITAIN
FREMANTLE OVER-
SEAS RADIO,
18 PARK ST.
PARK LANE.
LONDON, W1.

STUDIOS & OFFICES
BROADCASTING HOUSE
TOBIN STREET
HALIFAX, N. S.

October2, 1947.

Mr. Holland Shepherd,
The Halifax Amateur Radio Club,
First Ave.,
Fairview,
Halifax Co., N.S.

Dear Mr. Shepherd:

This is to acknowledge your letter
of September 19, 1947, concerning the programme
interference experienced throughout the broadcast of
the Track and Field Meet held at the Wanderers Grounds.

During the past six months we have
repeatedly reported interference to Mr. R.I. Harris,
of the Department of Transport, due to the broadcast
activities of a number of persons in the Halifax
area. On many occasions this interference was so
strong that our operators in our Main Control Room
were unable to cue up recordings on our speakers, or
conduct voice level tests on our announcers.

During this same period, we received
hundreds of telephone calls from annoyed listeners,
complaining about the interference on our signal.
To say that this experience was annoying is to state
the case mildly.

On the Saturday afternoon in question,
it so happened that I was at the microphone at the
Wanderers Grounds, when the interference blanketed
out our attempted coverage of the field events.

I am unable to agree with your state-
ment that the remarks were derrogatory. On the contrary,
I quite assure you that in view of the situation, they
were decidedly restrained. There was no slight intended

"Please address reply to the Company, attention of Official or Department concerned."

144

Mr. Holland Shepherd -- 2

Oct 16th, 1947.

toward your amateur fraternity, and I trust that you
will appreciate the re-action of professional radio
men, who had voluntarily given up a Saturday afternoon
to provide a public service to their listeners.

Yours very sincerely,

G.J.Redmond,
Station Manager.

GJR:mps

Mr. Gerald Redmond
Station Manager 'CHNS'
Halifax, N.S.

Dear Sir:-
May I thank you for your reply to the letter
of the Halifax Amateur Radio Club, which has been digested,
with appreciation, by our executive.

Your statement that no slight was intended to the
amateur fraternity was received with approval by our members,
and I have been instructed to thank you for your apology.

At the same time, the executive wishes me to
express to you our deepest regard and sympathy for yourself and those other 'professional radio-men' who so
generously, had 'voluntarily' given up their Saturday
afternoon to provide so valuable a 'public service' to
your listeners.

Yours very truly,

H. Shepherd Secty.

HARC Files – The stains on these old letters are rust from the staples.

Interference was found anywhere and everywhere back then and of course the amateur radio community was to blame.

HARC had a new Interference Committee to handle the BCI problem in 1949. Many hams, especially those just starting out were using war surplus equipment that was inclined to produce problems. The majority of the broadcast receivers available and especially the television receivers when TV first became popular were quite susceptible to interference problems. One of the local radio inspectors gave a talk at one of the club meetings on the problems they were experiencing, so this committee would have been a welcome asset to anyone who became accused of creating this interference.

The first TVs for sale were advertised in QST in late 1948. The top of the line was a 16 x 12-inch screen of 192 square inches that sold for $695.00. A ten-inch direct view (whatever that was) sold for $295.00 and a 7-inch model sold for $169.50. A new Chevrolet car sold for around $1000.00 at that time so TV was expensive.

Doug Johnson, VE1OM, stated CHSJ-TV Saint John, New Brunswick, was the first TV transmitter to operate in the VE1 area in his May, 1954, report. The first program I saw was "Hop along Cassidy" on a set in a store window at Berwick, Nova Scotia. I believe Boyd Burgess, VE1ID, or Bill Woolard, VE1BT, were involved with this store. They both may have been running the store at the time. Bill Woolard, VE1BT, and Bill Bligh, VE1BC, were close friends and became interested in amateur radio at the same time in Berwick.

So many of the old newspaper clippings on file at HARC are in such bad shape I could not scan and use them. I could not resist this one of VE1LG on the left and VE1MA on the

145

right. They both were so well known and appear in numerous clippings of HARC and both attended nearly every Hamfest. Fred was the oldest ham in attendance at most of them. He did not make it to 100 years of age but was not far from it.

AVID FANS—Two avid radio "hams", Fred K. Bath (left) and his son, Harry, are shown spending an evening at the "rig" of the former, as they converse with other wireless fans in widely spaced parts of the world. The two Middleton men live in adjoining houses and each has his own station. (Photo by Wetmore).

The ARRL made a script for TVI available in 1953 that could be used via radio and television programs to explain TVI, the causes and cures. This was a very big problem and a very big headache to one and all back then. My father-in-law was a good example. His TV antenna was such a mess it is a wonder he received anything, but all the interference was caused by the amateur radio world around him. There was no arguing the point.

The amateur radio community also had ITV and this broke down to Interference Television from lousy TV receivers made from manufacturers that knew better. This produced all kinds of weird sounds and signals especially on the 80-meter band.

HARC members Bill Bligh, VE1BC, and Cliff Short, VE1AW, had a grand old time in 1952 working TV DX they called it. Apparently, they each had a pretty good antenna trying to pick up as many stations as possible. One wonders if there was ever an award created for that.

AMATEUR TELEVISION

The editorial and much of the November, 1953, issue of QST was on colour TVI. This same issue included a detailed description on the building of an Amateur Television Transmitter (ATV). Amateur TV made its debut in the VE1 call area in 1960. Orest Chaban, VE1AFQ, assisted in the preparations but was unable to attend the actual transmission. (Mike Goldstein, VE1ADH, does not remember Orest as having a call but this is listed in the 1963 callbook. The entry in QST was listed with the VE1AFQ call sign only and Mike remembers Orest as part of the gang so I believe this to be accurate).

Canadian Broadcasting Corporation

Halifax County Vocational High School, Bell Road, Radio Television Class 1959

Back Row Left to Right: Bernard Johnson, David Hamlin

3rd Row: David Hayes, Grant Pattison, Ken Duggan Gerard Gagnon, Orest Chaban VE1AFQ and Ronald Burbidge

2nd Row: Leonard Grant, Frank Myers, Don Colp, Mike Goldstein VE1ADH and Wilfred Boudreau

1st Row: Murray Hutchinson, Robert Schultz VE1IF

Jack Leahy, VE1ZZ, Mike Goldstein, VE1ADH, Jean Bilodeau, VE1IJ, and Bob Schultz, VE1IF, made history in this area when they were successful in an Amateur TV transmission between Rawdon and Blomidon on 440 mc despite heavy rain. Jean Bilodeau, VE1IJ, was better known in this area as VE0NI the amateur operator in *HMCS ST LAURENT.*

Page 63, QST, August, 1960

This is Bob Shultz, VE1IF, and the ATV transmitter that made the above record.

MIKE VE1ADH EXPERIENCES

Mike, VE1ADH, gave me this bit of interesting history:

"Jean, VE1IJ, used to hang around with us constantly, when he was in port, and we had a lot of fun together. As I recall, it was Jean who helped me put up a second-hand 3-element 10-meter beam (scrounged from Aaron Solomon, VE1OC), on the roof of my parents' house. Unfortunately, Jean managed to put his foot through the roof slightly, and that project was followed quickly by a massive rainstorm, that flooded the house (all but my ham gear), and I was banned from the roof for years to come!

Speaking of Jean and the 10-Meter beam, I have a story for you ... that winter, I used a long wire for 75/80M, and that beam on 10M. That was a year of fantastic sunspot maximum, and I only used to come down off 10M for the Maritime Net, on 75M, in the evenings.

> My buddy, Andy White, VE1AEW, came over one afternoon during a fantastic snowstorm, to chase DX with me. The beam was on a "zip-up" pole, and we used the "Armstrong" method of rotation.
>
> I'd hear a DX station in Africa. Andy would throw open the window of my bedroom shack, look out into the storm, and turn the beam to the correct heading. I'd chase the DX, listen some more, and give Andy a new heading, and he'd repeat the process.
>
> At the end of the afternoon, Andy went home with galloping pneumonia, and I went to change the antenna to the long wire, for 75M ... only to discover we had been on the long wire all afternoon, and all Andy's efforts in biting the snowstorm had been for nothing! Andy was just over six feet three, and a Judo student of note, and I never did tell him."

Mike, VE1ADH, was transferred to Ontario and became VE3GFN. I remember working him in 1978. How could one forget "Good for nothing"! Gary Lloyd was with me and said he would love to have that call. I managed to talk Gary into becoming VE1BZD two years later. He eventually became VE1JB.

JACK LEAHY VE1ZZ

PREFERS SIMPLE METHOD—In spite of all the complicated and confusing equipment Jack Lahey, above, of Tuft's Cove, still relies on the Morse code and a key to do the greatest part of his work in the "radio ham" field which he has developed into a permanent job at the Naval Research Establishment in Dartmouth. (Conrod photo.)

HARC Files

1954

Brit Fader, VE1FQ, stated in his Bulletin for June, 1980, that Jack, VE1ZZ, had 221 countries worked on 80 meters only.

James D. Cain, K1TN, did a one-page article on Jack Leahy, VE1ZZ, in the February, 1996, issue of QST and calls Jack a "Ham's Ham". He is that for sure and is still at the same QTH in 2008. In 1996 Jack had worked 264 countries on 160-meters but was well equipped to handle all bands well from his many homemade antennas. This is a quote from that article: "Peter Lovino, KA1BQ, who last year was participating in a contest operation from EA9IE in Northern Africa, struggling to hear North America on 160 meters. "Every now and again VE1ZZ would drop by and ask 'How are things going?'" Peter says it was as if a local had asked the question on 2 meters through the repeater!" I remember that part of the story from when I first received that QST in 1996.

YL & XYL'S

Eleanor Wilson, W1QON, was the YL Editor of QST in the 1950's and she recorded this in her YL News and Views article in July 1953:

"Who are an YL and what is an XYL? Confused? Occasionally we have difficulty interpreting how other people interpret these terms, so we'll tell you what we mean when we use 'em.

Originally, the male operator referred to his girlfriend sitting by in the radio shack as his "young lady" – abbreviated by the cw boys to "YL". A wedding band, it was thought, transformed an YL into an ex-young lady or "XYL". In the early days there were so few women with amateur licenses that the terms referred generally to unlicensed YL's and XYL's.

Today, in this column, "YL" is used to denote any licensed feminine amateur radio operator – whether she is nine or ninety, bachelor girl or wedded spouse. "XYL" refers only to the unlicensed wife of a male amateur. These terms are consistently used as defined, even though some may puzzle why girls with multiple harmonics are referred to as YL's and not XYL's.

If you feel these terms are misnomers, you are not alone. Marriage, we blissfully hope, does not make a gal an "ex-young lady". And we all know some YL's who admit to fifty or sixty summers. Literally speaking, both terms are not completely appropriate. Like so many things that "just grew", the terms have stuck with us – or should we say, we are stuck with them, until some better ones are uncovered. True, anything is better than "OW" – perish this thought! "OG" is slightly less painful but rather unimaginative.

So, what say, YL's and XYL's are you happy being so called or can you think of something better?

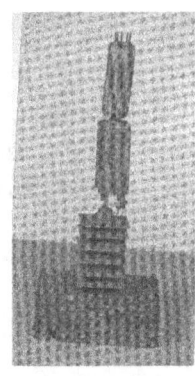

Page 89, QST, June 1964

This is a Deluxe 88 Filter

The photograph did not reproduce very well

One OM's reply to that were SLO and MLO – Single Lady Operator and Married Lady Operator. We are lucky that did not catch on but one can only wonder what other replies she received.

A lady visiting a W1 Field Day site in 1956 noted that all the hams were enjoying themselves and wanted to know if lady hams were called "sows".

Every so often someone will ask me the correct meaning of the sign 33 as used similar to 88 and 73. According to the June, 1980, issue of QST on page 71, Clara Reger, W2RUF, created that as a greeting from one YL operator to another YL operator and it means "Love sealed with friendship".

There is an excellent long article of five pages in the May, 1940, QST titled "The YL's Unite". This is the story of the Young Ladies Radio League (YLRL). Ethel Smith, W7FWB, from Wenatchee in the state of Washington founded this league. It hardly managed to get started when they had 71 paid up YL members from

30 states, Alaska, Canada, Puerto Rico and Hawaii. This article does not mention the term OW or XYL. They were all known as YL and at the time they were after international recognition and ARRL affiliation. Ethel held the call W7FWB, W3MSU and K4LMS before she became a silent key in 1997. Her biography plus a photo appears on page 73 of the April, 1997, issue of QST.

Page 72 QST December, 1961

They had branch units in: Cleveland, Cincinnati, St-Louis and New York City by July 1941. The October, 1941, issue of QST has another excellent five-page article on the YLRL. This article starts off stating their motto is QRV – I am ready. It states the No. 1 Unit is New York City and the article has a lot of photographs. They are all rather young-looking girls and one is Dot Willett, W8UDA, who is blind and teaches Braille for a living. Yes, a few were married and some were married to amateur radio operators. Although the girls do not appear in the language of the day when describing anything connected to amateur radio they were definitely there. They were referred to as YL and OW as late as 1942 so it took some time for the XYL acronym to become popular. There was a photograph and article by Jean Hudson, W3BAK, in the April, 1942, issue of QST. Jean was the World's Code Speed Champion (Class E) at the Chicago convention in 1933 when only 9 years old. The girls were definitely into amateur radio back then.

Page 62 QST January, 1959

By early 1960 QST was listing 25 Nets and Round Tables, both Phone and CW for the YLRL.

SNOWSTORM 1960

NOVA SCOTIA SNOWSTORM FEB 3 / 1960

ON WEDNESDAY FEB 3RD. ,THE WESTERN SECTION OF NOVA SCOTIA
WAS HIT BY A TERRIFIC SNOW STORM ,WHICH DROPPED UPWARDS OF
THIRTY INCHES OF SNOW ON THE LEVEL , AND CRIPPLED POWER,
TELEPHONE , AND TELEGRAPH COMMUNICATIONS BETWEEN BRIDGE-
WATER AND YARMOUTH AND BRIDGEWATER AND MIDDLETON .
    DUE TO THE STORM MANY OF THE BOYS WERE SNOWBOUND AND
UNABLE TO GET OUT TO WORK ,AND AS A RESULT, THERE WERE QUITE
A NUMBER THAT SHOWED UP ON THE 75 METER BAND , SOME AS EARLY
AS 7.30 OXXXXXXXXXXXXXXXXXX IN THE MORNING .
AN EMERGENCY NET WAS ESTABLISHED WITH THE FOCAL POINT BEING
LIVERPOOL , SINCE IT WAS THE AREA HARDEST HIT BY THE STORM .
    VE1TN GOT ON THE AIR FROM LIVERPOOL AND OPERATED FROM
HIS HOME QTH, DURING THE TIME THE POWER WAS ON AND HANDLED
TRAFFIC FOR THE TELEPHONE COMPANY, CANADIAN NATIONAL
TELEGRAPH COMPANY AND THE NOVA SCOTIA POWER COMMISSION ,
AND CLEARED HIS TRAFFIC TO MIDDLETON , BRIDGEWATER AND
HALIFAX . LIVERPOOL LOST THEIR POWER FOR SEVERAL PERIODS
DURING THE THREE DAYS , AND ONE PERIOD THEY WERE WITHOUT
POWER FOR NEARLY EIGHT HOURS . DURING THESE POWER BLACK
OUT PERIODS, VE1TN OPERATED FROM THE MOBILE RIG OF VE1US ,
AND WAS ABLE TO KEEP TRAFFIC MOVING AND KEEP LIVERPOOL IN
TOUCH WITH THE OUTSIDE NXXXX . WORLD .
    MANY STATIONS FROM OTHER PARTS OF THE PROVINCES AND
ADJOINING PROVINCES CHECKED IN OFFERING THEIR SERVICES ,
WHICH WAS MUCH APPRECIATED .
STATIONS WHO ACTIVELY PARTICIPATED IN THE EMERGENCY WERE :
VE1TN ,UC, VN , ABJ ,PA ,MA ,KC ,FQ ,DW, NZ ,ABF ,LY ,GX ,AFU ,WL , ADH,
IR ,QM ,BO ,ABB ,SE ,AFD ,MO ,LG, FV, AAR
VE1BJ, OPERATING OUT OF SHELBURNE ON TEN METERS PASSED
TRAFFIC THROUGH K9QET FOR RELAY TO VE1BO, WHO PUT IT BACK
ONTO 75 METERS FOR LOCAL DELIVERY .
    VE1DW OPERATING OUT OF YARMOUTH, PASSED SOME TRAFFIC TO
THE HALIFAX AREA FOR THE WEATHER OFFICE AND ALSO FOR TCA .

HARC Files

We left Halifax just before this snowstorm in *HMCS SWANSEA* and we heard all about it via CHNX the six-megahertz transmission of CHNS. We were listening to it in the beautiful warm weather of the Gulf Stream. The coldest temperature on record for Halifax had been recorded shortly before this and is the reason I remember it so well.

MOON BOUNCE

The other big news in 1960 was Moon Bounce. Amateur radio operators were bouncing signals off the moon and achieving some amazing distances on 1296 megacycles.

RADIO COMPLAINTS

In September, 1960, the Canadian Director of ARRL, Noel Eaton, VE3CJ, stated that the Department had received 25,188 complaints total and only 92 involved amateur radio operators. These were key clicks, over modulation, out of band transmissions and so on, as follows:

Interference to standard broadcast stations	10
Interference to TV reception	59
Interference to both TV and broadcast stations	19
Interference to other services	4

This was all of Canada so Noel felt the amateur radio operators were not so bad. Brit Fader, VE1FQ, stated in his Bulletin for June, 1976, that the Department of Communications had received 30,290 cases of interference in 1975 and only 150 were attributed to amateur radio. One has to agree that the radio amateurs had a very good record.

The VE1FO call sign has been the club call sign since 1946 and has been the call sign the club used at field day 1946 and each field day since. I am convinced that Doug Smith and HARC are the only stations to have been assigned the VE1FO call sign.

HARC CLUB RECEIVER 1947

HARC boasted a new receiver for the VE1FO club station in 1947. It was a BC-348 like the one above and still held by the club in 1974. This was the companion receiver to the BC-349 transmitter that was fitted in many aircraft during World War II. This equipment was known as war surplus and was readily available after the war. QST ran a number of  articles on how to convert various pieces of this equipment to operate within the amateur radio bands. QST also ran the address where one could purchase copies of the manuals to various pieces of this equipment.

Page 12, QST, October 1970

Many receivers back then had an intermediate frequency (IF) of about 455 kilocycles. One could obtain an outboard unit called a Q multiplier that improved the selectivity of these receivers. I remember Heathkit sold one in kit form. The BC-348 had an IF of 915 kilocycles and the June, 1948, issue of QST described in detail how one could build one for this receiver called a "Q5-er". They must have been a popular ham receiver at the time.

THE NORTHERN STATIONS

Art Crowell, VE1DQ, was using a Q5-er in 1948 for his contacts with VE8NB and VE8OE on the Hudson Strait. These skeds with VE8NB and VE8OE terminated when the two operators moved back to this area in late 1948.

HARC Files

VE8NB was Mickey McWilliams who I am sure sailed in as many ships over the years as I did. He started as a naval telegraphist during the war sailing in the corvette *HMCS ARVIDA* with international call sign CGQF. He retired from Halifax Coast Guard Radio in Ketch Harbour and held amateur radio call sign VE1HE during retirement.

VE1FO ON AIR FROM N.S. TECHNICAL COLLEGE 1968

Bill Gillis, VE1NR, made this statement in his Maritime News item in the August, 1968, issue of QST: "VE1FO is on air from the Nova Scotia Technical College and his first permanent set-up since 1939". The VE1FO club station went on the air in June, 1968 but the equipment was moved to field day shortly after. It was stated at the September, 1968 meeting that the VE1FO station had not been used since field day. The October meeting stated the VE1FO station would soon be back in operation and that only those with phone licenses should be allowed to operate it. Wes Street, VE1EK suggested at this meeting that the VE1FO station be manned every day. The H.A.R.C. is one of the few Amateur Radio Clubs in the country that has been fortunate enough to have had access to excellent facilities in which to house a Club Radio Station. The VE1FO Club Station has operated in all the club meeting locations since 1968, and this included an outdoor tower at each location.

ATLAS AND SWAN TRANSCEIVERS

It is stated on page 23 of the December, 1979, issue of QST that over the years Atlas had built 18,000 transceivers. This will give one an idea of how many units there really are when you read an advertisement. Actually Herbert G. Johnson, W6QKI, designed, founded and built the Swan line of transceivers. The first models were assembled in his garage. This increased to the point Swan was building 400 transceivers a month. Frank Milton, W6BZN, donated his Collins KWM-1 to the *BOUNTY* as station VE0MO in 1962. He replaced that KWM-1 with a Swan and claimed he liked it just as well if not better than the Collins. Frank lived at Oceanside, California, the home of the Swan factory. He was in charge of the sound department at Metro Goldwyn Mayer at the time. They were the motion picture company that had Leo the Lion growl at the beginning of each movie. Herbert Johnson, W6QKI, was so successful with the Swan he went on and founded the Atlas line of transceivers according to page 81 of the April, 2000, issue of QST.

The VE1FO club station, a Swan 500CX, gave them a lot of trouble in the early 1970's. This transceiver was purchased by a motion at the monthly meeting on October 16th, 1970, for $734.85. There was talk of sending it back to the factory to be checked out but a cold solder joint was found and cancelled that idea. The Swan transceiver kept blowing the finals, a pair of 6LQ6's, from being tuned improperly. The 6LQ6 is a beam power pentode and was designed and built as the horizontal sweep tube in a television receiver. The 500CX had two and they had to be mated so that they performed as near identical as possible. A good description on using television sweep tubes as power amplifiers can be found covering several pages beginning at page 11 in the February, 1980, issue of QST.

These tubes were easily damaged by the inexperienced when the transceiver was tuned. The handy part was they were fairly cheap and every TV repair shop around would stock and sell them. The Swan 500CX first came out in 1970, the year this one was purchased, as the last of that line of transceivers. The power output was 480 to over 500 watts depending on whose figures you quote, and this is a powerful transmitter considering the size. If one could find a pair of matched 6LQ6's today they would be quite expensive. This would make restoring one of these transceivers expensive. The Swan 700CX was several years after the 500CX and it came with 8950 finals that proved to be much better.

Page 4, QST, August 1970

This is a Swan 500CX mounted under the dash of an automobile for mobile use.

The VE1FO Swan 500CX was used in a CQ DX contest held out in Hammonds Plains in January, 1974, and performed very well. They used it on a dipole at 105 feet and on a Monocone antenna. They reported that the Monocone was a terrific omni-directional antenna. They picked up 918 contacts for 463,760 points in this contest.

In March, 1974, Brit Fader, VE1FQ, Harley Grimmer, VE1MX and Dick West, VE1AGX had taken the 500CX to the High School at Tantallon and found it worked very well and that it was a good radio location. They showed the students all bands and worked one station in Chicago for 45 minutes. Dick Grantham, VE1AI, also mentioned that the 500CX still had the same 6LQ6 finals in it that were used on field day in June, 1973.

SWAN TRANSCEIVER REPLACED WITH HEATHKIT

This 500CX received a lot of hard use. It was used every field day and for any number of other contests and various occasions. The club agreed to sell this transceiver to Dick West, VE1AGX, at the club meeting on October 16th, 1974, for $500.00. It was replaced exactly four years to the day by a Heathkit SB102. The Heathkit had a more rugged tube, the 6146B for the finals. The minutes of the meeting simply state the Heathkit was purchased from Oakley who had built it. One assumes this is Oakley Peck, VE1UC, from Bear River who was and is a very good technician. The club paid $500.00 for the SB102.

The VE1FO club station was still running the Heathkit SB102 in November, 1979. It was approved at the November 21st monthly meeting for the transceiver to receive some general maintenance and a complete set of new tubes.

David MacKinnon, VE1ALO, has a Swan and it may be the one I have tried to describe that served HARC so well for four years. It sounds very good the few times I have heard it. Dave is one who is blessed with a very good CW bug fist (using his Vibroplex semi-automatic CW key). One you could copy and enjoy all day.

VE1FO SHACK BROKEN INTO

The VE1FO radio shack had been broken into at least once at the Nova Scotia Technical College. They had the lock on the door changed several times. There was also talk of closing the station for good although it was believed that the VE1FO station had worked 60 countries by March, 1973. Most if not all of these countries had been worked by the Swan 500CX.

HARRY PHILIPS VE1AGW OFF THE AIR

Harry Phillips, VE1AGW, was not home when his neighbour arrived with a TVI complaint in January 1962. He promptly dropped Harry's transmitter and receiver out the window where they landed 30 feet below on the ground. Harry was off the air for quite some time. Harry probably took his neighbour off his Christmas list as well. This was a very big problem years ago. It was so bad that when K5YEE moved and was putting up an 80-meter dipole in 1962 he had 4 TVI complaints before he was finished putting up the antenna.

TVI COMPLAINTS

I ran into the same problem in 1975 when I moved to Sambro Head to operate Halifax Coast Guard Radio. I had two 40-foot towers and put one up on either end of the house to simply see what would happen. All I had was the two towers. I did not have a station. One of the neighbours blamed me for messing up their TV so I took along the Department of Communications phone number and went for a visit. It went nowhere because he was a retired navy chief and a friend of my father another retired navy chief. Darn it, we got along the "finest kind" after that. The various amateur radio organizations did not get the ridiculous tower complaint that I expected.

In 1975 Mr. Benno Friesen, Conservative Member from Surrey-White Rock, B.C., in the Federal Government, brought up the problem of a constituent who had $800.00 worth of stereo equipment that he could not use because a ham neighbour kept interfering. The communications minister at the time was Gerard Pelletier and he told Friesen to get this neighbour mad enough to use profane language over his ham station so D.O.C. could shut it down. That would have been a big help, and from the one who should have been doing all he could to get those manufacturing this stereo equipment to clean it up. The manufacturer knew better. It would not have taken that much to clean it up when they made it.

One could get themselves into a real mess and everyone should be aware of Jack Ravenscroft, VE3SR, and his legal case in Ontario. Jack was put off the air, fined and had to pay legal costs. The detail on this can be found on page 66 of the June, 1986, issue of QST. The verdict to put Jack off the air came down on April 9th, 1986, and Jack appealed this ruling. The appeal was filed on May 6, 1986. Jack filed another appeal on February 28th, 1987, and was told his case would be heard in October, 1987. Jack was very frustrated and wrote ARRL a letter describing things. A DX station that was on the air at the time would have been his 360th country had he been able to contact it. The world's amateur radio community came to his rescue with financial assistance. The legal costs for all of this cost a fortune. Poor Jack was told in the fall of 1987 that he would be lucky to have his case heard in January or February, 1988. His was a civil case and there were many criminal cases that took priority. Jack was simply waiting with no amateur radio to operate.

The decision came after Jack's case had been heard on January 28th and 29th, 1988. This decision is found in Canadian Newsfronts on page 77 of the April, 1988, issue of QST. It is much too long to record here but Jack was back on the air and had 90 days to fix the neighbours appliances to a standard approved by DOC. He also had to pay the neighbour $5,000.00 for the inconvenience of losing these appliances while they were being fixed. The equipment today is better designed but can still create a headache if one is not careful.

HARC RECEIVED P.O. BOX

HARC received Post Office Box number 663 in April, 1950, that lasted until 2010. Of course, one had to add the postal code of B2T 2T3 when they were created in 1973. This may have been the address of Ron Hart, VE1MZ, and he turned it over to the club when he moved. He was using that address on his license. Ron became W9IVP in 1955. In 2010 the HARC address became P.O. Box 8895, Halifax, Nova Scotia B3K 5M5. This address made it more convenient to the club at 3380 Barnstead Lane.

Yes, it was a different world back then. The United States post office announced in the June, 1950, issue of QST that all amateur clubs should list all the local amateur radio call signs with their local post office. They should be listed with their proper QTH (address) in order for the post office to deliver all QSL cards they received addressed to call sign and city only.

HALIFAX AMATEUR RADIO WIVES' ASSOCIATION

The Halifax Amateur Radio Wives Association was formed in 1946 by the wives of the HARC members to assist in the 1946 Hamfest at Bedford. There were 13 members when it was formed and the same 13 were still active members in 1959. There were 22 paid up members when they wrapped up the season with their annual dinner in 1950. Not all the members were amateur radio operators. Sid, VE1WJ (Mrs. Doug Johnson, VE1OM), Evelyn, VE1OW (Mrs. Bill Bligh, VE1BC), Sara VE1IV (Mrs. Ron Hart VE1MZ) and Helen, VE1YL (Mrs. Wes Street, VE1EK), were members.

Page 58 QST January, 1962

According to the Maritime Division news for February, 1948, there were five of the 30 members with licenses only. Another held call sign VE1YW but I have been unable to locate her name. The call sign is not listed in the 1949 Callbook. Jean Hughes (Mrs. Neil Hughes, VE1YZ) has the VE1YW call in 2008 and has held it for years, but Norm Brooks had it for a while before she was assigned the call. Norm was VE1NEB in 2008 and became a silent key in 2012.

When I first learned of the Halifax Amateur Radio Wives Association, I felt that the group became the Sparkettes but not so. The Sparkettes were all licensed YL amateur operators. The formation of the Maritime Sparkettes was announced by Christine Weeks, VE1AKO, in the Maritime News section of QST for April, 1966. Meetings were held at 1900 GMT on 3740 kHz. This was soon changed

QST page 60 February, 1961

to every Thursday at 0930 local time on 3770 kHz with the last Thursday of the month at 1330 local time. There were members from all the Atlantic Provinces and newcomers were welcome. VE1AQI was net control but I did not know her. The Maritime News section of the June, 1967, issue of QST states that VE1OZ was operating VE0NP and he and XYL, VE1AQI, were transferred to Ottawa. Therefore, VE1AQI had been here a short while only.

An OM Speaks

I got heart palpitation –
I worked my first YL;
I asked her for a schedule
And she said, "All very well."

There were hours when we chatted
And letters sent galore;
For months we kept this up,
Then I could wait no more.

Said I, "My dearest maiden
What time for you is best
For me to come a-callin'
My eyes on you to rest?"

She quickly set the date
And I traveled far to see
This wondrous lovely lady
Who QSO'd with me?

Oh, dear, is there no justice
For hams who have such faith?
I came into her life
Just thirty years too late!

W8CKH (November, 1933)

Page 71, QST December, 1975

According to the Maritime News for December, 1968, VE1AHV was president and Alma Hills, VE1MY, was secretary of the Sparkettes. They were meeting then on 3770 kHz but failed to record the time. They stated that new or junior YL's could call in on CW.

This is in the YL News and Views article and the caption reads: "Maritime Sparkettes who attended the 1975 Atlantic-Canada Radio Convention were: seated Chris Haycock, WB2YBA, 1975 YLRL president; standing l-r, Margie, VE1YU; Gernen, VE1BCB; May, VE1AMB; Jean, VE1YW; Joan, VE1APL; and Judy, call not given. (Photo courtesy of WB2YBA)

Louise Ramsey Moreau, WB6BBO, was editor of the "YL News and Views" section of QST in January, 1969. She gives some interesting history of the YL activity in that issue. She states Eleanor Wilson, W1QON, was the first editor and that she held the position for 11 years. She stated the YL News and Views section was first published with the January, 1952, issue of QST. She also goes on to state that there were 2 American YL's only in 1910 and by 1916 there were 6 American YL's and at least one in England.

The ARRL YL News and Views section has been renamed simply YL News and does not appear in QST every month. This section is still called YL News and Views in the Radio Amateurs of Canada publication The Canadian Amateur in 2008.

THE FIRST CANADIAN YL GROUP

The formation of the first Canadian YL group, the Trilliums in Ontario, a couple of months previous to the formation of the Sparkettes must have had some influence on the YL's with the VE1 calls. Chris, VE1AKO, was a very pleasant operator with a good sense of humour. When I first met her, she was one of the net control operators of the Maritime Net on 3750 kHz each evening at 1900. Her OM was Ted, VE1AGM, who was engineer in *BLUENOSE*

II and operated a Collins KWM-2 as VE0MY. She and Ted enjoyed a trip to attend the Midwest YLRL convention, according to the July, 1969, Maritime News. This convention was held at Toronto that year and involved all the American and Canadian YL's with around 50 attending.

Nova Scotia Government

VEOMY

The December, 1969, Maritime News stated that the Sparkettes had created an award in memory of Bert Whittaker, VE1RT. This was for the YL that handled the most traffic annually between January 1st and July 31st for each year. Bert, VE1RT, became a silent key on January 16th, 1968.

There appeared to be lots of news on the VE3 Trilliums in YL News and Views but very little on the Sparkettes. They did list the Sparkettes Executive for 1975:

President – May Jones, VE1AMB
Vice President – Betty Howell, VE1AMS
Secretary Treasurer – Alma Hills, VE1MY

Alma Hills, VE1MY, and husband Bill, VE1KK, moved to British Columbia in 1986. Alma became VE7EKV and Bill VE7EKW. Most if not all of their three children were living out west.

This is the QSL card of Bill and Alma Hills and their station in 1958. This is the station I remember so well from visiting their home.

Another YL radio club, the Canadian Ladies Amateur Radio Association (CLARA) was formed in 1967. There was very little information on any of these Canadian YL clubs in QST. The YL News and Views column for August, 1987, mentions the fact that CLARA would celebrate its 20th anniversary on September 11th to 13th, 1987. CLARA was and is the national YL Canadian club with members from every province and territory.

There was a CLARA Hamfest held in Toronto from September 26th to the 28th in 1997. A description of this Hamfest is recorded on page 95 of the December, 1997, issue of QST. It is mentioned in this that CLARA was formed with the help of the Trilliums and the Maritime Sparkettes. It states the Trilliums were closed at this Hamfest and all remaining assets of the Trilliums were turned over to CLARA. It does not mention what became of the Maritime Sparkettes but one assumes a similar fate and what was left of that group, if anything, was turned over to CLARA as well.

According to the Maritime News Section in QST for January, 1947, it stated the code class was making good progress with the Dit and Dah Club. So, they were at least trying to make an amateur operator out of some of the ladies who were members of the Halifax Amateur Radio Wives Association. They were still holding code classes in 1948. The Dit and Dah Club held a banquet in 1948 that was enjoyed by one and all.

Helen Hubley, VE1YL, received her licence in 1933 and married Wes Street, VE1EK, later that year. This was 13 years after the phrase YL was created in 1920. Helen retained the VE1YL call sign for the rest of her life. She was one of, if not the first YL in this area. There were three ladies in a class of over 30 students with the Halifax County Amateur Radio

Association during the winter of 1926 – 1927 but I have found no record of one receiving a license.

In a newspaper article by C. G. Robinson that Brit, VE1FQ, has marked as 1935 (but appears to be 1936) stated there were 30 members in HARC. It goes on to state that there were two YL members in the club at that time. He goes on to state that they were just as enthusiastic about radio as the other members. But unfortunately, it did not list their call signs or names. One would have been Helen, VE1YL.

There was an excellent article on Iris White, VE1AYL, at Glen Falls, N.B., in The Evening Times-Globe, Saint John on Saturday, May 12th, 1962. Iris stated she was first licensed on March 23rd, 1938. She said at the time Helen had the VE1YL call and Evelyn had the VE1OW call in Halifax so she requested and received the VE1AYL call for "A Young Lady". She also stated she was one of the first three amateur operators in the area to receive a three letter call sign.

The first ARRL YL/OM annual contest was held in January, 1950. Top score OM was to receive a gold loving cup and the top score YL was to receive a silver loving cup and if either won it three years in a row they were able to keep the cup. The rules were rather lengthy and because I found no VE1 involvement I will not record them. Carl Evans, W1BFT, was the first to win a cup three years in a row so was able to keep the cup. According to the June, 1953, Maritime News VE1ABT in New Brunswick was very active in this contest but I am unable to locate the name for this call sign or any record in this contest.

According to Doug, VE1OM, in his report for August, 1954, VE1ABT was Doreen and her husband was VE1PF and they recently had a baby. At the same time, she came in 2nd in Canada and 4th in North America in the CW portion of the YL/OM contest. It does not appear to be of any record for HARC.

The February, 1966, issue of QST had a photograph of the Ontario Trilliums Amateur Radio Club. The caption of this identified all the women in the photo by call sign and all were VE3 calls. It also stated that this was the first YL group formed in Canada. From the records I have been able to locate I would agree with that statement.

A "Campfest" was held at Beaverbank Labour Day weekend 1966 that was put on by a few members of HARC and The Maritime Sparkettes. Moe Lake, VE1PX, was one who put a lot of work into this "Campfest". Janet Barwise, VE1ARB, won the code copying contest. Janet enjoyed CW and handling traffic with her Heathkit DX-100 when I first met her years ago. She holds call sign VY2XX in 2008. Janet became a silent key on February 15, 2018 at age 84. Maritime News for February, 1967, noted that she had passed her advanced license and that she had handled 20 messages that month. Janet's husband is not a ham radio enthusiast.

The Halifax Amateur Radio Wives Association was first known as the Dit and Dah Club. The first name of the group was found too cumbersome and was changed to the Halifax Amateur Radio Wives Association. We have seen it listed as the Halifax and Dartmouth

Amateur Radio Wives Association. They had an annual party in June 1950. They were commended for the social and charitable work they had been doing. They had a paid-up membership of 22 members at that time. They had a banquet and social evening at Lakelodge in the summer of 1951. Apparently, this group more or less disbanded in 1952 and was reformed in 1953 under the name, the Ladies Dit and Dah Club. The December, 1956, Maritime News states the Ladies Dit and Dah Club was going strong and had been for nearly ten years. The name "Ladies Dit and Dah Club" sounds like something Evelyn, VE1OW, would create. Evelyn Bligh, VE1OW, received her license and the VE1OW call sign in early 1937 about the same time HARC received the VE1MK call sign.

In the minutes of the monthly meetings for HARC during 1976 they make mention of the ladies' auxiliary of the club. One assumes this was simply a group of the ladies that were involved with the members of HARC. At least I have found no record that the Amateur Radio Wives Association was still active.

Brit, VE1FQ, gave more detail on this in his Bulletin for January, 1976. "The Ladies Auxiliary of the H.A.R.C. is sponsoring a Bingo at the Royal Canadian Legion Building, 5837 Cunard Street on Sunday February 1st." In the February Bulletin he stated Shirley Trites, VE1AZY, was the leader of this Bingo and it received a very poor turnout.

There were about 430 amateur radio operators in Nova Scotia in 1959 and over 15 were YL operators. The Halifax Amateur Radio Wives Association was felt to be unique in 1959 and as far as was known was the only one of its kind in Canada. This association held monthly meetings in the homes of the various members in rotation and each season ended with an annual dinner and an election of officers for the coming year. The association was very active and took on a number of projects, such as supplying Christmas packages to needy families, or furnishing bundles for a local welfare home, etc.

Calling CQ

When I call CQ while using the key,
At least one station will come back to me.
For on cw a guy makes a choice
By strength of the signal and not by a voice.

I call CQ on the ten-meter 'phone,
I twist the dials – wear my thumb to the bone.
I try it again. I yell and I yell,
Do they answer me? No! They call a YL.

At last, I got mad and said to my wife,
Come here, sweetie pie, and help save my life.
Put out a CQ on this old pile of junk,
Let's prove this YL biz is true or the bunk.

She calls a CQ in her sweet-mannered style,
Ye Gads! Hear the answers – I didn't touch the dial.
They gave her reports of R5 . . . S9 . . . plus.
Not in all my experience have I heard such a fuss.

She kept it going for quite a long time.
Oh! Boy! Am I happy, someone's on the line?
I said, sweetie pie thanks a lot for the help,
Now give me the mike, let me give him a yelp.

Now I know band conditions can change mighty fast,
But I can't understand how her contact can last;
For soon as she's finished and I take the mike,
QRM blanks me out and the guy has to hike.

I'm going back to cw where all CQs hit,
Where I get many answers and don't have to sit
And holler and shout and call to beat hell
Then hear the guys answer a sweet-voiced YL.

But on second thought, I think I will stay
On ten-meter 'phone and enjoy every day.
Now why should I crawl up on a shelf?
I'm going to answer some YL's myself!

- G. E. Hoffstetter, W9JC

Page 28, QST May, 1950

QSL MANAGERS

John Roue, VE1FB, joined the Navy during World War II and decided to make it a permanent career in 1947. In 1957 Commander John Roue managed to get back his old VE1FB call sign. John had turned the QSL card manager job over to Brit Fader, VE1FQ. Art Grant, VE1EP, did not know why the change was made. Brit stated he held the job from January 1st, 1939 until January 1st, 1987.

U.S. amateurs are reminded that the postage on QSL cards mailed to Canada is two cents — not one cent as so many believe. A card carrying insufficient postage is subject to delay at the Canadian end while the mailman goes through the embarrassing formality of collecting the extra postage from the VE addressee.

Page 116 QST July 1947

The Canadian Divisional news section of QST magazine terminated with the termination of the amateur radio license at the outbreak of World War II. But the list of Canadian QSL Managers continued in each issue of QST and Brit Fader is listed as the VE1 QSL Manager all through World War II.

After the war Brit spent his annual vacations delivering QSL cards to those in the VE1 call area. A lot of amateurs do not bother to collect these cards and it is difficult to deliver them. The Editor of QST in August, 1953, stated that the QSL manager for the W2 call area had a stack of unclaimed QSL cards that was 41 feet high.

The difference between a drunk and an alcoholic is simple. The drunk does not have to attend the damn meetings and this same factor applies to amateur radio. The majority of those who get an amateur radio license simply want to learn radio and nothing more. They have no desire to be part of any group, meeting, or anything like that. There are those who have to be in charge and create various rules for the rest to abide by. The QSL card was created at the very beginning of the amateur radio hobby. Simply mail a post card to the one you worked to confirm the contact. This went on to create the QSL Card Bureau. There are incoming bureaus where one receives a QSL card and outgoing bureaus where one can send a QSL card. Without the bureaus it is very expensive and with the bureaus it is simply expensive. There are those who will do anything to get a card to confirm a contact, but the majority of amateur radio operators simply participate only when it is absolutely necessary.

That stack of unclaimed QSL cards that were 41 feet high at the W2 QSL Bureau must have been 41 feet long and recorded as high simply to try and impress someone. They claim there were a lot of good DX cards in that pile. It not likely occurred to anyone that those to whom those cards were addressed did not care if they ever received them. All of these QSL bureaus now have cards lying around since Pontius Pilot was an air cadet. Well, not quite that long, but they are definitely decades old. There must be some reason for retaining them but for the life of me I cannot think of one. As a commercial radio operator, we retained our logs for three months. The one who did up this month's month-end report stored the logs for the month with the others and removed the logs three months old. They were burned in a burn barrel for that purpose until the invention of the shredder. After that they were shredded and the paper recycled. If a log involved an accident it was retained for three years, then destroyed. Why the amateur radio QSL bureaus retain these cards over three years is above and beyond my comprehension.

Jack Helmbold, K6INM, stated on page 24 of the October, 2006, issue of QST that the W6 QSL Bureau has an awful lot of these unclaimed cards as one can imagine. He goes on to state that there are some really good DX cards and that it is a shame because those who send these cards go to a lot of trouble and expense. We all understand that but he went on to state that the W6 Bureau destroys these cards after a year or so. I did not know there was a bureau that destroyed the unclaimed cards. I felt that just maybe the cards were treated as unclaimed mail and that no one could destroy that. I had often felt that the unclaimed cards should have been returned to the sender stamped held at the bureau for a certain period, unclaimed and returned. We have the signal KN to indicate we want to communicate with the station called only. Maybe a signal should be created to indicate that we will not send a QSL nor do we want one for the contact.

COMPUTER LOGGING PROGRAMS

There have been various types of logging programs created for computers. The one most popular with the members of HARC is known as DX4WIN. Howard, VE1DHD, has given courses on this program for the members that are interested in 2009. Their web page makes the following claim; "DX4WIN is an easy to use, yet powerful logging program for every ham. It has been designed for the serious and the casual DXer. It has a number of features that make operating in a contest fun from the DXers perspective, and if you use a contesting program, DX4WIN can import your log after the contest."

Another that became available on September 15th, 2003, is ARRL's Logbook of the World (LoTW). This is described as an electronic alternative to collecting QSL cards for awards purposes. They state: "ARRL's *Logbook of the World* (LoTW) system is a repository of log records submitted by users from around the world. When both participants in a QSO submit matching QSO records to LoTW, the result is a QSL that can be used for ARRL award credit. To minimize the chance of fraudulent submissions to LoTW, all QSO records must be digitally signed using a digital certificate obtained from ARRL. Obtaining such a certificate requires verification of the licensee's identity either through mail verification (US) or inspection by ARRL of required documentation (non-US). Software developed by ARRL can be used to convert a log file (in ADIF or Cabrillo file format) into a file of digitally signed QSO data, ready for submission to LoTW".

Therefore, it appears as though it is getting easier to send and receive contact confirmation. One can still use the old QSL card via the various incoming and outgoing bureaus. A few of those QSL card collectors claim amateur radio is losing a lot of tradition with the loss of the QSL card.

VE1FQ QSL BUREAU

The address to the VE1 QSL Bureau was changed to the same address as HARC in 1958. The QSL Bureau had been using Brit's home address on Henry Street up until that time. Brit lived there with his mother until he married.

HARC has not only retained Brit's VE1FQ call sign but has the QSL bureau, and has it named the Brit Fader Memorial QSL Bureau. This is a fitting tribute to a well-known and hardworking HARC member. Bob Burns, VE1VCK, was presented with a lifetime membership in the HARC in 2008 for his excellent work in running this QSL Bureau. Bob had to give it up in 2007 and it was turned over to Tom Caithness, VE1GTC.

ARRL INCOMING MAIL

By February, 1970, the ARRL had their own pickup truck making 3 mail deliveries each day to headquarters. There it took 4 girls and one supervisor 3-1/2 hours to sort, open and date stamp each piece of incoming mail. At that time ARRL was mailing around 100,000 copies of QST around the 20th day of each month. It was and is a busy place.

SEVERAL FM REPEATERS IN OPERATION

The October, 1969, Maritime News was happy to report several FM repeaters in operation in Nova Scotia and New Brunswick and that several VE2 and VE3 stations had checked in while in this area on vacation that summer. They were still calling the 2-meter band the FM band at that time. The January, 1970, Maritime News gave more detail and stated there were FM repeaters at Moncton, Halifax, St. John's and Truro. They went on to state that FM repeaters were planned for Saint John and Sydney.

The Maritime News for January, 1977, states that there were 25 active repeaters in the Maritimes. It did not take long once it got going. The minutes for the monthly club meeting on September 17th, 1975, have this statement; "Yarmouth Repeater is now 01-06, whatever that means".

At the time they were installing these first repeaters each identified by keying the call sign of the repeater in Morse code. This created the digital Morse code generator and a homebrew model was described in detail starting on page 11 in the June, 1970, issue of QST. This was a big improvement, especially for the various beacons around the country. Up until that time these beacons were keyed via a rotating wheel with the Morse characters carved around the edge of the wheel so that they closed a mechanical switch to make the keying transmission of one or two letters. This of course was not amateur radio. This was the various beacons along the coast for ships and all the beacons found all over the country for aircraft. The amateur radio world improved the commercial radio world again.

Back then the first FM 2-meter rigs were "Rock Bound". Crystal controlled on both transmit and receive. The store-bought rigs were usually 6 or 12 channels and one had to buy two crystals for each channel, one for the receiving section of the channel and the other for the transmitting section. My first rig was 12 channels but I did not have all the channels because I did not need them. They were great rigs to operate because when one was on the correct channel that was all there was to it. Very simple and I am a simple person who enjoys simple things.

While at an amateur radio club meeting in Seattle, Washington, in 1962 we were told that if we were having a QSO and the repeater drowned us out, the repeater had priority. The reason for this was rather simple. The first repeaters were on 146.34/146.94 and 146.94 was the simplex or direct frequency. Simplex frequency 146.52 was created shortly after that.

The first repeater in Canada was one in Toronto that identified with the VE3RPT call sign and went on the air in 1966. This was four years after that meeting I attended in Seattle in 1962.

Harrison was advertising what was probably the first Icom transceiver in the August, 1970, issue of QST. It does not state that it is Japanese and it makes no mention of Icom. The transceiver was simply called an IC-2F. This was a 2-meter FM transceiver that had six channels and came with crystals for two channels – 146.34/146.94 and 146.94/146.94. The power input to this transceiver was 20 watts that gave a 10-watt output. QST gave this transceiver one of its detailed technical descriptions starting on page 48 of the January, 1971, issue. In this article they called this radio a Varitronics – Inoue IC-2F. There is a photograph of Tokuzo Inoue, JA3FA the founder and president of ICOM on page 19 of the September, 2000, issue of QST. So that is where the name Inoue came from!

By 1972 amateur radio was trying hard to get all the new repeaters that were showing up on 2-meters to operate on standard frequencies. There was also a lot of discussion on auto patching; making phone calls through these repeaters.

In September, 1976, I came up past Halifax in a large ore carrier. My cabin was up behind the bridge and I had my Larsen 5/8ths magnetic mount up on "Monkey Island". That is the roof of the bridge and was roughly 110 feet above the water. A 450-foot steel ship in salt water makes a prefect transmitting platform. I wanted to let Joan and our two boys know I was going by but could not use 146.34/146.94. When I keyed it for the Halifax repeater, I got both the Halifax repeater and the Charlottetown repeater. They both were on the same frequencies.

REPEATERS LINKED

Don Welling, VE1WF, stated in his Maritime News for March, 1984, that several Halifax amateur radio operators had worked stations with a W4, W5 and W6 prefix via the VE1LHR repeater and links to 10-meters. Today these various repeaters are linked by various means so that it is not unusual to have someone in England call in on one of the repeaters in this the Halifax area. It does not take long for things to change when a group of amateur radio operators put their minds to it.

Canadian Newsfronts in the October, 1986, issue of QST gave the statistics for amateur radio licenses in Canada on April 27th, 1986. This record stated there were 556 repeater licenses in Canada and 28 were in Nova Scotia.

The May, 1991, issue of QST describes the first of the integrated repeaters where one could operate several repeaters at once by linking them together.

1976 HAMFEST

This was another Hamfest held in 1976 and is self-explanatory.

As host club, The Halifax Amateur Radio Club, welcomes you to attend Hamfest '76, The Maritime Amateur Radio Convention for 1976, at St. Marys University Convention Centre, Halifax, N.S., August 20, 21 and 22, 1976.

Available for the Convention are: Live-in accomodations, complete dining facilities, free parking, private rooms - single or double, apartments - 2 bedroom with 2 beds per room, bath, shower, etc., supervised swimming pool (professional life guards).

Situated in Downtown Halifax, the prices are extremely reasonable: Single room (private) $9 per day; Double room (private) $15 per day.

You may extend your visit up to two weeks with all the above facilities: Single room, 7 days $45; Double room, 7 days $79; 2 bedroom apt. with kitchen, 7 days $135.

For your convention kit, write or telephone Leo G. Perry VE1AMI, Box 663, Halifax, N.S.; telephone 902-463-6668.

Two complete programmes - one for OMs, XYLs and YLs (Amateurs). One for XYLs and Jr. Ops sponsored by the XYL's Auxiliary.

Convention Center is in a High Rise Complex, all under one roof!

HARC Files

NOVA SCOTIA AMATEUR RADIO ASSOCIATION

The Nova Scotia Amateur Radio Association (NSARA) was incorporated and received its by-laws on April 17th, 1957. This association was formed by 14 amateurs from Nova Scotia at the Maritime Section A.R.R.L. Hamfest held at Bathurst, New Brunswick, in 1956. Dr. Leo Doucette, VE1FH, of Cheticamp was elected president and Hugh Corkum, VE1VN, Lunenburg was elected Secretary. There were only

430 amateur radio call sign holders in the province at the time and within six months 150 of them were members of this association.

The NSARA is involved in many activities providing support to Amateur Radio in Nova Scotia.

- Liaison between Provincial and Municipal levels of Government

- Emergency Measures Preparedness and Communications Training

- Amateur License Plate Program

- Contesting to promote Amateur Radio

    - Worked all Nova Scotia Counties (WARC)

    - VHF Century Club (VHFCC)

    - The NSARA Contest

- School Business

- Frequency Co-ordination (VHF – UHF)

- Assistance with provincial linking

- Sponsoring of the Nova Scotia Packet Internet Gateway (VE1NSG)

- Co-sponsoring of the Internet Radio Linking Project (IRLP) Voice over Internet World Wide

- Yearly Hamfest and Picnic, CW Contest, Fox Hunt, Home Brew Competition, and Distress Message Copying

ILLEGAL USE OF AMATEUR CALL SIGNS

I believe those who make up an amateur radio call sign and somehow get on the air will be with us forever. I do believe a lot of this garbage has moved over to the land of computers and the Internet. But this was very bad and involved a lot of illegal VE1 call signs during the late 1950's. There were several notices to report any known instances of this. I know my VE1BC call has been a popular call for these characters several times since I was assigned it in January, 1975. Dave Oldridge, VE1EI, tried to contact one several years ago that was using my VE1BC call. Dave felt he was in the area of Fredericton, New Brunswick, but was unable to contact him.

STOLEN EQUIPMENT

Each and every issue of QST lists any amount of stolen amateur radio equipment. One fellow showed up at an amateur radio flea market in the U.S. and wanted to know what a certain transceiver was worth with the serial numbers removed. He was told at least two years so disappeared in a hurry.

BOY SCOUT EXERCISE

The HARC station VE1FO/1 was kept busy at the Lord Nelson Hotel in Halifax during the Maritime Boy Scout Convention in 1959. VE1FO/1 was also active during the Boy Scout Jamboree on the air. This activity has continued to 2008 when Sam, VE1YVN, and Doug, VE1LDL, participated in the 2008 Boy Scout exercise on the air.

DXCC

Two HARC members were in direct competition with each other trying to get the high spot on the DX roster. Art Grant, VE1EP, and Doug Johnson, VE1PQ, each had over two hundred countries confirmed in 1959. It would likely be impossible to record all the HARC members over the years who have earned DXCC. This is the oldest contest in amateur radio and probably the most popular amateur radio contest.

Page 101, October 1992 QST

The all-time DXCC leader recorded in the November, 1989, issue of QST was George de Grenier, W1GKK, of North Adams, Massachusetts. If George is not the all-time leader still, he is well up on the list of all time high scores. He had 372 countries confirmed then. Confirmed, not worked, and this was over a period of 50 years so some of those 372 are now deleted countries. He did not run an elaborate station it was simply a modest set-up and he made these contacts simply from pure patience and persistence.

GENERAL RADIO SERVICE/CITIZEN BAND RADIO SERVICE

The Citizen Band radio service in the United States is known as the General Radio Service in Canada. These frequencies were the old 11-meter Amateur Radio Band. I do not have the date the United States created this service from this band, but the ARRL stated on behalf of the Canadian government in January, 1959, that the band would remain for amateur radio use only in Canada at that time. That at least gave our illustrious leaders time to create the General Radio Service in Canada. The General Radio Service (GRS) was activated in Canada on April 1st, 1962, and a portion of the old 11-meter band was retained for amateur use in Canada at that time. It is interesting to note the GRS label did not catch on in Canada and it remained simply CB.

The local CB club was known as the "Kingfisher CB Club" here in Halifax. They claim it was a much better club than HARC because it had more members and more activities such as dances and so on. A member of HARC would on occasion attend a CB club meeting and on occasion members of the CB club including the Kingfisher president would attend a monthly HARC meeting.

The HARC minutes for their monthly meeting May 19th, 1976, stated that the local DOC office in Halifax had issued 10,000 CB licenses and CB appeared to be taking over. The minutes two months before on March 17th noted that the CB crowd was using high power stations with fictitious amateur radio call signs outside the old 11-meter band and had been heard by various members of HARC.

The CB crowd was getting rather brazen and rather self-important when it was noted in the minutes of the June 15th, 1977, monthly meeting of HARC that the local CB clubs wanted a list of all amateur radio call signs issued in the Halifax area. They wanted this list in order to see what was available.

The editor of QST had a lengthy article in the September, 1977, issue concerning the illegal activity of CB. It really was getting very bad at the time. With human nature as it is one has to admit it was true to form. How could the various authorities expect anything different? The CB operators were leaving the assigned CB frequencies and using illegal amplifiers and there did not seem to be any end to what was taking place. HARC had tried to get the CB crowd to take an interest in amateur radio. They had a film on loan from ARRL and Brit Fader, VE1FQ, showed this to various CB clubs around Halifax and as far away as New Glasgow. At least some of those who had participated in this CB activity became amateur radio operators.

HARC supplied a radio equipped vehicle to the police on special occasions such as Halloween. A police officer would ride with the amateur radio operator in his vehicle and they would go on patrol. The Kingfisher CB club took this over from HARC once they were well established. This is rather hard to believe today with insurance and everything the way it is, but back then the automobile was still considered a miracle of the modern age. The automobile has since grown to the point it has taken over our lives and owns us. There is nothing that costs so much, creates so much pollution and is worth so little so fast. And people will simply go without eating in order to own one. Let's face it there are 15 times the people on this mud ball called earth than there should be.

Brit Fader, VE1FQ, kept close tabs on the Halifax and area CB clubs. He gave a detailed description of their activity at the monthly meeting held on October 19th, 1977. He stated at that meeting that the CB club had held a symposium at St. Mary's University that was set up for 600 and only 150 turned up. He went on to state that there were 600,000 licensed CB operators in Canada and 30% more than that were unlicensed.

It is recorded in the minutes of the January 18th, 1978, monthly meeting that the city was getting fed up with the many CB stations and wanted some assistance from the hams with this matter.

It was noted at the monthly HARC meeting on April 18th, 1979, that HARC paid the $13.50 for the CB license for the EMO trailer. That is unusual, a Ham Club buying a CB license, but it will give one the cost of a CB license.

The FCC in the United States put an end to the CB license in 1983 and from then on one did not require a CB license but they still were held accountable under the various FCC rules and regulations.

DOUG VE1UY HOME

HARC member Doug Conrad, VE1UY, was home for a vacation in 1959 from his duties as a ship's radio officer according to the Maritime News section of QST. Doug was the radio officer in the Liberian vessel *RIO SACRAMENTO* with international call sign 5LTT. Doug was the only Canadian in the vessel. She had German officers and Cayman Island crew and he said they were a great bunch to sail with. Doug had flown to Japan and joined the vessel while still under construction. Doug held call sign VE1ZL for several years and was known as VE1 Zima Lulu. We felt that described him better than the standard phonetics. Doug's XYL Dorothee held call sign VE1CBA until Doug managed to obtain his old VE1UY call again. Dorothee was also a ship's radio officer. She was serving in the German merchant navy when she and Doug met. Dorothee now holds the VE1ZL call sign.

1959 HAMFEST

This is the detail on the 1959 Hamfest and thanks to HARC member Murray Alary, VE1ALS, for providing this detail.

Murray Alary, VE1ALS

HALIFAX AMATEUR RADIO CLUB
1959 MARITIME SECTION A. R. R. L. CONVENTION

EXECUTIVE

Ray Wilson, VE1WL, President HARC
Don Bain, VE1LZ, Chairman Convention Committee
Norm Weedmark, VE1QV, Secretary-Treasurer

COMMITTEES

Publicity	Finance
Brit Fader, VE1FQ	Gerry Snair, VE1TA
Cyril Boudreau, VE1RJ	
Program	Registration
Doug Johnson, VE1OM	Aaron Solomon, VE1OC
Doug Johnson, VE1PQ	
Ladies	Display and Advertising
Ada Crowell	Wes Street, VE1EK
Accommodation	Posters
Jack Mather, VE1LY	Tom Clahane, VE1SP
A. R. R. L. Booth	P. A. System
Binks Fisher, VE1AFN	Dave Morrison, VE1GV
Ralph Beach, VE1WD	
Hidden Transmitter Hunts	Convention Station, VE1FO
Brit Fader, VE1FQ	Mike Goldstein, VE1ADH
Cliff Short, VE1AW	Bob Shultz, VE1IF
	Andy White, VE1AEW

The assistance of others not listed is acknowledged:
Members of the Halifax Amateur Radio Club.
Members of the Dartmouth Amateur Radio Club.
Members of the Halifax and Dartmouth Amateur
Radio Wives Association.

Murray Alary, VE1ALS

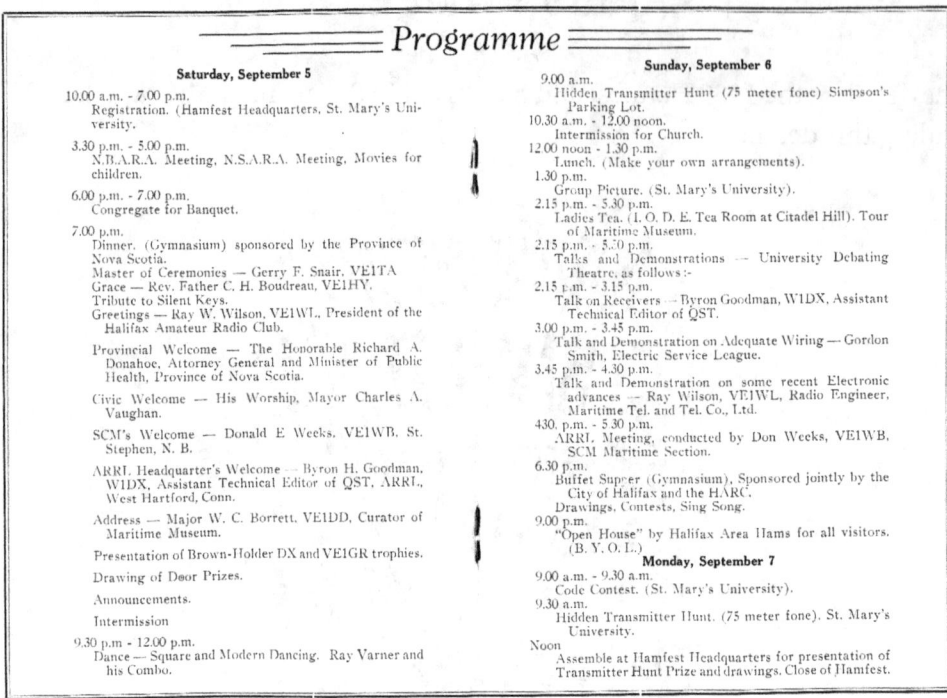

=== *Programme* ===

Saturday, September 5

10.00 a.m. - 7.00 p.m.
Registration. (Hamfest Headquarters, St. Mary's University.

3.30 p.m. - 5.00 p.m.
N.B.A.R.A. Meeting, N.S.A.R.A. Meeting, Movies for children.

6.00 p.m. - 7.00 p.m.
Congregate for Banquet.

7.00 p.m.
Dinner. (Gymnasium) sponsored by the Province of Nova Scotia.
Master of Ceremonies — Gerry F. Snair, VE1TA
Grace — Rev. Father C. H. Boudreau, VE1HY.
Tribute to Silent Keys.
Greetings — Ray W. Wilson, VE1WL, President of the Halifax Amateur Radio Club.

Provincial Welcome — The Honorable Richard A. Donahoe, Attorney General and Minister of Public Health, Province of Nova Scotia.

Civic Welcome — His Worship, Mayor Charles A. Vaughan.

SCM's Welcome — Donald E. Weeks, VE1WB, St. Stephen, N. B.

ARRL Headquarter's Welcome — Byron H. Goodman, W1DX, Assistant Technical Editor of QST, ARRL, West Hartford, Conn.

Address — Major W. C. Borrett, VE1DD, Curator of Maritime Museum.

Presentation of Brown-Holder DX and VE1GR trophies.

Drawing of Door Prizes.

Announcements.

Intermission

9.30 p.m - 12.00 p.m.
Dance — Square and Modern Dancing. Ray Varner and his Combo.

Sunday, September 6

9.00 a.m.
Hidden Transmitter Hunt (75 meter fone) Simpson's Parking Lot.

10.30 a.m. - 12.00 noon.
Intermission for Church.

12.00 noon - 1.30 p.m.
Lunch. (Make your own arrangements).

1.30 p.m.
Group Picture. (St. Mary's University).

2.15 p.m. - 5.30 p.m.
Ladies Tea. (I. O. D. E. Tea Room at Citadel Hill). Tour of Maritime Museum.

2.15 p.m. - 5.30 p.m.
Talks and Demonstrations — University Debating Theatre, as follows:-

2.15 p.m. - 3.15 p.m.
Talk on Receivers — Byron Goodman, W1DX, Assistant Technical Editor of QST.

3.00 p.m. - 3.45 p.m.
Talk and Demonstration on Adequate Wiring — Gordon Smith, Electric Service League.

3.45 p.m. - 4.30 p.m.
Talk and Demonstration on some recent Electronic advances — Ray Wilson, VE1WL, Radio Engineer, Maritime Tel. and Tel. Co., Ltd.

4.30 p.m. - 5.30 p.m.
ARRL Meeting, conducted by Don Weeks, VE1WB, SCM Maritime Section.

6.30 p.m.
Buffet Supper (Gymnasium), Sponsored jointly by the City of Halifax and the HARC.
Drawings, Contests, Sing Song.

9.00 p.m.
"Open House" by Halifax Area Hams for all visitors. (B. Y. O. L.)

Monday, September 7

9.00 a.m. - 9.30 a.m.
Code Contest. (St. Mary's University).

9.30 a.m.
Hidden Transmitter Hunt. (75 meter fone). St. Mary's University.

Noon
Assemble at Hamfest Headquarters for presentation of Transmitter Hunt Prize and drawings. Close of Hamfest.

Murray Alary, VE1ALS

There are probably a few copies of that group photograph taken at 1:30 PM Sunday September 6th packed away in a dusty old shoe box someplace. It is unusual to not find the Bligh's, VE1BC and VE1OW mentioned. Ada Crowell was the treasurer of the Halifax Amateur Radio Wives Association at that time. A record states she was the XYL of VE1YL. Helen Street was VE1YL at that time and no Crowell is listed with a Y call sign in 1949 or 1963. Art Crowell, VE1DQ, became a silent key in 1963. Therefore, unless someone can identify Ada Crowell, we do not know who she was.

RECORD CODE SPEED U.S. ARMY SERGEANT

The U.S. Army claimed in 1960 that they had a Sergeant that could send 30 words per minute with either hand, 18 words per minute with the right foot and 16 words per minute with the left foot. That speed with the left foot is no big deal. I am sure most of us have worked any number of those.

OSCAR - SATELLITES

The first Orbiting Satellite Carrying Amateur Radio (OSCAR) made history when it first flew on December 12th, 1961. I heard of this but was so busy trying to learn radio at Radio College of Canada in Toronto that I paid little attention to it. This first one simply orbited the earth transmitting "Hi" in Morse code. I met a few of those responsible for this in California the next summer. It was all the news and a lot of excitement within the amateur radio world that one can imagine at the time. The December, 1986, QST has a complete story on the first Oscar on its 25th anniversary.

There is no end to the improvement that can be made on these satellites. The one launched on June 15th, 1988, was known as AMSAT phase 3C and became AMSAT OSCAR 13. This one was launched into preliminary orbit by the European Space Agency's Ariane-4 rocket and had communication capabilities unheard of a few years before. The apogee or the highest point reached by this OSCAR was nearly 22,000 miles with the resulting DX potential. This OSCAR had two modes; Mode S, with 436-MHz uplink and 2401-MHz downlink frequencies, and Mode JL, with 145-MHz and 1290-MHz uplink, and a 435-MHz downlink. QST made a request to share your station design if you communicated via this OSCAR.

19 months after the launch of OSCAR 13 six more were sent aloft on January 21st, 1990. A photograph of the Ariane V-35 rocket that carried these six aloft appeared on the front cover of the April, 1990, issue of QST. These six became OSCAR 14 through 19 inclusive. This made it possible for a hand-held transceiver in North America to talk to a hand-held transceiver in Europe.

Amateur radio received a lot of excellent publicity on December 4th, 2008, when the high school students of Quispamsis, New Brunswick, managed to communicate with the space station going overhead. Teacher Valerie Conrod and The Loyalist City Amateur Radio Club had been 3½ years organizing this event. They had a microphone placed in the school gymnasium and 20 students took turns walking up to the microphone and asking the station a question. The audio was very good and both the reception and transmission could be clearly heard by all in the room. There were several hundred students and their families in attendance for this science class. This was played over several news broadcasts by the Canadian Television network (CTV). The club call sign VE9LC was used. It was a very interesting experience for one and all, and I am sure the students who took part and those who witnessed this event will cherish the memory all their lives.

The news media gave four college students from Humber College in Toronto excellent coverage via television and newspapers on February 2nd, 2009. They had made history when they contacted the space station 440 kilometers overhead. They did it the hard way by building everything from scratch. Their professor Mark Rector said they did not know a thing about radio when they started. The four were Patrick Neelan, Paul Je, Kevin Luong and Gino Cunti. They spoke to Sandra Magnus on the space station. Sandra was using the International Space Station Amateur Radio Club call sign NA1SS. The college students were using call sign VA3JUV. This was the call sign assigned to Gino Cunti. It had taken the students 18 months to put it all together and see it work. This would not have been possible without amateur radio and it is a shame so little credit was given to amateur radio.

Making radio contact with one of these satellites, space shuttles, space stations and so on is a complete science within itself today. At least I would have to study it for some time before I tried to make contact. If not, I am sure I would simply be QRM for those that knew what they were doing.

ITU AMATEUR RADIO STATION

The International Telecommunication Union (ITU) Amateur Radio Station opened in 1962 in Geneva, Switzerland, to promote amateur radio in all nations.

NEW COUNTRIES 1960'S

There were a lot of new countries created during the 1960's. A lot were former protectorates of other nations that gained their independence. Africa and the West Indies were two areas that saw a big increase. This gave concern toward the viewpoints of the various uses of the HF spectrum. Some of these small nations were leaning towards large broadcast facilities on HF.

Several small nations like Jamaica received their independence from the larger nations like Great Britain. It was a while before their amateur radio stations were assigned a call sign with a two-character prefix followed by a digit and then letter suffix. For example, when Jamaica received her independence her amateur stations had been assigned VP5xx as a call sign. On receiving their independence, they were assigned 6YAxx as a call sign and two years later in 1964 it became standardized. In other words, one with 6YABC was now 6Y5BC. This same fact applied to several of these small nations and a few still use the two-character prefix, the letter A and a one, two or three letter suffixes. It is also worthy of note that Jamaica and the Cayman Islands were one and the same with VP5xx call signs until Jamaica gained her independence. The Cayman Islands remained a British protectorate and received call signs with the ZF prefix from then on.

TOURIST BUREAU QSL CARDS

In 1962 Nova Scotia amateur radio operators could obtain Nova Scotia Tourist Bureau Post Card QSL Cards via NSARA. I have seen the tourist bureau QSL cards in use back in the 1930's and for various years since then. It is a good way to advertise the province to the traveling public.

AMATEURS HANDLED ELECTION RETURNS

There was a bad ice storm in Nova Scotia on April 8th, 1963, that disrupted communications for an election. Various Nova Scotia amateur radio operators received high praise, including members of HARC, for handling some election returns to various points within the province.

Amateur radio operators all over the province stepped into the communications breach during Monday's freak electrical-snow storm. Liverpool and area was particularly hard hit with telephone communications down. Election results and traffic for the hard hit power companies was handled by the amateurs. Typical of the many ham operators operating on that day was A. F. "Wiggy" Wigglesworth of Liverpool, shown above operating station VE1US. (Morton Photo).

HARC Files

A. F. "Wiggy" Wigglesworth became a silent key in 2008.

ZIP AND POSTAL CODES

The United States received their zip codes in 1963 and the ARRL started advertising and requesting these codes in the September, 1963, issue of QST. The first Canadian postal code showed up ten years later with the July, 1973, issue of QST. "Shep" Shepherd, VE3DV ex VE1RR, recorded it after his address on the Section Communications Manager Masthead. The rest of the Canadian postal codes appeared in the SCM address list in January, 1976, with the exception of Nova Scotia and British Columbia. Aaron Solomon, VE1OC, first listed his postal code as SCM in the May, 1976, issue of QST but British Columbia did not list theirs.

CANADIAN LICENSE NUMBERS 1960'S

Noel Eaton, VE3CJ, the Canadian Director of ARRL provided these figures.

CANADIAN LICENSE FIGURES

Below we present, through the courtesy of Director Eaton, amateur station license figures by regional offices, as of March 31, 1964, with comparison for earlier years:

Region	1964	1963	1962	1961
Vancouver	1398	1415	1150	1280
Edmonton	1073	986	939	912
Winnipeg	1201	1193	1118	1087
Toronto	3907	3742	3417	3192
Montreal	1890	1773	1692	1586
Moncton	1161	1073	1016	953
Shipboard VE0	10	26	15	21
TOTALS	10,640	10,208	9347	9031

Page 81 QST, August 1964

These numbers had increased to a Canadian total of 18,015 on April 30th, 1978. The same total in 1977 was 16,357 and in 1976 were 15,346, as one can see a steady increase. 2,091 of those 18,015 amateur licenses in 1978 were here in Atlantic Canada. That would include VO1, VO2 and VE1.

I would like to think that the big increase in the VE0 prefix in 1963 was from my use of the VE0MO call sign in 1962, but I have no idea what caused it. There were 62 naval ships in commission in 1960 if that is any help. There was definitely a big increase that died rather quickly as one can see.

The April, 1979, issue of QST stated that the total number of amateur radio licenses of all classes in the United States at the end of 1978 was 356,336. In the January, 1985, issue ARRL stated they had a gain of 18,800 licenses but a loss of 19,644 of those who did not renew their license, therefore the number of amateur radio operators in the United States was on a decrease.

ASSISTANCE FOR PARADE AND YACHT RACE

A big event for HARC in 1967 was to provide communications for the Natal Day Parade. The CBC (Canadian Broadcasting Corporation) amateur radio club assisted HARC with communications for the Marblehead to Halifax yacht race in 1967 as well. Unfortunately, this yacht race web site does not describe the history of these races back that far.

CLUB LICENCE AND LETTER RE STATE OF WAR

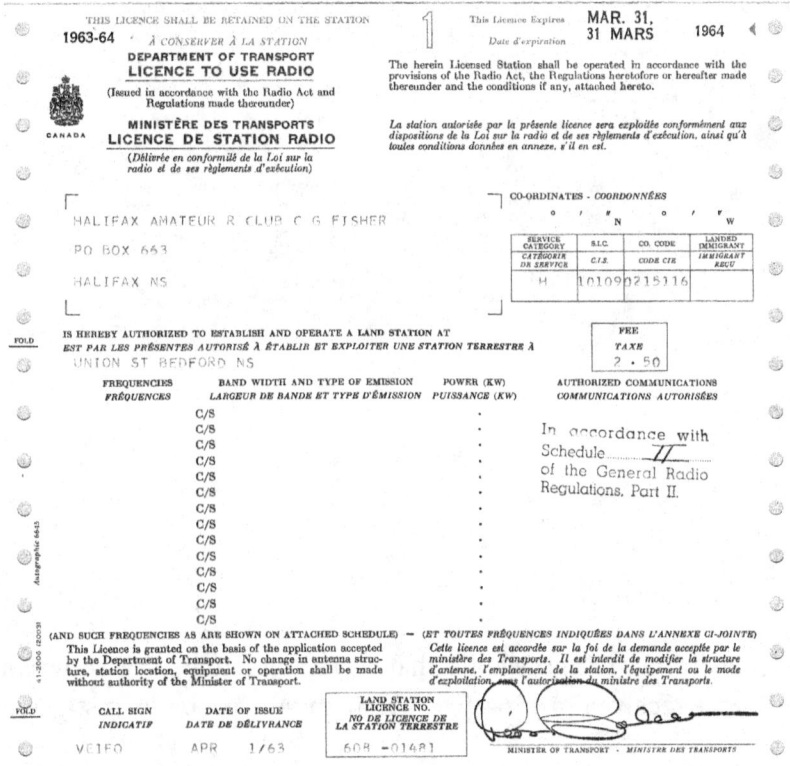

AIR SERVICES
TELECOMMUNICATIONS AND
ELECTRONICS BRANCH

CANADA

DEPARTMENT OF TRANSPORT
OTTAWA

Dear Sir:

     Should it become necessary for Canada to declare a state
of war, or should this country come under attack by an enemy your
Amateur Experimental Station Licence will be automatically suspended
and all on-the-air amateur activities shall cease immediately.

     In order to disseminate information to the public the
Government has empowered the Canadian Broadcasting Corporation
to develop an Emergency Broadcasting Plan within the framework
of which all Private Commercial Broadcasting Stations (including
"standard band", short wave, FM & TV) will, if the situation
warrants, broadcast official bulletins. It should be recognized
that this is the most practical way of so informing the majority
of the public. In remote northern areas such bulletins or official
instructions will be disseminated by appropriate military or government radio stations. Therefore in order to ascertain if such a
situation exists it will be your responsibility to pay strict
attention to bulletins of national importance and if you become
aware by this means or by any other means whatsoever, that war has
been declared or this country is under attack you shall take
immediate steps to deactivate and render inoperative your transmitting equipment.

     However, if you are enrolled in Civil Defence for radio-
communication purposes your equipment may be kept in operating
condition to be available as required, under appropriate authorization.
In such instances your station shall be operated for Civil Defence
purposes only and it may be required to be moved to an appropriate
Civil Defence communication site.

     Your station shall not be reactivated under the terms of
your amateur licence until authority to do so is received in writing
from this office.

                   Yours truly,

(F.G. Nixon)
Director,
Telecommunications and
Electronics Branch.

HARC Files

This is the 1964 club license with the letter that was stapled to it. This shows one the license and letter each license received in 1964. The Vietnam War probably instigated the letter.

MARBLE HEAD YACHT RACE, NATAL DAY PARADE & CAM VE1RO

HARC provided the communications for the Marblehead, Massachusetts, to Halifax Yacht Race in 1965 under the direction of Don Bain, VE1LZ. HARC provided these communications for several years when an amateur operator was placed on various boats and on shore to relay the position of each yacht as it entered Halifax Harbour. HARC handled this in 1975 and were thanked in the minutes of September 17th, 1975. HARC handled the Marblehead race in 1977 and this was noted in the August, 1977, Maritime News of QST. This exercise was taken over by the Kingfisher CB Club who had assisted HARC during some of these races.

The minutes of the monthly HARC meeting held on January 19th, 1977, state that the race officials for the Marblehead Yacht Race to be about 20 hours held around July 10th, 1977, requests HARC supply the communications. They claim they had CB but prefer HARC.

The minutes recorded for the monthly meeting of the October 19th, 1977, state that they had received word from the Nova Scotia Yacht Squadron that they were very pleased with the communications handled by HARC. This was the fastest of only 20 hours on record and some of the American yachtsmen said they could not get over how fast they received the times. They claimed they could never get their times that fast at Marblehead. HARC was to handle the communications again for the 1979 Marblehead Race.

Don Welling, VE1WF, stated in his Maritime News of October, 1980, on page 97 of QST that 4 HARC members had participated in the Natal Day Parade. The four were Dick Grantham, VE1AI, Brit Fader, VE1FQ, Don Bowers, VE1AMC and Cam Maillette, VE1LCR. Cam was actually VE1RO but the Lions Club sponsored a radio station for him complete with the call sign VE1LCR for "Lions Club Radio".

In Don's Maritime News for May, 1981, he states Cam had received an award for outstanding public service and he was made a life member of the Armdale Fairview Rockingham Lions Club.

ARRL DECREASE

ARRL noticed a decrease in the Amateur Radio population. They claimed the FCC was examining 33,000 prospective licensees per annum up to 1964. They claimed that in 1965 they examined 20,000 only but no one seemed to know the reason for the decrease. Some were blaming it on CB but the CB licenses were way down as well.

VE1 CAN BE 3C1

HARC members could replace the VE1 prefix of their call sign with 3C1 to celebrate the centennial celebrations that took place in 1967. Canada was 100 years old that year and held the 3BA-3FZ block of call signs issued by the International Telecommunication

Union (ITU) at that time. The VO calls could replace theirs with 3B and the VE calls across Canada could replace theirs with 3C. Canada gave up this block of call signs in 1969 and the International Telecommunication Union reassigned it to several nations immediately.

NATIONAL CONVENTION MONTREAL 1967

The June, 1967, issue of QST did not have field day on the front cover. It had an elaborate drawing advertising the National Convention to be held at Montreal. This included the Canadian symbol celebrating one hundred years as a Canadian nation. This symbol was a stylized maple leaf with 1867 on the left side and 1967 on the right of the stem. The first time the national ARRL convention had been held outside the United States. The YL News and Views section on page 96 was mostly Canadian with a lot of photographs of Canadian YL operators. I knew several. The VE1 YL's were Alma Hills, VE1MY, Sandra Carr, VE1WR, and Betty Howell, VE1AMS.

CALL SIGN CHANGES

There were a lot of changes in call signs both in Canada and the United States and this prompted several listings under the heading "Who the devil is who". These lists gave the new call and showed the previous call of the holders that had made a change. In Canada a lot of three letter calls had been changed to two letter calls.

FIRST JAPANESE EQUIPMENT

The first Japanese equipment advertisement I found in QST was one for a Yaesu FTDX-400 in the January, 1968, issue. This was the only Japanese equipment advertised. It was sold by Spectronics in California. In the March, 1969, issue both Harrison and Spectronics are advertising this same station. The November, 1969, issue of QST has a long technical description of the Inoue FDFM-2 the first Japanese FM 2-meter rig to appear. The FTDX-400 had been replaced with the FTDX-560. The inside back cover of the January, 1970, issue of QST has a full page Yaesu advertisement that calls it the "F" Line by Spectronics. By this time Spectronics has both a California store and an Ohio store. Henry another amateur radio store advertises the first Kenwood station on page 5 of the October, 1970, issue of QST. The Japanese invasion was on and what a shame to see all those beautiful American stations and companies disappear. Ten-Tec appears to be the only one left in 2008. Ten-Tec had disappeared as well but in 2021 they are trying to get started again in Dayton, Ohio.

CANADIAN FEE QUADRUPLED

CANADIAN FEE QUADRUPLED

The Canadian Government on April 1 raised the amateur license fee from $2.50 annually to a new rate of $10; amendments are $6.00 each (e.g., change of address). Needless to say, the immediate reaction of VE amateurs was one of outrage and dismay.

Early in April, high officials of the Department of Transport (DOT) met with officers of the League, the Radio Society of Ontario and Radio Amateur du Quebec, Incorporated, to discuss the issue. The amateurs were told that the change was necessary under Government policy; any reduction would require action by the Cabinet.

Accordingly, each of the three groups is filing a formal brief with the Minister of Transport; all clubs in Canada are urged to follow suit, with copy to the local Member of Parliament and to ARRL Canadian Director Noel B. Eaton, VE3CJ.

Page 71 QST, June, 1968

Every amateur in Canada was encouraged via every means possible, including club bulletins, maritime news, and so on, to write their Member of Parliament. The ARRL submitted a brief on this to the Department of Transport on behalf of Canadian amateurs. They printed this brief in the September, 1968, issue of QST and it consisted of three pages.

The May, 1969, Maritime News states that the $10.00 license fee remains but the $6.00 amendment fee is cancelled as of April 1st, 1969.

The total number of Canadian licenses was released in the June, 1970, issue of QST. They had been expecting an increase of about 800 Canadian amateur stations but received a decrease of 600 stations. 93 licenses were lost in Atlantic Canada alone. The number dropped from 1399 to 1306 licenses. ARRL blamed this on the $10.00 license fee increase and the fact this fee had been $2.50 only up to and including the 1968 fee. When one considers the amount of money involved in the average station it is

"YOUR RADIO CLUB NEEDS YOU, BUT IN FACT YOU NEED IT MORE"

Page 78, QST, September, 1968

hard to believe that fee could affect the numbers by that much. DOC increased the license fee from $13.00 to $20.00 per annum on April 1st, 1985.

THE AMAZING WORK OF BRIT VE1FQ

The more research I did on this project the more amazed I became with the amount of work accomplished by Brit Fader, VE1FQ. I found this on page 72 in the December, 1980, issue of QST. I was going to simply scan and place it on here but it deserves more than that so I retyped it.

"Those QSL Bureaus...

This month we take a quick look at our Canadian QSL bureaus. There are nine individual bureaus in Canada, one in each VE call area and one in VO, Most Canadian amateurs know that you leave several self-addressed, stamped envelopes on file at your individual bureau. When an envelope won't hold one more card, it's sealed, dropped into the mail and sent along to you.

What few Canadian amateurs realize is that before cards arrive at their individual bureaus, they're probably passed through the CRRL (Canadian Radio Relay League) Central QSL Bureau in Halifax, NS. This bureau is operated by Brit Fader, VE1FQ. Brit also operates the VE1 QSL Bureau, and has been doing so since 1937. With his name at the top of the *Callbook* list of Canadian QSL bureau managers. Brit was always receiving cards for every part of Canada. In 1977, When Len Sumner, VE3DOR, retired from the Central Bureau, Brit officially took over.

The CRRL Central QSL Bureau is quite an operation. Here are a few figures:

Year	Cards Forwarded	Postage Costs
1978	276,149	$ 822.21
1979	319,211	1036.18
1980 (first 10 months)		
VO	12,320	26.40
VE1	38,840	0.00
VE2	86,760	101.40
VE3	111,860	255.70
VE4	10,580	27.40
VE5	10,050	29.70
VE6	28,560	84.15
VE7	53,460	180.26
VE8	3,150	12.25
VY	1,450	8.50
Total	357,030	$ 725.76

What's really amazing is that Brit personally handles and sorts every QSL card by himself. Brit says it's really no work, that he's a mail clerk by trade, and that sorting cards is second nature to him. Still 357,030 are the first 10 months of this year alone and a lot of cards!

Of course, there's even more sorting done at the nine individual bureaus. The two largest, in VE3 and VE7, are operated by clubs, the Ontario Trilliums and Burnaby ARC. Other bureaus are run by individuals. Few amateurs can properly appreciate the hours of work put in by bureau managers and their volunteer assistants to get those QSL cards out to you.

Who pays for the operation of these bureaus? Fortunately, bureau personnel don't charge for their time! Costs of operating all VE-VO QSL bureaus are covered by the League."

Amazing! What in the world would a pile of 357,030 QSL cards look like? It is nice that HARC has the QSL bureau named for him but the more one learns of Brit the more you feel it is not enough. Maybe rename the club the Brit Fader Amateur Radio Club. Brit was QSL manager of the CRRL Central Bureau here in Halifax for a period of three years. He then turned the job over to Andy McClellan, VE1ASJ, at Saint John, New Brunswick, on February 1st, 1981. This was an incoming QSL bureau. Andy and the Kennebecasis Valley ARC handled this Central Bureau. Andy kept everyone up to date via the Maritime Net that met every evening at 1900. He often came on stating the number of cards they had handled. This bureau handled 400,000 cards in 1980 alone. Would that equal the 41 feet of the W2 QSL Bureau? I doubt it. Andy wanted everyone to know he was on the job and Don Welling, VE1WF, recorded his record in his Monthly news. In December, 1981, he stated Andy, VE1ASJ, had shipped 74,190 cards. One assumes that was the month of November, 1981, and why would one count the number of cards? It is recorded in Canadian Newsfronts for March, 1982, that Andy, VE1ASJ, had shipped 468,745 cards in 1981.

Andy, VE1ASJ, created the CRRL Central outgoing QSL bureau in 1983 and Don Welling, VE1WF, took this over in 1984. 60,640 QSL cards were handled via this outgoing QSL bureau in 1984.

Brit, VE1FQ, had to relinquish the incoming QSL bureau on January 1st, 1987, due to failing eyesight. Andy, VE1ASJ, took this over as well until it was taken over by HARC and named the Brit Fader memorial QSL Bureau.

HARC CLUB CREST

There were six crests submitted by six members of HARC for the club crest at the club monthly meeting held on June 21st, 1968. The one submitted by Bazil Nowe, VE1ATF, was the one chosen. Bazil has had the VE1ATF call sign all these years and still had it in 2009.

The HARC monthly meeting on October 17th, 1979, claims the Nova Scotia School of Art and Design is working on a club logo. The Art Class at this school presented their ideas to HARC members at the monthly meeting held on November 21st, 1979, and it was stated it would be sometime in the new year of 1980 before a decision would be made.

One of the members at the HARC monthly meeting held on March 19th, 1980, claimed the logo that was accepted looked like a piece of stovepipe. Therefore, this became the stove-pipe logo. The April 17th, 1980, monthly meeting stated the proposed logo was delayed and that further data was required. The last minutes I have was dated June 18th, 1980, so I have nothing further on that logo.

The HARC logo in use in 2008 was created in 1988.

Howard Dickson, VE1DHD, added the 75th and the statement "Celebrating our Past Anticipating our Future 1933 to 2008" for the 75th anniversary celebrations.

IMPORT DUTIES REMOVED ON EQUIPMENT

TARIFF ITEM REWORDED

Finance Minister MacEachen's November 12 budget contained some good news, for amateurs at least. Tariff item 44534-2 was reworded to permit duty-free entry of amateur transmitters, receivers, transceivers and related equipment with provisions for WARC bands or with general-coverage receive. The item was not expanded to include antennas, presumably to protect Canadian companies that do manufacture these. CRRL, and particularly CRRL Counsel Bob Branca, VE3VW, deserves much of the credit for this success. Bob not only suggested the rewording that was adopted; he pressed for the rewording through extensive correspondence, telephone contacts, and personal visits with Tariff Board and Department of Finance officials in Ottawa.

Page 62, QST, January, 1982

On page 64 of the November, 1969, issue of QST it was announced that ARRL and the Canadian Director, Noel Eaton, VE3CJ, had been trying to get the import duties removed from American equipment imported into Canada by Canadian amateur radio operators. Ron Hesler, VE1SH, gives more detail on this in Canadian Newsfronts found in the November, 1979, issue of QST. This is ten years later and it took a lot of work to see this happen.

This took eleven years of hard work before these import duties were removed on October 28th, 1980. These duties were removed from amateur radio transceivers, transmitters, receivers, and transverters, assembled or in kit form. Also included were linear amplifiers, VFOs and power supplies designed for use with this equipment. However, federal tax still applies to these items.

THE CREATION OF THE DEPARTMENT OF COMMUNICATIONS

The Department of Communications was created in 1970 and took over all the radio duties that had been held by the Telecommunications and Electronics Branch of the Department of Transport.

At this time all amateur radio licenses were issued for a five-year period at a cost of $10.00 per year and if the amateur moved to another province, they had to have the receipt for the current year. A description of the antenna had to be given to DOC or Civil Aviation if the antenna was within two miles of an airport. If the antenna was more than two miles from an airport but more than 50 feet high one had to notify the Civil Aviation.

HIGH WINDS

We have been doing it all wrong because Harris, W1BU, states in a lengthy article titled "If your antenna didn't blow down last winter, it wasn't big enough". With the winds we get around here the antenna does not have to be very big to get blown down.

THE AVERAGE AGE OF OPERATORS

The editorial on page 9 of the February, 1971, issue of QST was on average age and the number of amateur radio operators in the United States. The average age worked out to 41 years old and the number of amateur operators had remained the same for the past six or eight years but they did not state a figure. This survey had been conducted by several companies and ARRL.

USS PUEBLO

There was an interesting 4-page article on the capture of the *USS PUEBLO* starting on page 55 of the February, 1971, issue of QST. She was the spy ship captured by North Korea that some claim the navy wanted captured so they could obtain some new monitoring equipment. Most of those on board were of the communications technician specialty and I had served with several while in the equivalent trade in the Canadian navy. There were three amateurs on board; K1SCQ, WA6XKE and K7RSM and the article was by K1SCQ after they were released. The thing I found interesting was that the North Korean Guards tried to communicate by the Q code by simply sounding out the letters in Morse code. Apparently, the reply was often a string of dits. The USS PUEBLO is still in commission and still held in North Korea in 2021.

ARRL TO DATA PROCESSING

The ARRL went to data processing and a computer in October, 1972, in order to handle the 100,000 copies of QST. The first of these to be posted by computer were the copies mailed in April, 1973. This was most of them because some still had some glitches that needed ironing out.

AGE LIMIT TERMINATED

The age limit for a Canadian to sit for an amateur license was terminated in 1972. One had to be 15 years of age for many years before that.

3.7% LICENCE INCREASE

There were 12,607 amateur radio licenses in Canada on March 31st, 1972. This gave an increase of 3.7% over 1971.

CLUB MEETINGS CHANGED TO WEDNESDAY

The HARC club meetings were changed from Friday night to the third Wednesday evening each month in 1972. The first Wednesday evening meeting was on September 20th, 1972. This change had been approved at the meeting on June 16th, 1972.

REQUEST FOR PHONE PATCH TO ENGLAND

HARC will often receive a request for one thing or another. A lady in Halifax wrote HARC and requested a phone patch back to England in the early 1970's. The phone service today makes this a thing of the past. Not only that, but the United Kingdom has never permitted phone patches. I confirmed that with Robert, G4PYR.

UNIQUE HARC DEMOSTRATION

There were any number of requests for assistance in obtaining an amateur radio license from various clubs and organizations. The air cadets at RCAF Gorsebrook requested a demonstration with plans to set-up a club station for the cadets. Barry Hyndman, VE1XW, was the Technical Committee Chairman at the time, May, 1971. He took the club station, a Swan 500CX over to Goorsebrook and set it up with their antenna already in place. During the demonstration he contacted a Canadian warship off the coast of England. The warship was in the 1 AM time zone but as fate would have it one of the air cadets had a brother in that ship. They dragged this brother out of his bunk and up to the radio room. The two brothers had a great chat. One could not experience a better demonstration if it had been planned and practiced because Murphy would have stepped in and ruined it. Unfortunately, the minutes of the meeting record the ship's call sign simply as VE0 so there is no way to identify the ship.

HALIFAX SCHOOL FOR THE BLIND

The Halifax School for the Blind amateur radio station went "on the air" in 1973. A number of Halifax area amateurs including members of HARC were instrumental in getting this station up and running. Jim Shand, VE1ASN, spoke on the CNIB (Canadian National Institute for the Blind) Amateur Radio Club during the November 21st, 1969, monthly meeting of HARC. This is a quote from the minutes of that meeting:

"Camille, VE1RO, read a letter in Braille re this matter stating how the blind operators can be licensed that club assistance helped assemble the equipment and stated that various clubs, companies, etc., have assisted in various ways. Equipment specially assembled is for CW and includes Braille instructions and directions."

Jim, VE1ASN, went on to state that he had been working on the scheme and had offered club assistance to NSARA to help the CNIB ARC and had offered financial assistance from the club. Jim was HARC president at the time. It was also mentioned that a personal sponsor was required for each blind operator. Camille, VE1RO, thanked the club and Jim, VE1ASN, suggested the club assist other blind operators.

The 1972 Annual Club Report to ARRL stated that Dick Grantham, VE1AI, Don Bower, VE1AMC, Murray Alary, VE1ALS, Harley Grimmer, VE1MX and Doug Johnson, VE1OM were sponsors for blind amateur radio operators.

Camille Maillette, VE1RO, was a very good CW operator and I mentioned this to the Station Operations Supervisor at Halifax Coast Guard Radio one time when they needed CW operators. We could not figure out a way for Camille to read the traffic he would have to transmit so it went no further. Camille was a great traffic operator on the various amateur radio CW traffic nets and gave an excellent description of them at the February 20th, 1970, monthly meeting. Cam was a telegraph operator before he lost his eyesight.

At the HARC monthly meeting on May 19th, 1972, Dick West, VE1AGX, stated he had six "White Caners" pass the D.O.C. amateur radio exams and that they needed more sponsors. Barry Hyndman, VE1XW, stated at the monthly meeting on June 16th, 1972, that he had six pass their D.O.C. exams and he also read a letter that he as president of HARC had written each "White Caner" congratulating them on passing their exams and offering them free HARC membership. It was also stated at this meeting that NSARA would supply the sponsors for each "White Caner". The D.O.C. made it mandatory that each have a sponsor to help them install their station and teach them how to operate and maintain it properly.

Cam, VE1RO, stated at the monthly meeting held on September 20th, 1972, that 8 adults and 6 students had their licenses and most had their call signs. All were "White Caners" and asked for help in getting a station for the school that already had a radio shack wired and an antenna. It was moved by Doug Johnson, VE1OM, and seconded by Cyril Boudreau, VE1RJ, that HARC write D.O.C. advising them that HARC will sponsor the CNIB amateur radio station.

LOG TERMINATION FOR 2 METERS

It was not until late 1974 that the FCC in the United States relaxed some of the log keeping requirements for amateur radio. Canada followed soon after and this meant it was not necessary to keep a log for 2-meters especially while mobile. That was a bit of a pain and one would have to write up their log each time they stopped while mobile. No doubt some did it while mobile. No doubt a few of those did not notice that tree, ditch, light post, etc., coming at them while writing up their log. A few have no doubt been operating that big station in the sky since.

HARC HELP HANDICAPPED CHILDREN

At the HARC monthly club meeting on August 16th, 1975, the club agreed to help the camp for handicapped children with money and also agreed to help repair their public address system.

BILL BORRETT VE1DD RECEIVES PLAQUE

Bill Borrett, VE1DD, was presented the Canadian Broadcasters Award Plaque for long service in the broadcasting industry according to the Maritime News for July, 1976.

ARRL NO LONGER GO PORTABLE OR MOBILE

ARRL announced in the December, 1976, issue of QST that U.S. amateur operators no longer have to announce when they are going to go portable or mobile.

OPERATING IN CANADA

Q. Does my U.S. amateur license allow me to operate in Canada?

A. U.S. amateurs of General class or higher may operate in Canada; however, prior permission must be obtained from the Canadian Department of Communications (DOC). DOC form 16-52 is used to request permission to operate in Canada and is available from ARRL hq.

Q. Can Canadian amateurs operate in the U.S.?

A. Yes. Again, prior permission is required. Canadian amateurs apply to the FCC on FCC form 410, which is also available from ARRL hq.

Page 73, QST, September, 1977

The "good old days" could be a bit of a pain when it came to the paper work at times.

VE1AGH & VE1FQ VISIT HMCS PROTECTEUR

QST Congratulates

□ Bertus Backer, VE1AGH and Britt Fader, VE1FQ, who were honored recently by the captain and crew of HMCS *Protecteur* for relaying more than 200 messages back home while at sea. They received engraved plaques and some new equipment for their shacks. Both men live in Lower Sackville, NS.

Page 74, QST September, 1977

Not only did both live in Lower Sackville, both were active HARC members. Internet and email access by all crewmembers has eliminated this.

CF1ISH

Halifax ham operators getting new call sign

Ham radio operators around the world will be picking up a new call sign out of Halifax near the end of the summer.

It's CF1ISH. Technically, it should be read as CF-one-ISH but it doesn't take much imagination to read it as seafish.

The signal is appropriate because it is the Halifax Amateur Radio Club's way of telling the world that Halifax is the host of the World Fishing Exhibition.

"We are delighted that the ham radio operators are taking part in this way," Fisheries Minister Dan Reid said Wednesday.

During the eight days of the exhibition, from Aug. 31 to Sept. 7, the 80 members of the Halifax Amateur Radio Club will be telling their fellow hams around the world what is going on.

Normally, the Halifax club's

call is VE1FO. But when the club heard about the exhibition, it applied to the federal department of communications for the seafish call.

"We wanted to go ahead right away to give a bit of advance notice about the exhibition," says club vice-president Mike Pothier, "but the department limited us to the period of the exhibition."

The club is notifying ham radio magazines of the project and plans to man its radio station "for as many hours as possible" while the exhibition is on, taking calls from hams who know there's a new call sign — or prefix, as the hams know it — on the air.

Ham radio operators try to collect these prefixes as part of their hobby.

HARC Files

I remember this exhibition; this is where the new fisheries patrol vessel *CAPE ROGER* was on display. She was fitted with a Marconi Commander D marine radio station. *CAPE ROGER* must have been one of the last vessels built at the Pictou shipyard. *CAPE ROGER* was assigned international call sign VCBT. It was unusual to see a government ship without the CG prefix; Canadian Government.

OPERATING FROM THE ISLANDS

The two main islands around Nova Scotia that see a lot of amateur radio activity are St. Paul's Island in the Gulf of St. Lawrence off the Northern tip of Cape Breton Island and Sable Island south of Nova Scotia and south of the Canso Strait.

Sauli, VE1AIH, and Eric Mills, VE1AST, made an expedition to Sable Island during the early summer of 1979. They made 4,200 contacts on all bands and the 20-meter band was the heaviest. This is from the minutes of the HARC meeting on June 20th, 1979.

According to the November 21st, 1979, monthly meeting of HARC Dick Grantham, VE1AI, Harley Grimmer, VE1MX, Walter Rawle, VE1AWS, Clive Bagley, VE1AMZ and Don Bowers, VE1AMC, had been recent visitors to Sable Island. According to Aaron Solomon, VE1OC, in his Maritime News for February, 1980, they made 7,000 contacts and many were 6-meter cross band contacts.

According to Don Welling, VE1WF, in his Maritime News for February, 1981, Dick Grantham, VE1AI, was planning another DX expedition to Sable Island for December. Was that December, 1980 or 1981?

The Maritime News section of the Canadian Division for August, 1978, by Aaron Solomon, VE1OC, stated that an HARC group had made an expedition to St. Paul's Island off the northern tip of Cape Breton Island. This group made 6,000 contacts and was composed of five HARC members: Dick Grantham, VE1AI, Sauli Arosankari, VE1AIH, Mike Pothier, VE1AJP, Don Bower, VE1AMC, and Harley Grimmer, VE1MX. They used special call sign VY0CA.

A slide show was shown at the HARC monthly meeting held on July 19th, 1978, of this expedition. The weather had been very rough with the antennas receiving much damage but they managed 2,750 SSB contacts and 2,920 CW contacts. All contacts were made with 200 watts. In other words, no linear amplifier was in use. They managed to work 115 countries and all 50 states that make up the United States of America.

Another group of amateurs were on St. Paul's in 1982 using call sign VE1SPI and Dick Grantham, VE1AI, was one of the operators. Dick told me he has made around six expeditions to St. Paul's and three to Sable Island.

The DOC assigned CY9SAB permanently to Sable Island and CY0SPI permanently to St. Paul's Island for all future DX expeditions in 1983. Both islands appear to have used every call sign imaginable. Sable Island was assigned CY0X during the expedition in the summer of 2008. It is rather obvious that assigning these places a permanent call sign so one and all would know exactly who they are is a wasted effort.

Andy McClellan, VE1ASJ, headed up an expedition for St. Paul's in September, 1983, according to Maritime News for December, 1983. Andy was assigned call sign VE9DX in 2008.

The Breton DX Group with call sign VE1DXX made an expedition to St. Paul's Island in 1988. There were four members of this group; VE1AL, VE1XT, VE1AWG and VE1BHR all from the Cape Breton area of Nova Scotia. They took along a friend of one of group with call sign W5KNE. They stated that DOC had assigned call sign CY9SPI to St. Paul's but they managed to do some arm twisting and DOC gave them permission to use CY9DXX, the suffix of their VE1 group call sign. So much for the permanent call sign whatever it was. This group left a detailed description of this expedition complete with photographs covering 4 pages of QST starting on page 14 of the March, 1989, issue.

WV2B and WA2UJH operated from St. Paul's Island from the 9th to the 12th of July, 1993. WV2B/CY9 was the call sign in use for this expedition. There was little detail on this expedition as recorded in the October issue of QST.

There was another expedition to St. Paul's Island in Mid-August of 1993 and the complete details with several photographs appeared in the June, 1994, issue of QST. This expedition used call sign CY9CWI and the description is by Fred Archibald, VE2SEI, one of the members of the expedition. Fred is now a member of HARC and holds call sign VE1FA.

St. Paul's Island was invaded by the amateur radio world at least twice in 1997 according to QST in the September and November issues. The expedition from June 27th to July 3rd was using call sign CY9AA and the expedition in early August operated with call sign CY9SS. There is a detailed description plus a number of photographs of the CY9AA expedition in the August, 1998, issue of QST.

Islands on the Air (IOTA) is an organization that gives one an excuse to visit these islands at the end of July and tries to work as many stations as possible. IOTA was created in the mid-1960s by a British short-wave listener, Geoff Watts. It became popular the world over and the awards program was taken over by the Radio Society of Great Britain (RSGB) in 1985. The Amateur Radio Lighthouse Society (ARLHS) is another organization that gives one an excuse to visit the islands that still have a lighthouse or had a lighthouse at one time. IOTA is an annual event for a 24-hour period whereas with ARLHS one can operate whenever they like for as long as they like.

INTERNATIONAL LIGHTHOUSE LIGHTSHIP

The International Lighthouse – Lightship Weekend (ILLW) was formed in 1999 and is another organization that usually holds an event in August each year that gives amateur radio operators an excuse to operate. One has to find the owner or the one in charge and obtain permission to visit these places. Many are now privately owned. All of these organizations have the various locations numbered in one form or another.

ISLANDS ON THE AIR

Some members from HARC put Bon Portage Island on during the IOTA period in 2007 using call sign VC1W. Fred Archibald, VE1FA, was one in this party and records this event on page 27 and 28 in The Canadian Amateur (TCA) for May/June 2008. They were on the island from July 25th to August 1st and were able to use the VC1W call sign for the IOTA portion only. They then reverted to the HARC call sign VE1FO calling it portable for the duration of the visit. These are just a few of the many expeditions that have been made to these islands over the years using a variety of call signs.

The Nova Scotia government claims that the province of Nova Scotia has over three thousand coastal islands. There are 15 times the people on earth that should be here. There is always someone who is convinced they are in charge of us in one way or another. Therefore, obtaining permission to visit each of these islands would be a monumental task. An award for working all of these islands would be a challenge for everyone. Learning someone in the W9 call area, for example, worked all of these islands would be hard to believe.

LIGHTHOUSES

There are nearly 200 lighthouses, lights and sites where a lighthouse once stood at one time or another around Nova Scotia. The ARLHS keeps an accurate up to date list. An award for working each of these sites would be another real challenge. This is another project obtaining the permission to visit each site would be as monumental a task as obtaining the award for working each site.

This history exercise has proven to me that those who obtain high scores on the various awards in the world of amateur radio do not own elaborate stations. They are simply ordinary people with the ordinary run of the mill station. They persevere with simply a lot of persistence and patience.

HALIFAX CLUB INCORPORATED

The Halifax Amateur Radio Club was officially incorporated in July 1978.

RICHARD CLATTENBURG VE1AYN LOST AT SEA

HARC member Richard Clattenburg, VE1AYN, was lost at sea from *CSS DAWSON* with international call sign CGBV during the summer of 1978. This was recorded at the monthly meeting held on August 16th, 1978. It was also recorded in the Silent Keys section of the January, 1979, issue of QST.

DAVE OLDRIDGE VE1BFV TRIP WEST

HARC member Dave Oldridge, VE1BFV, went from Halifax to Vancouver and back during the summer of 1978. He made contact with Halifax from nearly every town he stopped at going out and coming back. Brit, VE1FQ, recorded this in the minutes of the monthly meeting held on November 15th, 1978. Dave told me that one of those he visited was CW only and he worked him on phone listening to his CW right up to his door.

HARC DUES INCREASE

According to the monthly HARC meeting held on November 15th, 1978, the dues were increased to $10.00 per annum and $5.00 for students, senior citizens and associate members. The club was having a bit of financial trouble and this was to alleviate the problem.

VE3KAY

The best record ever in Canadian amateur radio is the record of Kay Clark, VE3KAY. Kay was both blind and deaf but some amateur radio operators worked with her and managed to see her pass her amateur radio license. She copied Morse code by feeling the code vibrate a metal container. It is hard for you or me to imagine what that must have been like, to be unable to see or hear and then be able to feel and communicate both locally and around the world through the vibrations. The VE3 regional office had already assigned the VE3KAY call sign but they told the holder to pick a call sign and gave Kay that call. This was unheard of at the time.

AMATEUR RADIO'S
SECOND "HELEN KELLER"

Kay, VE3KAY, received her license and also a special notice of commendation for her achievement in getting that license at the ARRL National Convention in Toronto in June. Both blind and deaf, Kay is the second YL able to pass the test and become one of the "Helen Kellers of Amateur Radio." The other gal is Mary Lou Stockill, WB6SSZ. While Mary Lou is able to hear a single note of the audio spectrum, Kay reads the code through sensing the vibrations of the dots and dashes with her fingers. She is a member of CLARA and will be active on all the amateur cw frequencies.

Though deaf and blind, Kay is now very active on cw as VE3KAY. Supporting her efforts were gals from both the Ontario Trilliums and CLARA. *(W2EEO photo)*

Page 74, September, 1977, QST

On page 57 of the November, 1978, issue of QST it states that Kay, VE3KAY, was presented with the Diamond Jubilee Award for opening the world to deaf and blind amateurs.

RUSSIAN OVER THE HORIZON RADAR

The Russian government was testing an over the horizon radar using the high frequencies of the radio spectrum. Its pulsing transmission was rather annoying and could be quite loud at times. The amateur radio community was encouraged to write various government officials and complain of this interference.

The following is from League Lines, page 10, QST June, 1979:

> "Do you remember the "Russian Woodpecker" that wideband pulse-type signal causing harmful interference on the amateur (and other) bands, generally between 7 and 21 MHz? Chances are you don't have to think back very far, because it's still with us! Despite numerous complaints to the FCC by the users of several radio services in the U.S. and despite the fact that the U.S. State Department has been involved in the matter for nearly three years, the interference persists. The   next time you experience the interference, don't give up in despair. Write the Watch Officer, Monitoring Branch, FCC Washington DC 20554."

This request continued for several months and on page 78 of the December, 1979, issue of QST is the letter Senator Barry Goldwater, K7UGA, sent to the President of the United States and to Congress regarding this interference. On page 65 of the January, 1980, issue of QST it states that CRRL had written the Canadian Prime Minister regarding this woodpecker interference in November, 1979.

The answer to this letter to the Prime Minister is recorded on page 68 of the March, 1980, issue of QST. The answer came from the Minister of Communications who stated their direction-finding equipment proved that the transmission was coming from Riga, Latvia. The minister went on to state that they had contacted both the U.S.S.R. and the International Telecommunication Union at Geneva, Switzerland.

HARC made up a form letter so one and all could use if they desired for this purpose. We in the commercial radio world simply carried on working each other through the noise. I have no knowledge of the commercial radio world making a complaint re this racket, and a good racket it could be at times.

157 KHZ

I found this rather interesting. 157 kHz is not that far from 136 kHz.

> • **Canadian lowfers report first 136 kHz QSO:** Larry Kayser, VA3LK, and Mitch Powell, VE3OT, report that, despite poor to medium conditions, they successfully completed the first two-way QSO in Canada on 136 kHz at 1400 UTC on July 22. The distance was 431 km (268 miles). The pair used very slow-speed CW—QRSS—where dits are 3 seconds long and dahs are 9 seconds long! VA3LK and VE3OT have received special letters of authorization for LF testing and evaluation. Frequency range is 135.7 to 137.8 kHz, emissions permitted include CW, FSK and BPSK at a bandwidth of up to 3 kHz. Powell says VA3LK is operating on 137.710 kHz and he is on 137.780 kHz. More information on LF Amateur Radio is available on the Radio Amateurs of Canada Web site, http://www.rac.ca/infodx.htm.—*Mitch Powell, VE3OT/RAC*

QST October 2000 71

The part I find hard to understand is why such slow code speeds. 157 kHz was one of our CW frequencies at the Aeradio stations I operated. We rattled along at our regular CW speeds via Vibroplex bugs. This would have been around 25 WPM and maybe higher on the downhill runs. I was at VFT2 Teslin, Yukon Territory, and I worked VFM8 Mayo, Yukon Territory and VFU6 Coppermine, North West Territories, with no problem. I was roughly the same distance from Coppermine as I was from Edmonton. We were using an RCA N26 transmitter and the RCA AR88 receiver if I remember correctly. We normally worked VFD Watson Lake, Yukon Territory, or VFD4 Dease Lake, British Columbia. Watson Lake was one of the few Aeradio stations without a digit in the call sign. The East Coast Canadian Aeradio stations were on 160 kHz and their stations lasted longer on that frequency than our West Coast stations. These frequencies were replaced by landline teletype at the western stations. Mitch Powell, VE3OT, operated these same northern aeradio stations at one time.

PLEXIGLAS PADDLES FOR VIBROPLEX BUG

If your "Bug Fist" is in agony try this:

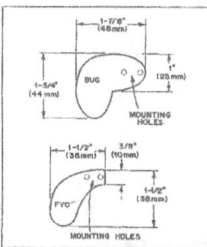

PADDLES REDUCE FATIGUE

In using my bug keyer, I found that the arrangement of the paddles was conducive to muscle tiring. After a little head scratching and a hint from W9DU, I produced a new set of paddles that reduced the fatigue. As may be seen in the photograph a set of L-shaped thumb and finger pieces have been applied to the bug.

The new paddles are made of 3/8-inch (10-mm) thick Plexiglas. Dimensions may be varied to suit the individual operator. To accommodate a pair of bevel-head machine screws for mounting the paddles, I drilled the thumb (dit) piece and both drilled and tapped the finger (dah) piece. For keyers requiring less operating force, such as the FYO, 1/8-inch (3-mm) plastic may be used and shaped as drawn. — *James J. Di Spirito, Jr., WB9TCT*

Patterns for making fatigue-reducing paddles for a bug (top) or a W8FYO keyer (bottom).

These Plexiglas paddles make sending with a bug easier.

Page 36, QST September, 1978

That is a Vibroplex bug in the photo above. They were manufactured and sold from their office in New York City for years. This was moved to Portland, Maine, in 1979 and was sold and moved again to Mobile, Alabama, in 1995. Mitch Mitchell, W4OA, owned the company in Mobile and sold it to Scott Robbins, W4PA on December 21st, 2009. One can still purchase a full line of their equipment in 2012. Scott resigned from Ten-Tec and moved the Vibroplex Company to Knoxville, Tennessee.

Vibroplex

This is the deluxe version of the standard Vibroplex semi-automatic transmitting key. This was the most popular model of these semi-automatic keys although during the heyday of Morse, both landline and radio, everyone and anyone built these keys. The reed or long thin piece one can see with the round weight attached in this photograph was often a hack saw blade in the "homebrew" or home built models.

These are the trademarks of the Vibroplex Company and have been since it was created. This is where the term bug originated and has been the name given these semi-automatic keys for years. Bug is much shorter and easier to use than semi-automatic. Horace Martin is the one who designed and built these Vibroplex keys in New York. Tom French, W1IMQ, produced The Vibroplex Collector's Guide. It is a most interesting 87-page document he produced in 1990. I received my copy on June 19th, 1991. It was $14.95 U.S. and with shipping and handling it came to $22.00 Canadian.

Two famous trademarks of the Company:
the Bug logo and the name coined by Horace Martin.

Vibroplex Collector's Guide page 39

VIBROPLEX REVISITED
◈ Two *QST* articles — "A Lost Dit of Vibroplex History" [Feb 2009, pages 58-59] by Brian R. Page, N4TRB, and "Vibroplex — The Company and its Classic Key" [Jan 2003, pages 48-49] by John Ceccherelli, N2XE — both state that Vibroplex's first or original manufacturing facility was located in Norcross, Georgia. This is not correct. The first Vibroplex factory was located at 53 Vesey Street in New York City. The United Electrical Manufacturing Company (UEM) was incorporated in New York on February 17, 1904. Horace G. Martin was the Vice President and General Manager of UEM; he and Edward Buchanan were two of seven directors of the corporation. Initially, UEM manufactured the Autoplex, a semi automatic electro-mechanical key invented by Martin. In 1907, UEM moved to Norcross, Georgia.
Edward Buchanan had lofty plans for UEM in Norcross that included the production of an automobile called the Nor-X. After the collapse of the brokerage firm A.O. Brown & Co. in August 1908 and subsequent failure of UEM (Albert Brown was one of four subscribers of UEM), Martin tried to personally obtain financing for the Nor-X, but was unsuccessful. In order to support his wife and five children, he found work in Atlanta as a telegrapher. His father-in-law was a high ranking officer with Western Union in Atlanta. Martin continued to assemble and sell his Vibroplex on the side while working as a telegrapher and moved to New Jersey in early 1910.
JOHN CASALE, W2NI, ARRL Life Member
Troy, New York

Page 24, QST May, 2009

GERRY HARRIS VE1AAC WX NET CONTROL

Aaron Solomon, VE1OC, recorded in his Maritime News for September, 1978, that HARC member and former president, Gerry Harris, VE1AAC, was now the Net Control of the Weather Net. That has not been a bad run for Gerry because he is still the Net Control for the Weather Net in 2008. That should add up to 30 years in anyone's arithmetic. The weather net had been in service for 18 years when Gerry took over as net control. Gerry wrote a brief history of the Maritime Weather Net. The weather net was started by Bert Whittaker, VE1RT, on September 26th, 1960. The weather net has its own personal call sign, VE1MWX. Gerry's history can be found on the West Cumberland amateur radio club web site. Gerry Harris, VE1AAC became a silent key in 2021.

ARRL CRRL MEMBER TOTALS

The editor of QST stated in his message for the month of October, 1978, that the ARRL membership stood at 170,000 members. On the next page of the same issue a request was made for volunteer examiners to examine visually impaired candidates who wanted to sit for examination for an amateur radio license. This program must have worked out so well that it was the basis of the volunteer examination program the FCC has been using since 1984. In the Canadian Newsfronts section of this same issue it stated the total CRRL membership was 6,911. In Canadian Newsfronts for June, 1981, it stated the CRRL membership as of March 3rd, 1981, was 6,572. So, there was a decrease in that three-year period. In 1983 Canadian Newsfronts stated a membership of nearly 5,000 members and that would be a further reduction. At the time they felt it very good with economic conditions the way they were but one will note they did not record the exact number.

According to Canadian Newsfronts the Canadian statistics as of April 27th, 1986, was 22,870 amateur operators in Canada and that 1,105 of them were located in Nova Scotia. These were official DOC figures.

DOC CREATES NEW LICENCE

DOC CREATES NEW AMATEUR LICENSE CLASS

The Canadian DOC has taken final action to create a new class of amateur license. The new Amateur Digital Radio Operator's Certificate will require passing an examination equivalent to the Advanced Amateur, *plus* a new exam on digital techniques, packet radio, and microprocessors, *without* a Morse code requirement. Privileges are limited to 144 MHz and up, with pulse modulation reserved for holders of the new license. Present Advanced Amateurs can get their licenses endorsed by passing the new exam.

All Canadian amateurs will be permitted to use "packet radio" between 220.1-223.5, 433-434, and 24,000-24,010 MHz; however, 221-223 and 433-434 MHz will be set aside exclusively for packet radio in Canada. Narrowband, low-power pulse modulation will be permitted at 145.5-145.8 and 434-434.5 MHz. The rules also have been changed to permit fm with up to 15 kHz deviation at 144.1-146 MHz in addition to 146-148 MHz, permitting Canadians to use the new U.S. repeaters in the 144.5- to 145.5-MHz segment.

While a dramatic departure from tradition, the new regulations are less startling and have much less impact on present operations than the original DOC proposals (May 1978 *QST*, page 59). More details will appear next month in "Canadian NewsFronts." — *David Sumner, K1ZZ*

Page 56 QST November, 1978

Canadian Newsfronts on page 61 of the December, 1978, issue of QST describes this license in detail. The editor of QST asked if the U.S. amateur operators should petition the FCC for this identical license in his editorial in the June, 1980, issue. John de Mercado, VE3LBA, was the only one really "gung-ho" about this license and one of the few who bothered to write and pass it as I understand the news at the time. The U.S. had no interest in it. The

ARRL stated in the February, 1991, issue of QST that the reason this license flopped was because the written exam was too demanding.

MORSE CODE TERMINATED

Canada went to the codeless basic exam in October, 1990. The FCC terminated the requirement for a code exam on February 23rd, 2007. This would appear to be the last nation to terminate the code requirement.

CANADIAN ARMED FORCES RADIO SYSTEM

There is a lengthy description in Canadian Newsfronts on page 67 of the January, 1979, issue of QST describing the formation of the Canadian Armed Forces Radio System. The Canadian Armed Forces were created 11 years previous to this in 1968. This new system was based on the American system and the amateur radio operators communicated with the various military units around the world outside the amateur radio bands. The military was quite impressed with a trial experiment with this system on a 13 MHz frequency.

WORLD ADMINISTRATIVE RADIO CONFERENCE 1979

The QST editorial for October, 1979, was all about the promised removal of CW from the amateur radio licenses. The World Administrative Radio Conference was held from September 24th until November 30th, 1979. The first three days were spent in choosing a suitable chairman for this event. There were 13,000 proposals to be handled at this conference and nearly 11,000 were dealing with frequency allocations. There were two nations at this conference that were in agreement with canceling the CW requirement for amateur radio and at least 15 opposed to this move. The big reason for retaining CW was that it would create problems with the reciprocal licensing agreements with the nations that retained CW. The ARRL wanted those amateurs opposed to the removal of CW to write the International Telecommunication Policy at the Department of State in Washington.

THE THREE R'S IN SCHOOL

Mayday!

Newcomers to Amateur Radio often wonder why Morse code is still required for attaining a ham license. I can tell them why.

Disaster struck on what was supposed to be our "dream of a lifetime" fishing trip. Three of us took off in a Cessna 180 float plane bound for a small lake in Manitoba's north country, 260 miles away.

At first, luck was on our side. Within two hours, each of use had caught trout weighing about 25 pounds. Our rocky fishing point was narrow, however, and we were having trouble keeping our lines untangled. To ease the congestion, Chuck fired up the Cessna and taxied to the other side of the lake to fish.

That's when our luck started to change. Dave's new rod cracked during a cast, his line broke, and half of his rod sailed into the water. Then came the bugs. An exotic mixture of black flies, mosquitoes and other insects began devouring a rare dish — us. We summoned Chuck back across the lake with the insect repellent, which was in the plane. Right when we thought things couldn't get any worse, they did — much worse.

As Chuck started the plane to taxi over to us, an unseen current gripped the aircraft and pushed it out into the middle of the lake. In an effort to break free of the current, Chuck applied more power. With a sickening crunch, the plane struck and stuck to rocks hidden beneath the foaming water. Eventually, Chuck managed to pry his plane loose and taxi to shore to check the damage. A hole gaped in the right float which was quickly filling with water. A crude patching job failed to close the hole, and the waterlogged plane would not take off. In one final, desperate attempt to lift the damaged plane out of the water, the left float was punctured by another rock. We were trapped! Our secluded fishermen's paradise had become a watery prison.

Chuck immediately began calling "Mayday," but the only response was silence. We built a signal fire and alternated calling "Mayday" at 30-minute intervals hoping to contact a commercial flight on the polar route. Still, our calls got no reply. The elt (emergency locator transmitter) was screaming its electrifying siren on 121.5 MHz, the same frequency as our vhf radio. I turned off the elt after each distress call to listen for replies.

Just as the light and our hopes were dimming, we got a response. A Pan American pilot answered and accurately copied our position as 59° 20' north by 97° 40' west. Later another pilot radioed that a rescue helicopter was on its way.

When the chopper pilot estimated that he was 40 miles south of our location, we refueled the signal fire so he could see it on his approach. Convinced that our rescue was near, we began packing our gear. We stowed away the adf (automatic direction-finding receiver), but left the vhf gear in the plane for any last-minute communications.

After more than an hour of billowing smoke and nerve-racking waiting, the helicopter had not shown up. I decided to use the vhf unit to see if anything was wrong. Before I could get to the rig, it began squawking. An airline pilot was asking if 57° 20' north by 97° 40' west was the correct location. I couldn't believe it! The search was taking place two degrees south of us; approximately 120 miles away.

I grabbed the microphone to send the correct coordinates. The pilot reported that he was receiving only a carrier, no voice. I tried again, but got the same result. Apparently, when I removed the adf from the plane I had broken a wire under the panel in the audio circuit. We were without communications. Or were we?

I put the microphone on my knee and hit the button with my finger in the familiar staccato of Morse code — SOS SOS SOS DE CF IXJ CF IXJ CF IXJ HW CPY BK. The airline pilot said to someone else, "That guy down there sure knows his Morse code. He's telling us something, but none of us knows Morse code." My heat sank. What good was code if the guy at the other end couldn't copy it?

As I had done so many times on the ham bands, I began to send very slowly while the pilot looked up each character in his flight manual. Finally, he confirmed our correct position. The next morning, a Canadian Forces Hercules search and rescue plane zeroed in on us. A few hours later, a helicopter arrived and, thanks to Morse code, we were transported back to civilization. — *Jim Prentice, VE4JI, The Pas, MB* QST

Page 21 QST November, 1979

We often hear of the three R's taught in schools; reading, writing and arithmetic. I have been convinced for years that there should be three more subjects just as important and require a passing grade equal to the three R's. Touch typing is a must for anyone in today's world and should be taught from the beginning in every school. The automobile used to be the miracle of the modern transportation age but has grown to the point it now owns us. There is nothing so expensive, that creates so much pollution and loses its value so fast, and yet there are those who will go without eating to own one. The proper operation of one is a must for everyone and those who cannot operate one properly should be weeded out at an early age. The third one is just as important as the other two and this story on Mayday is but one of many examples.

I remember this incident very well. American made CF-IXJ was serial number 32480 and had been imported into Canada in 1956 so was getting along towards 25 years old if not older. The Cessna 180 was as much "work horse" in the Canadian north as several other aircraft models. They were equipped with skis, wheels or in this case floats. Like so many "work horse" models of aircraft they seem to go on forever.

There have been so many incidents such as this Mayday over the years one could fill an encyclopedia with the details. Prisoners of war have spelled out torture blinking their eyes in Morse code and therefore had communication with the free world. There are two excellent medical stories in the July, 1990, issue of QST. One a patient had communication from eye blinking and another where a husband communicated with his wife by tapping Morse code on her wrist. This was their only means of communication. Pilots have changed the sound of their engines and communicated with Morse code when all else failed. I was with a machine gun regiment of the reserve army in high school and could easily transmit Morse code with one of those machine guns. I would like to see someone modulate that with phone.

One does not hear the term Hot Box with the railroads like they used to. The railroad cars may have been improved to the point this is very rare today; like all the propeller and

205

rudder problems that plagued ships years ago. A hot box is a wheel and axle on a railcar smoking from lack of lubrication. This was/is a very real problem and could cause a serious fire. One amateur radio operator was driving along parallel to a train years ago and spotted a hot box on one car. He stepped on it and pulled up alongside the engine and transmitted "HOT BOX" in Morse on his car horn. One of the engineers stuck his head out the engine window and looked back. On spotting the hot box he acknowledged receipt of this on the trains whistle and thanked him. This amateur was wondering how to get a QSL card for that contact and I should have saved this as written in QST some years ago.

One does not need to be a speed demon in Morse to make it a useful tool. If they simply could recognize the majority of the characters, they would have communication with someone who knew the code or one who could recognize as many characters as they could.

The Canadian Amateur titled this another reason for saving CW. I used to have a noon sked with Gary, VE1BZD years ago before he had his advanced license and became VE1JB. While transmitting to him with my old Vibroplex bug I glanced up when our youngest son Jeffrey comes in with 3 or 4 of his little friends in tow. They all were around 8 or 9 years of age. They gathered behind me and I could hear Jeff giving a good description of what was taking place. Towards the end of his description, I started to listen to what he had to say. He told them Gary and I were talking and the reason we were using that thing, my bug, was because we were telling dirty stories and did not want them to hear.

ASCII

The editor of QST December, 1979, was praising the FCC for requesting the use of American National Standard Code for Information Interchange (ASCII) by amateur radio operators at the World Administrative Radio Conference held at Geneva, Switzerland, from September 24th until November 30th, 1979.

AMATEURS EXAMINING AMATEURS

The DOC created a program of amateur radio operators examining those who wanted to be amateur radio operators in 1987. Dan Dawson, VE1JV, Dave MacKinnon, VE1ALO, and I examined a few right after this program went into effect. There had to be three advanced class licenses in order to conduct these examinations.

The FCC in the United States was in the process of creating a volunteer examination program for their amateur radio operators at this same time that went in place in 1984, three years previous to this.

FIRST AMATEUR IN SPACE

The big news in the amateur radio world for 1983 was the first amateur radio operator in space. Dr. Owen Garriott, W5LFL, went up in the spaceship *Columbia* and the complete detail of his proposed amateur activity was recorded in League Lines, November, 1983.

According to Dr. Garriott's radio log he worked the following VE1 stations:

Dave LeCain, VE1AFU
Gene Fougere, VE1BB
George Snow, VE1CAW   HARC
Cam LeBlanc, VE1CGY
Aaron Solomon, VE1OC   HARC
Bernie Bonner, VE1UT

One third of the total is HARC members. An excellent showing!

FLIGHT OF COLUMBIA
STS-9/Spacelab-1

Launched on November 28, 1983
and after 247 hrs, 47 min
landed at Edwards A.F.B. on December 8, 1983
* First launch of Spacelab (provided by the European Space Agency)
* Longest Orbiter flight to date
* First European crewmember
* First 'Payload Specialists' (non-career astronauts)
* First six-person spaceflight
★ First Amateur Radio station in space:
      W5LFL
Transceiver: modified Motorola MX-300 2-meter FM transceiver, hand-built by the Motorola Amateur Radio Club in Florida.
Antenna: directional ring radiator with cavity, designed to fit in the upper window of the spacecraft; built for NASA by volunteer employees of Lockheed.
Power: 4.5 watts
Mode: FM, CW (by keying carrier) All transmit and receive audio were tape recorded, which constitutes the station log.
Operating orbits: 40D, 56D, 82A, 71D, 91A, 96A, 97A&D, 110D, 111A&D, 112A, 113A, 129A, 130A, 134A, 134D, 135A&D, 144A&D, 145A&D, 146A, 149D and 150D.
Stations, 2-way contact: over 350
SWL: approximately 10,000 cards received
Countries: 23
Total operating time: about 4 hrs, 30 mins.

VE1CAW

Front

VE1CAW

Inside Front

W5LFL/VE1CAW

Space Shuttle Columbia

This confirms a two-way contact on 2 meters with Scientist/Astronaut Owen K. Garriott, W5LFL, operating in space from aboard Columbia on the flight of STS-9/Spacelab-1 between 28 Nov. and 8 Dec., 1983.

Owen K. Garriott, W5LFL

W5LFL
from the
Spacecraft
Columbia

VE1CAW

Inside Back

VE1CAW

Back

This is George Snow's QSL card. Thanks George – well done!

Owen Garriott's son, Richard, W5KWQ, is an astronaut like his father. Richard has his grandfather's amateur radio call sign. Richard was up in the international space station Soyuz TMA-12 from October 12th until October 23rd, 2008. He had a great time operating amateur radio and was amazed at the number he worked who had worked his father in 1983. Richard spent a lot of his operating talking to school children.

VE1AAC

WA4SIR was up in the space shuttle *Columbia* known as STS-35 from December 2nd, until the 11th, 1990. KB5AWP, N5QWL, N5SCW, N5RAW and N5RAX went up in the space shuttle Atlantis known as STS-37 on April 5th, 1991. This was an all-amateur radio crew in STS-37 and the detail can be read in the July, 1991, issue of QST. Their free time on these shuttles was spent in operating amateur radio. Talking to school students was a big part of their amateur radio operating. STS-37 failed in their attempt to contact the soviet space station *Mir* using call sign U2MIR on 2-meters. They felt confident that they heard each other but one can imagine the QRM these stations would create.

HARC member and former president Gerald Harris, VE1AAC, worked U4MIR and this is his QSL card to prove the contact. Well-done Gerry!

Gerry said he worked the station with an ICOM IC2AT and rubber duck ½-wave and had a video camera on his shoulder at the same time. He said it took 18 years to get the card.

VE1AAC

According to Ray Soifer, W2RS, Executive Vice President of AMSAT – NA in League Lines of QST for January, 1992, he found two soviet amateur operators operating from the same soviet space station *Mir*. Sergej, U5MIR, on 145.55 MHz asked him to shift to 145.51 MHz and work Aleksandr, U4MIR. He said this took place on November 23rd, 1991, and Gerry worked him on March 18th, 1992. That was quite a long time to be up there in a space station. This also indicates that each cosmonaut had the same MIR suffix in their call sign that was the name of the space station. They were simply issued the letter U and a digit as the prefix of the call sign. I'm the only one I know who would find that interesting.

The ARRL stated in the September, 1992, issue of QST that the FCC had just given their permission for any class of amateur radio license to operate the amateur station in a space craft. Up until that time the American astronauts had to hold an Extra Class Amateur License in order to operate from space.

There have been many spacecrafts go aloft since these first voyages. There appears to be unlimited room up there. There have been so many flights one begins to wonder if any have come close to running into each other. According to the National Aeronautics and Space Administration (NASA) the first incident of two space craft colliding occurred on Tuesday February 10th, 2009, over Siberia. NASA went on to state they had been expecting a collision sooner or later. The largest space museum is the one at Huntsville, Alabama, where one can learn a lot about these flights. Today one can make radio contact with a number of these spacecraft as they go over.

U.S. BURIES NO CODE LICENCE

The February, 1984, issue of QST has a couple of interesting statements. The first is that the editor in his editorial states that the FCC has buried the no-code license idea once and for all.

This was recorded in League Lines for April, 1984:

> "Wayne Green, W2NSD, filed a petition with the FCC February 10 requesting that all Amateur Radio operators be retested for their Morse code skill every two years. Green also proposed that the code requirements be increased in increments of 5 WPM at each retesting until the applicant reaches 35 WPM. Amateurs not able to pass this code test would be given an additional 60 days to improve their speed or turn in their licenses."

This was the reply as recorded in League Lines for May, 1984:

"The petition filed by Wayne Green, W2NSD, asking that radio amateurs be retested for Morse code proficiency every two years, until 35 WPM is achieved, was dismissed by the FCC on March 9, without a Rulemaking number."

That is rather interesting looking back from today. In case there is someone who does not know Wayne Green he was the owner and editor of CQ magazine. CQ magazine started with the first issue in January, 1945. Wayne also had the amateur radio magazine 73 that lasted 43 years from 1960 until 2003.

ALL CANADIAN HF FREQUENCIES PHONE

The other item in February, 1984, is the fact that CRRL and CARF were talking to DOC with the idea of making all HF frequencies phone with just a gentleman's agreement on the separation aspect.

MEXICO CITY EARTHQUAKE 1984

Mexico City was hit with a very bad earthquake in 1984 and amateur radio all over North America was kept busy with emergency traffic for quite some time. Bill Horne, VE1GL, and Serge Szpilfogel, VE1KG, were two HARC members who handled a lot of that traffic in this area.

HARC AND DARTMOUTH ARC NAVAL CADETS' REGATTA 1985

HARC and the Dartmouth ARC had a good time on September 23rd and 24th of 1985 providing the communications for a Naval Cadets Regatta at Shearwater.

RUN FOR LIGHT 1986

ASM Aaron Solomon, VE1OC, stated that several amateur operators from the Halifax and Dartmouth area participated in the "Run for Light" in his September, 1986, Maritime News. The Run for Light was a run to support Blind Sport Nova Scotia a division of Canadian Blind Sports Association.

ENJOYABLE SOCIAL 1986

Aaron, VE1OC, stated that the charter members of HARC and their XYL's attended an enjoyable social in his December, 1986, Maritime News.

WARC GIVES NEW FREQUENCIES 1979

The CRRL Canadian Newsfronts for April, 1979, state that Bud Punchard, VE3UD, had been selected as the amateur radio representative from Canada for the World Administrative Radio Conference (WARC). This was held from September 24th until November 30th, 1979. This WARC actually finished in early December. There were 2,000 delegates from more than 140 countries.

This WARC gave the world's amateur radio operators the 10.1 – 10.15, 18.068 – 18.168 and 24.890 – 24.990 MHz frequencies. This suggestion was first reported in December, 1974, so it took a lot of hard work to put it together and make it work. This also gave new amateur satellite allocations: 1260 – 1270, 2400 – 2450, 3400 – 3410, 5650 – 5670, and 5830 – 5850 MHz plus 10.45 – 10.5 GHz.

It took some time for DOC to give their permission to operate on these bands. The DOC gave their permission for the 30-meter band (10.1 – 10.15 MHz) on May 21st, 1982, for A1 and for A1 and F1 with an advanced certificate or license. DOC approved the use of the 17-meter band (18.068 – 18.168 MHz) and the 12-meter band (24.890 – 24.990 MHz) for amateur radio in Canada on July 29th, 1987.

The DOC gave permission for Canadian amateur stations to operate on full legal power on 160-meters, 1.8 to 2.0 MHz, on September 27th, 1985. The old "Loran A" navigation system had been replaced by "Loran C". "Loran A" had been shut down for some time before DOC got around to granting this permission. The amateur radio community had to share 160 meters with "Loran A" up until it was terminated.

VE1ASJ AND VY1CW QSO ON 160 METERS

Bill Richardson, VY1CW, Whitehorse and Andy McClelland, VE1ASJ, finally had a QSO on 160-meters in 1987. One would have to agree from Saint John in New Brunswick to Whitehorse in the Yukon Territory is very good DX on 160-meters.

CHU TIME SIGNALS

There is an interesting article with lots of detail on radio station CHU Ottawa on page 17 in the August, 1987, issue of QST. This is the Canadian station that transmits time signals and other information on the frequencies of 3,330, 7,335 and 14,670 kilohertz.

USSR CANADA POLAR BRIDGE EXPEDITION

The full details of a joint USSR Canada expedition known as The Polar Bridge Expedition or the Transpolar SKITREK can be found on page 62 of the June, 1988, edition of QST. This was a very interesting expedition involving nine Russians and four Canadians that crossed the North Pole from Cape Arktichesky, Russia, to Ward Hunt Island, Canada. All 13 skiers walked ashore side by side after a 91-day 2000-km trek on June 1st, 1988, according to QST in the August, 1988, issue. Communications was provided by amateur radio. There was a reciprocal agreement between Russia and Canada from October 30, 1987, until August, 1988.

Icom supplied their 761 HF transceivers for this expedition and the only unfortunate incident was the loss of the station at North Pole 28, the floating Soviet scientific station. Due to a huge breakup of the ice, a complete hut, clothes, antennas, as well as the radios, were lost. The expedition was prepared, as they had spread out equipment and supplies over the area. The group then relied on Soviet equipment to carry out their communications.

The only VE1 directly involved in this undertaking was Andy McClelland, VE1ASJ. Andy was one of the control operators at station CI8C located at Resolute Bay, Cornwallis Island. I had served with two of the members of the Polar Bridge Amateur Radio Network. Ron Belleville, VE3AUM, was one of my petty officers in the navy and Terry Keim, VE8TF, and I were stationed at Inuvik Aeradio VFA6. Terry became VE5TK when he retired. Terry became a silent key in February 2018. This expedition was mentioned at an HARC meeting with a brief description. I did not find a record of any HARC member making contact with this expedition.

SITOR TO AMTOR

God Bless Amateur Radio! Back during the 1970's several electronic companies created a telex over radio they called ship telex over radio and created the acronym SITOR. The Dutch Philips Company was felt to have the best design and that design went into production. If that was the best, I am glad I did not have to use one that wasn't. This best design was a nightmare but better heads were to prevail. I believe it was six American amateur operators that were given this equipment to try and make something out of it. They called it amateur telex over radio and created the acronym AMTOR. This certainly improved things for us in the commercial world of marine radio. SITOR was then a usable piece of equipment, but most of the stations using SITOR were shut down in 2009. SITOR is now obsolete. The AMTOR portion soon spread to the whole world of amateur radio. A detailed description of AMTOR is found in the November 1989 issue of QST beginning on page 26. As far as I know no member of HARC is using AMTOR and it appears to have run its course in the world of amateur radio.

ARRL REQUESTS CODELESS CLASS OF LICENCE

ARRL made a request to the FCC for a codeless class of license in 1989 according to the September issue of QST. The April issue of QST described this proposal to the FCC in detail. The American codeless license went into effect on February 14th, 1991, and the FCC dropped all code testing in February, 2007.

ARRL REQUESTS CANADA GET A
NOVICE AND TECHNICIAN CLASS

The ARRL was trying hard to get the Canadian amateur radio licenses changed to reflect the American licenses more with a Novice and Technician class license. There was a lot of discussion on this within the pages of QST and the meetings of HARC but it finally ended during the winter of 1977-78. The amateur radio operators did not want a novice class of license and the radio inspectors with the Department of Communications did not want the extra work involved to examine a technician class of license. Therefore, the Canadian license remained two classes of license; the Amateur License and the Advanced Amateur License, with the 6-month endorsement for the Amateur License to permit phone operation on 10-meters.

LOGGING REQUIREMENTS RELAXED BY DOC

In 1983 the DOC relaxed their logging requirements for amateur radio stations, but stated that one would require a log and QSL cards showing at least five CW contacts per week, in a six-month period, in order to obtain a 10-meter phone endorsement to their operator's license. Gary, VE1BZD later VE1JB, liked CW so well that in six months he presented three full log books to DOC for his 10-meter endorsement.

CRRL STATES DOC WANTS 3 CLASS OF LICENCE

The Canadian Radio Relay League Canadian Newsfronts for December, 1985, stated that DOC was in the process of changing the Canadian amateur radio licenses to three licenses similar to those in service in 2008.

CANADIAN AMATEUR LICENCE INCREASE 1984

The Canadian Amateur Radio Station License was increased from $13.00 per annum to $13.50 per annum in 1984.

Government of Canada
Department of Communications

Gouvernement du Canada
Ministère des Communications

RADIO STATION LICENCE

Issued in accordance with the Radiocommunication Act
and Regulations made thereunder

THIS LICENCE SHALL BE RETAINED AT THE STATION

LICENCE DE STATION RADIO

Délivrée en conformité de la Loi sur la radiocommunication
et de ses règlements d'exécution

LA PRÉSENTE LICENCE DOIT ÊTRE CONSERVÉE À LA STATION

CLASS OF LICENCE/CLASSE DE LICENCE		COMPANY CODE CODE DE LA CIE	LICENCE NUMBER NUMÉRO DE LA LICENCE
AMATEUR	THIS LICENCE SHALL CONTINUE IN FORCE UNTIL ▶ MARCH 31 1992 CETTE LICENCE RESTERA EN VIGUEUR JUSQU'AU 31 MARS 1992	020017647	663-0003759

ISSUED TO
DÉLIVRÉE À

SPURGEON GEORGE ROSCOE
BOX 1 SITE 5
RR 5 ARMDALE NS   B3L 4J5

SERVICE CATEGORY/CATÉGORIE DE SERVICE

AMATEUR - SERVICE - D'AMATEUR

TRANSMITTING FREQUENCIES FRÉQUENCES D'ÉMISSION	DESIGNATE BANDWIDTH & CLASS OF EMISSION LARGEUR DE BANDE NÉCESSAIRE ET CLASSE D'ÉMISSION	POWER PUISSANCE kW	AUTHORIZED COMMUNICATIONS/POINT-TO-POINT COMMUNICATIONS AUTORISÉES/POINT À POINT	RECEIVING FREQUENCIES FRÉQUENCES DE RÉCEPTION	CHANNELS VOIES TX-EM RX-REC

STATION LOCATION   ARMDALE NS
EMPLACEMENT DE LA STATION

CALL SIGN SIGNAL D'APPEL	DATE OF ISSUE/DATE DE DÉLIVRANCE	
VE1BC	APRIL 01 AVR 1991	

MARCEL MASSE
MINISTER OF COMMUNICATIONS/MINISTRE DES COMMUNICATIONS

SEE REVERSE SIDE–VOIR AU VERSO

This licence authorizes the licensee to establish and operate a radio station as described in the approved application, in accordance with specific items or conditions and applicable provisions of the Radiocommunication Act and its regulations. This authority should not be construed as approving the use of any antenna supporting structure which has not been approved by the Department of Transport from an aeronautical safety point of view. Except as provided in the regulations, no change in the apparatus or operations shall be made without the authority of the Minister of Communications, and the licensee shall notify the Department in writing upon a change of address.

The Department may, at a future date, require the licensee to install filters, tone coding devices, reduce the effective radiated power and/or antenna height as appropriate.

Service Category indicates the categories of service the station is authorized to perform.

In many cases licence fees are related to the number of transmit and receive channels. A code, used in the "channel" column, indicates the number of equivalent voice channels as given in the following table:

Cette licence autorise le titulaire à établir et faire fonctionner une station de radiocommunication comme décrit dans la demande approuvée, aux conditions précisées et conformément aux dispositions pertinentes de la Loi sur la radiocommunication et ses règlements d'exécution. On ne devrait pas interpréter cette autorisation comme une approbation d'un bâti d'antenne qui n'a pas reçu le feu vert du ministère des Transports du point de vue de la sécurité aéronautique. À moins d'indication contraire dans les règlements, aucun changement ne doit être apporté à l'appareil ni au mode d'exploitation sans l'autorisation du ministre des Communications et le titulaire de la licence doit aviser par écrit le Ministère de tout changement d'adresse.

Le Ministère peut obliger ultérieurement le titulaire de la présente à installer des filtres et des codeurs de tonalité, ainsi qu'à réduire la puissance apparente rayonnée et (ou) la hauteur de l'antenne, selon le cas.

La partie "Catégorie de service" indique les catégories de service que la station est autorisée à fournir.

Dans plusieurs cas, les droits de licence sont en fonction du nombre de voies de transmission et de réception. Un code dans la colonne "voie" indique le nombre équivalent de voies téléphoniques comme suit:

Channel Code	1 to 9	A	B	C	D	E	F	G	Other Letters H, I, J, etc.
Equivalent No. of Voice Channels	1 to 9	10 to 24	25 to 60	61 to 120	121 to 300	301 to 600	601 to 960	961 to 1200	Measured In units of 300 channels

Code de voie	1 à 9	A	B	C	D	E	F	G	Autres lettres H, I, J, etc.
Nombre équivalent de voies téléphoniques	1 à 9	10 à 24	25 à 60	61 à 120	121 à 300	301 à 600	601 à 960	961 à 1200	Mesuré par unité de 300 voies

For further information regarding your radio licence please contact your nearest Department of Communications District Office. Copies of the Radiocommunication Act and Radio Regulations may be purchased from Printing and Publishing, Supply & Services Canada, Ottawa, Ontario, Canada K1A 0S9.

Pour de plus amples renseignements, prière de communiquer avec le bureau de district du ministère des Communications le plus rapproché. On peut se procurer un exemplaire de la Loi sur la radiocommunication et du Règlement général sur la radio en s'adressant à l'Imprimerie du gouvernement canadien, ministère des Approvisionnements et Services, Ottawa (Ontario), Canada. K1A 0S9.

One can see that this is my Amateur Radio Station License for 1991 and the station was actually located at Sambro Head not Armdale. Armdale was the mail address. Had the station been close to an aerodrome I would have made a point of notifying the Department of Communications of the difference. This is both the front and back of this license.

This was my Amateur Radio Operator's Certificate and the reason it was dated February 20th, 1980, and not February 9th, 1962, is rather simple. Canada would issue an Amateur Radio Station License to one who held a Commercial Second-Class Certificate of Proficiency in Radio or higher. I had been operating amateur radio with my commercial certificates. I was doing a history project on marine radio at this time and wanted to mention this in that project. I asked the Department of Communications if I could have an Amateur Radio Operator's Certificate to use as an example.

My first commercial Certificate of Proficiency in Radio was dated February 9th, 1962. Canada is the only nation that I know that would issue an amateur radio station license to one who held a commercial certificate. This is one of the Canadian quirks in radio regulations that I agreed with. The word amateur is from some foreign word meaning "the love of" as I understand it. I doubt that I would have bothered to get involved in amateur radio had it not been for this provision. The very best operators are found in the world of amateur radio because they simply do it for the love of it. I can remember telling a number of commercial operators that they should visit an amateur station just to see how good it could be.

There are many quirks in the Canadian radio regulations that I do not agree with. If someone does not agree with me that is their right. But I do not agree with the special license and call sign for a ship. An amateur station in a ship should be identified as it is in the rest of the world; the operators own personal call sign. That way one knows who they are talking to. I also do not agree with these special prefixes from time to time. They make no sense to me and again one should use their own personal call sign and leave it at that.

DOC DID NOT LIKE SPECIAL PREFIXES

RAC ADVISES MEMBERS TO REJECT LICENSING CHANGES

Radio Amateurs of Canada has advised its members to just say no to Industry Canada's plan to drop Amateur Radio license fees and combine operator certificates and station licenses. RAC says the plan lacks provisions to adequately track call signs, station locations, and operator qualification levels. RAC also says IC—Canada's equivalent of the FCC—has been unable to answer specific questions about the plan.

"Industry Canada's proposal is viewed as another step by the Department to lessen its involvement in regulation of the Amateur Service," an RAC bulletin declared on June 5. "In recent years, Industry Canada has reduced surveillance and enforcement activities to a virtually ineffective level."

Expressing fears of "chaos" ahead, RAC concludes that eliminating the license fee and merging the license documentation "would lead to a further decline in the status of the Amateur Radio Service."

Late last year, the IC scuttled negotiations to delegate partial authority over Amateur Radio licensing in Canada to an arm's length organization associated with RAC. Canada has approximately 45,000 hams.

60 September 1998 QST

The DOC did not like the special prefixes that amateur radio operators requested from time to time for special occasions. DOC was trying hard to get rid of them in 1987 claiming they may be illegal. But they did not succeed and it appears as though they will be with us forever. In 1995 those with the VE1 prefix could substitute that for the XK1 prefix to recognize the 50th anniversary of the end of World War II. Each and every Canadian prefix was given a special prefix for that event.

We received it two years later via the letter and license below. Just before we received this letter and license RAC had asked Industry Canada to lower the code speed from 12 WPM down to 5 WPM according to page 85 of the April, 2000, issue of QST.

NEW AMATEUR LICENCE 2000

 Industry Canada   Industrie Canada
http://strategis.ic.gc.ca

Dear Amateur Radio Operator:

Effective April 1, 2000 a streamlined authorization procedure will be put in place for the amateur radio service. As a result, radio amateurs will no longer be issued a radio station licence on an annual basis or charged a licence renewal fee.

Enclosed is your new Amateur Radio Operator Certificate (wallet and diploma size). This certificate is your complete authority to operate amateur radio apparatus. It serves the function of both the traditional radio station licence and radio operator certificate.

No changes have been made to the regulations for operation of radio apparatus in the amateur service. However, amateurs are still required to notify Industry Canada of a change of mailing address.

If information appearing on your new certificate is not accurate, please provide us with the appropriate corrections by fax or mail to the address listed below. For corrections to your amateur qualification, please also forward copies of any documents confirming your qualification.

For more details, please refer to the documents concerning the streamlining of the authorization process for the amateur radio service available at Industry Canada's Internet site at the following address: **http://strategis.ic.gc.ca/radioamateur.e** or contact:

Industry Canada
Amateur Radio Service Centre
P.O. Box 9654
Postal Station "T"
Ottawa, ON
K1G 6K9

e-mail address: Spectrum.amateur@ic.gc.ca
Telephone: 1-888-780-3333 (Toll free)
Fax Number: 1-613-991-5575

Canada

Every amateur radio operator in Canada received this letter and the Certificate below in place of a station license on April 1ˢᵗ, 2000. This is self-explanatory.

Industry Canada / Industrie Canada

Certificate of Proficiency in Amateur Radio

This is to certify that

Spurgeon George Roscoe

has obtained the following qualifications:

Basic
12 W.P.M. Morse Code
Advanced

The certificate holder is authorized to operate amateur radio apparatus in accordance with the regulations made pursuant to the **Radiocommunication Act,** and to use the following call signs:

VE1BC

Certificate Number: 19809900716
Issue Date: 1 April 2000

Issued by

Canada

One can now have more than the one call sign as stated on this certificate and there is no way to keep track of these call signs except for the total number assigned. I also believe there are a lot of silent keys that are still on the books as well. Simply look up one who has been a silent key for a few years. Radio Amateurs of Canada (RAC) has a request on their web site to notify them of any silent key still listed. I should try it sometime just to see the mess I get myself into. It would never be a simple statement that they are a silent key and left at that. One would have to have proof of some description. There are a lot of incorrect addresses on the licenses that appear in rac.ca and qrz.com as well. If there were an annual renewal fee still in place it would eliminate both problems.

ONE AMATEUR PER 2,000 PEOPLE

In 1993 the world's amateur radio population was less than one amateur per 2,000 people. This has not likely changed since then but one has to admit the amateur radio population is an important part of the overall population. One would be hard pressed to find a group of people that does more to make life better for the world's population as a whole.

MANY INTERESTS GROUPS

There are many different interests within this amateur radio population and depending on the size of each interest group one could state more than 30 operating interests. If one decreased the size of each individual interest group there would be 50 or more. These figures are from the October, 1993, issue of QST. One has to admit this is a fascinating hobby with something within it for just about anyone.

RAC SPECTACULAR GROWTH 1993

Radio Amateurs of Canada (RAC) was reporting a spectacular growth in 1993. On October 1st, 1993, they claimed there were 41,014 Canadian amateur radio operators. They claimed 15,787 were new since the restructuring of the amateur radio service in April, 1990. They stated that 72% had the highest level of qualification and that even more encouraging was the drop in age of the average Canadian amateur radio operator. RAC stated that in 1987 60% were over 50 and in 1993 this percentage had dropped to 54%. They claim one cannot argue with city hall but one wonders what calculator came up with those digits.

NUMBER OF AMATEUR OPERATORS IARU

These are the number of amateur radio operators in 1994 according to the IARU. This was found on page 89 of the February, 1995, issue of QST:

1. Japan   1,300,000
2. United States   632,000
3. Germany   64,000
4. United Kingdom   62,000
5. Indonesia   60,000
6. Spain   47,000
7. Canada   44,000
8. Russia   38,000
9. Italy   30,000
10. Brazil   27,000

It looks like the IARU calculator found another 3,000 amateur operators here in Canada.

HARC & CAPITAL HEALTH

HARC has had a good close relationship with Capital Health in Halifax since 1993. In 2008 HARC placed a number of radio kits in various locations to be used in case of an emergency. HARC has had various exercises to ensure these units and those who operate them are ready for any emergency.

HARC ANNIVERSARY'S

Welcome!

On behalf of the Halifax Amateur Radio Club, I would like to welcome you to our 75th Anniversary Hamfest. Our remarkable record of continuous operation for three-quarters of a century deserves a big party, and we are glad you have come to help us celebrate. We have planned a variety of events that we hope will interest, entertain and excite each of you. Moreover, we have assembled an outstanding group of presenters for all parts of our program.

The **Maritime DX Forum** has grown steadily for the past four years as a separate activity of the Club, so we are happy to include it as a key part of the Ham Fest Program. The 75th Anniversary Planning Committee has also created a new event that is intended to complement the Maritime DX Forum with a program for those who are not as interested in HF DXing and contesting. The **Far Side of Amateur Radio** is a program to showcase emerging technologies of interest to Amateurs, including those in the VHF and UHF sphere. This will be the first year for this program and those of you participating will hopefully give us many ideas for the future.

Since Amateurs tend to be very outgoing people, there are several social events as part of the celebration. It is always great to meet those we have spoken with on air or to renew old friendships. In the Maritime tradition we will strive to show guests and visitors a very good time. The **Anniversary Banquet** will be an opportunity to formally acknowledge our birthday, to entertain you and to announce awards as well as to socialize.

The **Down East Flea Market**, usually held in May, is Saturday's central attraction and will draw people from many places in search of a bargain and personal contact with ham radio friends. During the flea market, there will also be a series of interesting talks for you to attend. In particular, there will be a unique opportunity after the flea market winds down to meet and discuss areas of mutual interest with Radio Amateurs of Canada officials, including RAC President, Dave Goodwin, VE3AAQ/VO1AU.

We hope that our Anniversary celebration will leave you with great memories to look back upon as well as an anticipation of the events we will organize in the future. Thank you for contributing to the success of this Anniversary party with your presence and enthusiasm. We hope you have a wonderful weekend!

73 - Bill Elliott, VE1MR
President, Halifax Amateur Radio Club & Chair of the 75th Anniversary Committee

Program of the 75th

HARC held their 75th Anniversary with a Down-East Hamfest over the weekend from Thursday, August 21st to Sunday August 24th, 2008. It was held in Roseria Hall at Mount St. Vincent University on the Bedford Highway in Halifax. The actual 75th date is March 16th, 2009, as one can see from the record in this exercise.

The Maritime News in May, 1963, states that HARC was celebrating its 30th anniversary and that three charter members were still with the club. The three were Cliff Short, VE1AW, Bill Bligh, VE1BC and Wes Street, VE1EK. If one considers the date of the 75th celebration a mistake it was made 45 years before this celebration.

Don Welling, VE1WF, states in Maritime News for January, 1983, that HARC will celebrate its 50th anniversary next March. It was a year early again. I was on the lookout for some clue for this discrepancy but found none.

Actually, the date March 16th, 1934, was simply the date that this club was renamed the Halifax Amateur Radio Club. There has been amateur radio activity and an amateur radio club of some description in Halifax since the beginning of radio. In the article Canadian Newsfronts for October, 1984, they congratulate HARC for 50 years affiliation with the league. From as near as I can tell the members of HARC were members of the American Radio Relay League before and after they renamed the Maritime Amateur Radio Association the Halifax Amateur Radio Club.

HAMFEST PROGRAM

Thursday, August 21, 2008

1600 to 2100 - *Hamfest Registration* with *Reception* beginning at 1900 (Cash Bar), Rosaria Terrace (second fl Rosaria Hall)

Friday, August 22, 2008

0830 Hamfest Registration, Rosaria Terrace
0915 – 1645 Maritime DX Forum – Vinnie's Pub (first floor)
0915 – 1645 The Far Side of Amateur Radio - in Board Room (second floor)

Maritime DX Forum	The Far Side of Amateur Radio
0915 Presentation: John Sluymer, VE3EJ, World Class Radio Competition and Contesting	0915 Presentation: Larry Hicks, VO1HL, Software Defined Radio
1030 Coffee Break in the lower foyer	
1100 Presentation: Jeff Briggs, K1ZM/VY2ZM, Understanding Marconi's 1901 Transatlantic Wireless Feat	1100 Presentation: Jim Cyr, VA1CYR, Geo-hamming
1215 Lunch, Rosaria Dining Hall: Soup, sandwiches, veggies & dip, sweets, coffee and tea	
1315 John Scott, VE1JS, About DXCC Card Checking	1315 Tom Caithness, VE1GTC, About VE1 QSL Bureau Operation
1330 Presentation: Dave Goodwin, VO1AU, and Garry Hammond, VE3XN, Canadians on the DXCC Honour Roll	1330 Presentation and discussion: Bill Elliott, VE1MR, Alternative aspects of VHF/UHF operations
1445 Coffee Break in the lower foyer	
1515 Presentation: Anne Santos, WA1S, The TX5C Clipperton Island Dxpedition	1515 Presentation: David George, VE1AJP Mak Music for your Ham Radio
1630 Closing words and door prize draw	1630 Closing words and door prize draw

1715 – 1815 *75th Anniversary Banquet Reception*, Rosaria Hall Dining Room
1830 – 2130 *75th Anniversary Banquet*, Rosaria Hall Dining Room

Saturday, August 23, 2008

Flea Market, Rosaria Terrace and the Rosaria Multi Purpose room (both on second floor)
0600 Vendors set up for flea market
1000 Flea Market open to buyers
Concurrent Presentations, Board Room (second floor)
1100 NS Power RFI presentation
1200 David Musgrave, VE1EDA – Paclink Radio for Emergency Communications
1300 Fred Archibald, VE1FA – A 1930s Radio Shack – A walk down memory lane
1400 Grand prize draw – FT-857D HF mobile
1405 Dave Goodwin, VO1AU/VE3AAQ – RAC President's forum with Atlantic Director Len Morgan, VE9M
End of flea market or 1500 - NSARA meeting (Annual General Meeting)

1800 – 2200 *Pizza, Beer and Bluegrass*, Vinnie's Pub, Rosaria Hall (first floor), Pizza & Cash Bar
Don Trotter, VE1DTR, and his bluegrass group *Sunrise*, will perform together with Frank Davis, VO1HP, from Newfoundland, and Richard Neville, Black Tickle, Labrador.

Sunday, August 24, 2008

1000 – 1200 *Farewell Brunch*, Rosaria Dining Hall (second floor), *The Canadian Country*: Juices, Fruit, Muffins, Buttermilk pancakes, Choice of sausage, ham or bacon, Scrambled eggs, Homefries, Coffee and Tea

Program of the 75th

TEACHING AMATEUR RADIO

Teaching amateur radio began before and with the formation of the various amateur radio clubs. There is no record of the first course taught at the Halifax Amateur Radio Club.

P. L. Whitman, 1AI, was president of the Halifax Wireless Association in 1923. Many new young members were developed by him. He donated a room and site for a club transmitter that he built. He must have been one of the first, if not the first to teach amateur radio in the Halifax area. The earliest record I have found of him is on the 1919 call sign list with the 1AI call sign. Apparently P.L. Whitman did not write an examination and obtain a license when this took place from the 1927 convention. At least I have been unable to find any record of this. The only record I have found since is that he was given a patent for a time dial in 1926. This consisted of two pieces of cardboard one placed over a clock face that would give the correct time in any of the world's time zones. He may have moved to the "Boston States" or "Upper Canada". There was a lot of that happening around here at that time. I find it hard to believe that one that active did not continue.

Richard Binns, 1EB, was building a receiver for the Halifax Wireless Club in 1923. This Club must have been one and the same as the Association. The amateur radio stations in Halifax in 1924 were: 1AI, 1AU, 1BQ, 1BV, 1DF, 1DT, 1EB and 1EF. The amateur radio stations in Dartmouth in 1924 were: 1AR, 1DD and 1DQ. I believe they all were not only members of this association but were also members of ARRL.

Bill Borrett, 1DD, was an official of some description with radio station CHNS in 1926. He and 1DJ started classes at CHNS for anyone interested in obtaining an amateur radio license. They also held a weekly broadcast on the ARRL and amateur radio each Wednesday evening.

There were ten new members in the fall of 1926 and I believe this would be ten new members to ARRL. At the same time a class of 17 was started by the Halifax County Radio Association. According to the January, 1927, report the Halifax County Radio Association had a class of over 30 students including 3 ladies.

Bill Bligh, VE1BC, held code practice transmissions from his station and gave a report to HARC at the November 27th, 1936, meeting. He stated at this meeting that he had received many letters from listeners outside this area. By 1938 Wes Street, VE1EK, Doug Smith, VE1FO, Brit Fader, VE1FQ and Cliff Short, VE1AW were all giving code practice sessions from their stations.

In the minutes of the HARC monthly meeting of May 16th, 1947, it was noted that the Department of Transport Radio Division had advised that Radio Clubs may give CW code practice on ten meters and higher frequencies with their club stations.

Chief Petty Officer John Powroz, VE1PW "Powerhouse Willie", and Sergeant Bill MacDonald, VE1WG, Halifax city police force was teaching a class of 30 students during the winter of 1962-1963. One of the students was John Powroz son. Moe Lake, VE1PX was teaching amateur radio classes at N.S. Technical in 1968.

The Halifax ARC and Dartmouth ARC were teaching 11 students from the Canadian National Institute for the Blind Amateur Radio Club. This was according to Walter Jones, VE1AMR, in his first Maritime News in the December, 1971, issue of QST. The HARC minutes of the monthly meeting on March 17th, 1972, stated that they felt they would have 10 to 14 "White Caners" (blind amateur radio operators) graduate soon from this class.

HARC ran a class in 1972 and the instructors were Dick West, VE1AGX, and Barry Hyndman, VE1XW with others as required.

Moe Lake, VE1PX, and Bill MacDonald, VE1WG were teaching classes in Halifax in 1972. They had 56 students when the class started at N. S. Technical. Moe managed to get seven students that passed their Department of Communications exams according to the minutes of the monthly meeting for June 16th, 1972, and that is about par for the course so to speak. It seems as though everyone has good intentions on the first day, but for some unknown reason simply loose interest and dropout as the class goes on. This was not only common with amateur radio it was the same with the commercial radio classes.

Jim Hicks wanted me to start a class so he could get his license shortly after I met him around 1980. I mentioned this to Dan Dawson one day. I did not know Dan had his amateur license. Dan wanted to know if he could teach theory with me. My theory was back in the days of tubes and not far removed from spark transmissions so needless to say Dan was a most welcome asset. Dan, VE1JV, and I started teaching in his basement in Williamswood. He is one of the finest radio theory instructors I have met. Jim Hicks became VE1AGK from our first class. I tried to be each student's first QSO (contact) and Jim was the first on March 13th, 1984, at 1605. A great thrill to work a former student for the first time. I note from my old logs that my first phone contact with him was one year later.

Dan, VE1JV, and I had a few members from the Ground Search and Rescue group in our classes. These members wanted their Amateur Radio License in order to use VHF hand held Amateur Radio equipment while they conducted their various searches. This was back when a 15-WPM Morse exam was part of the initial license. The termination of the Morse requirement has probably been a big help to this important group. Dan, VE1JV, and his XYL Louise, VE1BGD,

HARC Files

This is Dan Dawson, VE1JV, teaching an amateur radio class.

managed to get in a few searches with this group, the Waverley Ground Search and Rescue Group during 1988.

Adrien Blinn, VE1SZ, told me that we should apply to the adult education division of the Department of Education and in doing so wound-up teaching evening classes at the Harrietsfield School. We were quite successful but soon ran out of students. Dan went on one year without me before terminating in 1988 and concentrating on music and flying gliders at the Stanley Airport with his three children. I worked shift work at the time and used up a lot of my annual vacation in order to attend every evening class. It is a wonder my family did not disown me completely.

Lynn, VE1ENT, and Bill, VE1MR, started one class in the mid 1970's. Barry, VE1TRI, filled in as an instructor a couple of times. Mike, VE1NNX, has also taught classes for HARC. Clive Bagley, VE1AMZ, was teaching amateur radio with and for HARC when Dan and I were teaching on our own.

Clive had a class during the winter of 1973-1974 consisting of 30 students on January 16th, 1974, and his classes were two nights each week from 7 to 10 pm. This class cost each student $5.00, the dropouts to forfeit their $5.00 but any student that sat for the D.O.C. exams would get their $5.00 back. Clive also had a class going at the same location, the Nova Scotia Institute of Technology the winter of 1975-1976. Moe Lake, VE1PX, and Bill MacDonald, VE1WG, were hoping to start a class at that time. Clive Bagley, VE1AMZ, Dick Grantham, VE1AI and Mike Pothier, VE1AJP taught a class over the winter of 1978-1979. A Basic Amateur Radio Course was taught from September 1998 to January 1999 and the instructors were: Tom, VE1GTC, Dave, VE1RCN, Bill, VE1MR, and Al, VO1NO.

Howard Dickson, VE1DHD

Over the years the HARC has been fortunate to have had a large number of excellent and committed teachers for the basic Amateur Radio course. In the 1970's Clive Bagley, VE1AMZ, and Mike Pothier, VE1AJP, taught basic ham courses as stated. In the mid-80's Bill Elliott, VE1MR, started classes introducing the multiple instructor method. From the late 80's and well into the 90's Pearson Fryers, VE1SWL, carried on, managing the course assisted by a dedicated group of instructors. Tom Caithness, VE1GTC, took over in the late 90's and Scott Wood, VE1QD, managed the course from the beginning of the new millennium. Scott Wood has been successful in attracting some younger students into the hobby and now that there is no longer a Morse code requirement, this aspect of the program has been handled separately by Gary Bartlett, VE1RGB, who holds a weekly CW class at the Club station. Scott also introduced the concept of a winter self-study course where students are provided with a syllabus, the RAC Study Guide and unfettered access to their instructor by email. Although a far cry from the personal, hands-on approach that takes place in the more traditional fall course, this self-study approach has had surprising success. Barry Diggins, VE1TRI, took over the course from Scott in January 2008. One can see from this record that there has been a lot of effort to encourage anyone interested in amateur radio to get a license and enjoy this fascinating hobby.

Wayne Harasimovitch, VE1WPH has been the Basic Ham Course Instructor since Barry Diggins, VE1TRI.

MARITIME AREA NET MEETING 1986

These were the Maritime Area Net Meetings taken from the last Club Bulletin that Brit Fader, VE1FQ edited in November 1986. Local time:

Maritime Net	Daily	1900	3750 KHz	
White Cane Net	Mon-Sat	1745	3770 KHz	
Weather Net	Daily	0700	3770 KHz	
Old Timers	Sunday	0800	3750 KHz	
APN CW Net	Daily	2000	3635 KHz	
Lunenburg Net	Daily	2000	147.24 MHz	(+600)
Take 15 Net	Sunday	2045	146.85 MHz	(-600)
Swap Shop every 2nd	Tuesday	1930	3750 KHz	
Swap Shop	Sunday	2010	147.15 MHz	(+600)
DX Net	Sunday	2110	146.685 MHz	(-600)
40 Metre Net	Sunday	1000	7085 KHz	
PL Net	Daily	0900	3780 KHz	

As a comparison this is the list of nets that appeared in the 1964 issue of Atlantic Amateur Radio Newsletter, a publication Compliments of Canadian Admiral Corporation Limited.

Net Table and Frequencies
VE 1 RT

This table of Net times and frequencies is inserted in the Newsletter for everyone to keep it in mind, and try to avoid using the frequencies listed, while the Nets are in progress. PLEASE cut it out of the Newsletter, and use a piece of Scotch tape to stick it on the front of the rig, or somewhere where it can easily be seen:

WEEK DAYS

Maritime Weather Net	7.00 A.M. to 7.30 A.M.	3770 kcs
Maritime Traffic Net	7.00 P.M. to 8.00, or end	3750 kcs

SUNDAYS

Old Timers' Net	8.00 A.M. to 9.00 A.M.	3750 kcs
SKN CW Net	9.00 A.M. to 10.00 A.M.	3660 kcs
N.B.A.R.A. Net	9.30 A.M. to end of Net	3790 kcs
Ten Meter Net	10.00 A.M. to end	28300 kcs
Cape Breton Net	1.30 P.M. to end of Net	3730 kcs
A.V.A.R.C. Meeting	1.30 P.M. to end	3740 kcs
N.S.A.R.A. Meeting	3.00 P.M. to 4.00 P.M.	3750 kcs
A.R.P.S.C. Net	6.30 P.M. to 7.00 P.M.	3750 kcs

(This Net is for Members of the Amateur Radio Public Service Corps, which WAS known as A.R.E.C.)

Maritime Traffic Net	7.00 P.M. to 8.00 P.M.,	3750 kcs
	or end of Net.	

Newfoundland Tfc. Net

This Net is conducted on 3785 kcs, every night of the week, and its members have asked repeatedly that they be given a break for the half hour it is in progress.

It begins at 6.30 P.M., Atlantic Standard Time and is usually over in half an hour.

Please try to avoid crowding them. If others do it, we can't very well ask THEM, as it would be badly resented.

A. E. S. "Bert" Whittaker, VE1RT, Aspen, Guysborough County was the Editor of this publication. Bert received the Doucette Trophy in 1961 for the formation of the weather net.

In 1956 all CW operators were encouraged to hang out at 3535 kilocycles for what was to be a Trans Provincial Net. The frequency was to be monitored continually by VE1, VE2 and VE3 amateur stations. It was to be used for Civil Defense purposes mainly but it is not included on this list. Fritz Webb, VE1DB, was a regular on this net that they abbreviated to TRN. One assumes this stood for the Trans-Provincial Radio Net. Over this thirty-year period this net may have moved up the band 100 kilohertz and down-graded in size or area converted to the Atlantic Provinces Net (APN).

In his editorial "It Seems to Us…" in the October, 1965, issue of QST John Huntoon, W1LVQ, stated that the first instance of amateur radio emergency communication took place in a Middle West area left isolated by windstorm in 1913. This makes it clear amateur radio has participated in emergencies from the beginning.

The Acting Section Manager, Aaron Solomon, VE1OC, recorded the following statement in his Maritime News report for August, 1986: "50 years ago VE1GL organized the following for emergency traffic: VE1HJ, VE1FQ, VE1WC, VE3WW." In 1936 this would have been:

Bill Horne, VE1GL
Ralph Fraser, VE1HJ
Brit Fader, VE1FQ
John Doull, VE1FN – (VE1WC)
Walt Wooding, VE1ET – (VE3WW)

All five were very active members of HARC.

HARC has worked in close association with the Civil Defense authorities. The following appeared in the *Civil Defense Bulletin* issued by the Coordinator's Office at Ottawa:

"EMERGENCY COMMUNICATIONS – The first message transmitted over the emergency civil defense facilities in Nova Scotia was sent to Hon. A. B. DeWolfe at his home in Halifax from his Communications Committee Chairman, Mr. E. S. Campbell, [VE1QQ] on December 13 [1951] Two-way voice communications was carried on between radio transmitters in provincial headquarters and Halifax target area, control center. Radio contact was made also with amateur stations in the province, and out of the Province to St. Anthony, Nfld., giving an indication of the signal strength of the headquarters transmitters. Among those present at Provincial Emergency Headquarters for the test were Arthur Crowell, [VE1DQ] SCM Maritime, ARRL, and Squadron Leader Barrett, Director RCAF amateur radio network, both of whom are members of the provincial communications committee."

The Halifax Civil Defense net met each Wednesday at 7 PM on 3845 Kc. The ARRL had created the Amateur Radio Emergency Service (ARES) on August 15th, 1952.

Page 78, QST December, 2001

CIVIL DEFENSE EXERCISES

There are many records throughout the history of HARC where the club has participated in various exercises in order to train and be ready for any emergency. There was a Civil Defense Committee set-up in Halifax in 1951 and Art Crowell, VE1DQ, was chosen as a member to report directly to HARC. All the Civil Defense exercises were not as elaborate as the one in 1953. The armed services, police and utilities were all playing important parts in this one known as "Exercise Teamwork", which included a mock bombing of the city and anti-aircraft action by the local naval cruisers in the harbour.

Most of these exercises involved a fully equipped all band amateur station with the 80-meter band their main band of operations. A lot was made of the fact that some amateurs were mobile equipped and could move about with their automobiles. This was a good indication the cold war was in full force and shortly after this the big thing was to build a bomb shelter. I could never understand what one was to do when they crawled out of those holes in the ground after the bomb went off.

In 1954 a Civil Defense Convoy visited Halifax and special call signs were assigned these civil defense radio stations. Halifax received CJW201. There was a successful civil defense exercise titled "Operation Alert" according to the September, 1954, report by Doug Johnson, VE1OM. Those that participated in this exercise were Shep Shepherd, VE1RR, Wes Street, VE1EK, Doug Johnson, VE1OM, Fritz Webb, VE1DB, Doug Johnson, VE1PQ, Ray Wilson, VE1WL, Binks Fisher, VE1AFN, Ian McLeod, VE1GC, and mobiles Cliff Short, VE1AW, George Sandoz, VE1PT, and D. M. Copp, VE1NO.

Murray Banks, VE1GA, stated in 1957 that a new search team had been formed in the Greenwood, Nova Scotia area. This was known as "Land Search" and made up from members of the RCAF, RCMP, Fire Department and Amateur Radio. One can see that our present systems took a while "to grow" and did not happen overnight.

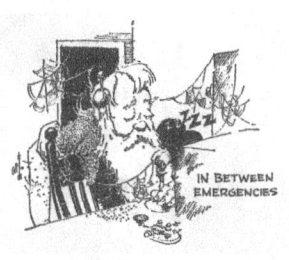

Page 43 QST, January 1965

HARC held a special Civil Defense course with 30 amateur operators in attendance in 1959. The instructors were Ray Wilson, VE1WL, Norm Weedmark, VE1QV and Ian Macleod, VE1GC. The Amateur Radio Emergency Communications (AREC) met at 1330 each Sunday on 3790 kilocycles after 1959.

The town of Marysville, New Brunswick, was cut off by a bad flood on February 4th, 1970, and this proved the big emergency that year. Emergency communications was set-up via amateur radio mainly on 80-meters. A lot of traffic was handled by various government officials including police and private individuals. At one point the Maritime Net closed and was turned over for use by this emergency.

The Civil Defense organization became known as the Emergency Measures Organization in 1975, and better known to the members of amateur radio by the acronym EMO. The EMO was actually first started during World War II in case Russia decided to attack the United States. This statement was made by Mr. M. Emmerson, Provincial Director of the provincial EMO at the HARC monthly meeting held on February 20th, 1974.

HARC Files

According to the June and July issues of QST in 1979 there were 50 Nova Scotia amateur radio operators who attended this seminar at Debert.

The EMO around Kentville, Nova Scotia is associated with the Kings County Amateur Radio Club, and that EMO holds the call sign **VE1EMO**. In 1970 the Kings County ARC had 12 members and 13 in a course to become amateur radio operators.

The EMO around Halifax and associated with HARC holds call sign **VE1QQ**. This EMO has held this call sign for fifty years. VE1QQ was the call sign of Mr. E. S. Campbell in 1959 and is listed in the 1963 Callbook as the Provincial Civil Defense organization. Mr. Campbell turned the VE1QQ call sign over to Civil Defense and replaced it with VE1AHQ. HARC used call sign VE1FO/1 as the call on these exercises from World War II until the Provincial Civil Defense was assigned the VE1QQ call sign. One had to identify each transmission with the call sign, the oblique stroke and the location of the transmitter when not at its registered location. VE1FO/1 stated it was the HARC station and was portable in area VE1.

COMMUNICATION — Amateur radio operators across Nova Scotia took part in a simulated emergency test of communications over the weekend. The test was designed to exercise emergency communications between the Emergency Measures Organization and zone headquarters in Sydney, Truro and Kentville and about 56 municipal emergency co-ordinators. Those manning temporary headquarters in the Halifax Amateur Radio Club Station at the Nova Scotia Institute of Technology included Gerald Hull, left, and Don Bower.   (Wamboldt-Waterfield)

Jan 29/79 H&M

HARC Files

CQWW CONTEST 1979

HARC Files

This is the EMO trailer on the front lawn of Bill Ash, VE1BBS, at 74 Quaker Crescent, Lower Sackville, for a CQWW contest on October 27th and 28th, 1979.

HARC Files

This is Bill Ash, VE1BBS, operating inside the trailer on his front lawn.

HARC Files

This is Dave Oldridge, VE1EI.

Dave was VA7CZ in 2008 and shortly before this photograph was taken, he was VE1BFV.

Dave came ashore from Sable Island to work at Halifax Coast Guard Radio VCS two years before this.

His move to VCS is recorded in Maritime News for September, 1977.

Dave VE1EI was made SEC (section emergency coordinator) for this area on October 1st, 1980.

Bill, VE1MR, bought that Ten-Tec Triton IV transceiver from Dave for his father.

The operators for this CQWW contest on October 27th and 28th, 1979, were: Sy, VE1BBO, John Gillis, VE1BFZ, Jim, VE1BJW, Peter, VE1BSE, Dave, VE1EI and Bill, VE1BBS. The coffee, logging and scoring was provided by Judy Ash the XYL of VE1BBS. They managed a score of 501,198 with that Ten-Tec Triton IV transceiver with a Wilson System 2 Beam at 25 feet. They had a dipole for 40 and 80-meters that did not work very well. They managed to fill a $2.99 Radio Shack CB/SWL Radio Station Log Book.

EMO & VEHICLES

Don Bower, VE1AMC, states that over the years he spent a lot of time in that old trailer. He recalled that it was brutally cold in the winter and that the inside of the trailer frosted over completely with ice, with icicles dripping from the ceiling. The metal chairs were particularly brutal. Summer nights were not much better, as the sides opened up for ventilation and access to the operators. The same accommodation was also provided for the bugs. They usually opted for stuffiness and heat as opposed to the bugs.

Aaron Solomon, VE1OC, states in his Maritime News for May, 1986, that HARC is the proud custodian of a 28-foot mobile communications van. Aaron also states in his news for October, 1986, that 37 Halifax metro amateur operators and the Liverpool Search and Rescue organization had spent 8 days searching for a 10-year-old boy. In Aaron's Maritime News for December, 1986, he states that 41 amateur radio operators received certificates of merit that participated in the Beaverbank EMO search for this missing boy.

The relationship between the local Amateur Radio Community and the EMO in the Halifax area has matured over the years and has been formalized by an official memorandum of understanding (MOU). The following Amateur Radio call signs have been assigned to these organizations for their use in Nova Scotia.

The following call signs are assigned to the Minister in Charge Emergency Measures

HARC Files

This is the Logistics Bus moving in on the 1994 Field Day Site.

HARC Files

This is the Control Bus moving in on the Field Day Site 1994.

HARC Files

This is a Search and Rescue 4-wheel drive pickup moving in on the 1994 Field Day Site.

Organization Nova Scotia: VE1EMO and VE1QQ. The VE1EMO call sign is sponsored by John Perkins, VE1FH, and the VE1QQ call sign is sponsored by Dave George, VE1AJP.

The following call signs are assigned to Emergency Preparedness Canada with Headquarters at Ottawa, Ontario: VA1EPC and VE1EPC.

The following call signs are assigned to Halifax Regional Municipality Emergency Measures Organization Amateur Radio, Dartmouth: VA1HSA, VA1HSB, VA1HSC, VA1HSD, VA1HSE, VA1HSF, VA1HSG, VA1HSH, VA1HSI, VA1HSJ, VE1HRA, VE1HRB, VE1HRC, VE1HRD, VE1HRE, VE1HRF, VE1HRG and VE1HRH as well as VE1HRM.

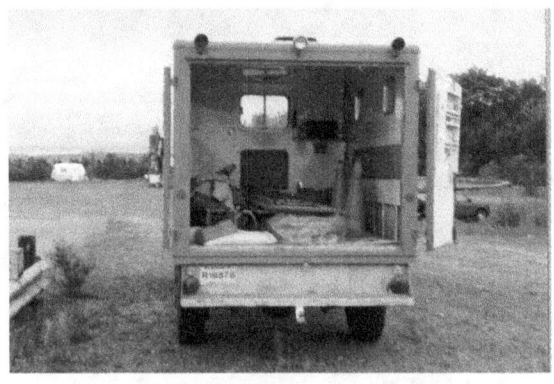

HARC Files

This is the interior of the 4-wheel drive pickup at the 1994 Field Day site.

There is also a highly productive collaboration between the Metro Halifax Amateur Radio community and the local Ground Search and Rescue group, with the latter making their command vehicles available for Field Days and EMO exercises. While the commitment to public service by HARC members has remained strong over the years, it is interesting to note how technology has advanced. The EMO trailer on the front lawn of Bill Ash, VE1BBS, was an early vehicle for remote communications. A Volkswagen van was another early vehicle and since then they have used four-wheel drive pickups up to and including former school buses.

These vehicles were usually painted international orange and can be seen many times in the photographs of HARC activities.

Howard Dickson VE1DHD

This is the EMO repeater trailer. This trailer was used at that time for a Multiple Sclerosis Bike Ride in July 2007. The trailer is equipped with a jack-up tower, Amateur Radio as well as EMO repeaters and an on-board generator.

Howard Dickson VE1DHD.

That is David Musgrave, VE1EDA, on the right and Tim Dunlap (the MS Bike Ride organizer) on the left.

David was the coordinator of the MS Bike Ride.

David Musgrave, VE1EDA, is custodian of the EMO trailer

HARC has provided communications for the MS Bike Ride event for years.

Dave George, VE1AJP, is custodian of the VE1QQ call sign.

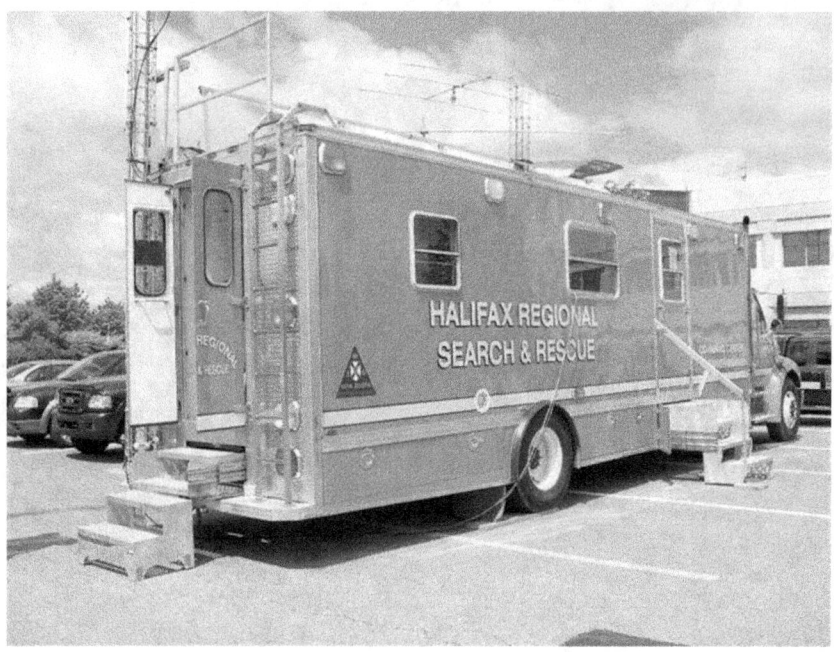

Lynn Bowser, VE1ENT

This is the Halifax Regional Search and Rescue Command bus.

It comes complete with the antennas to prove it.

They took delivery of this bus in 2005.

This vehicle cost over one quarter of a million dollars and is definitely an improvement over the Volkswagen van, the trailer on Bill's front lawn, the pickups and the school buses combined.

CHANGES TO 1946 CALL SIGNS

There have been more changes to the 1946 Canadian Amateur Radio Call Sign prefix. The Yukon Territory was assigned the VY1 prefix for their amateur radio call signs in 1977. Prince Edward Island was assigned the VY2 prefix and the VE9 experimental prefix was terminated. The VE9 prefix became the amateur radio call sign prefix of New Brunswick. This all took place in the early 1990's and unfortunately no official date was recorded for these changes. Some of those with the VE1 prefix retained that call sign and prefix after New Brunswick was assigned the VE9 prefix and Prince Edward Island the VY2 prefix. Some have saved their VE1 call sign and have a VE9 or VY2 call sign.

L. J. Shortly, VE8BR, made his first contact from the Yukon just before midnight on the day the call sign prefix was to change. He contacted K6UXO using VE8BR and as the QSO passed midnight Yukon local time he changed it to VY1BR but failed to record the exact date. One has to admit a contact to remember and this is recorded on page 58 of the September, 1978, issue of QST.

These Canadian provincial and territorial governments are simply cesspools of greed and discrimination and serve no useful purpose, but for some unknown reason we have to be blessed with them. Canada would be so much better with federal and municipal government only. On April 1st, 1999, Canada formed another of these known as the territory of Nunavut and their amateur radio stations were changed from VE8 to the VY0 prefix at that time.

About 1995 Canada began to issue amateur radio call signs with the VA prefix. These are the same as the VE prefix. There were so many VE3 amateur radio station licenses that this was done in order to obtain more amateur radio call signs, and this has spread to the other VE areas. There are now quite a number of call signs with the VA1 prefix in Nova Scotia.

A call sign no longer has an annual fee in Canada and is assigned for life. If it still had an annual fee there would not be so many call signs that are never used. Some have never been used since they were issued. I know of several that have the wrong address and some have QSL cards at the bureau for contacts that were never made. It has been a waste of time to try and correct this with those involved. Then again, I know just as many that have turned their station license, with call sign, back in and have no further interest in amateur radio.

One can now hold several call signs. Why one would want more than one is beyond me but some of those who have them apparently use them in different ways. One amateur radio operator living in the southeast section of Calgary in 2009 has three call signs. Two of them are two letter VE6 call signs and the third is a three letter VA7 call sign. This makes the whole call sign system a bit ridiculous. Why is a British Columbia call sign permitted in Alberta? Why is one operator permitted to hold two of the two letter call signs that so many amateur radio operators cherish? This Alberta operator has at least two vehicles and has a two letter call sign on each of them. This may be the reason he has these two call signs. I can think of no other reason.

The reason for all of this is rather clear from this research. Industry Canada wants the number of amateur radio licenses high so they can truthfully state so many licenses in Canada. Apparently, it does not matter that half or more of them are never used. One assumes that Radio Amateurs of Canada goes along with this scheme. The amateur radio community around the world has complained of the lack of interest in amateur radio and the high average age of amateur radio operators for years. This of course means a lack of valid licenses. This is the only thing I can think of that would make these numbers look very good in Canada, whether it is legal, ethical, or whatever.

Those not operating in the area of the prefix of their call sign may be doing it with a factitious address. I note in the 1949 Callbook that Dr. John Morton, VE1JM, was doing this. He was teaching at Carleton University in Ottawa, retained his VE1JM call sign and used that university as the address. He held two call signs in 1937; VE1JM and VE3ALK. He still held the VE3ALK call sign in 1963 but had relinquished the VE1JM call sign by then. Industry Canada told me that in order to hold my VE1BC call sign and live elsewhere I would require a Nova Scotia address. Every amateur call sign in Canada should be a Canadian call sign and a two letter call at that. One has a choice of 24 two letter prefixes and anything less than this is simply discrimination. A two-letter prefix from Canada's international blocks of call signs as required by law, then a digit and two letter suffixes of one's choice.

The VE0 prefix can now come with any suffix one desires and this has been the norm for several years. About the only thing one can more or less be certain about when working a VE0 call is that it is in a Canadian vessel of some description. Maybe! It may simply be a canoe on an inland puddle. When this call sign prefix was created it was to be used in vessels capable of transiting international waters.

This was recorded as part of DOC News page 62, QST, January, 1982:

> "DOC has revised its guidelines for VE0 call signs. In the past, VE0 call signs were issued as short-term endorsements on an amateur's station license. They were frequently recycled and rarely became identified with individual amateurs. DOC will now issue VE0 call signs in the same manner as other call signs, with a separate station license and an annual licensing fee. VE0 call signs will be reserved for amateur stations on ships that work primarily out of Canadian waters. Amateur stations on ships that work primarily in Canadian waters will be operated as mobile stations, in accordance with the provisions of the station license for the amateur's home station."

When one works a maritime mobile station today, the majority have no real relation to the vessel in which they are located. To be realistic one must have permission from the master or person in charge of the vessel and also have permission from the nation in which the vessel is registered. In other words, they should hold a call sign from the nation in which the vessel is registered. The majority of the maritime mobile amateur stations are simply passengers sailing in a "flag of convenience" vessel. I know that some of these "flag of convenience" nations do not permit amateur stations in their vessels.

Another way of looking at this subject is simply how much traffic has been passed via amateur radio stations fitted in ships that should have been passed via the ship's marine radio station. If one simply takes a look at all the greed possible and then simply imagines the possibilities, they can come close to what has actually taken place. I have no desire to record any of this known detail.

DAVE VE1ADH SHACKS

This is Dave's Shack in 1956 before he had a license.

This is Dave VE1ADH sometime in the 1990's

This is the VE1ADH Shack sometime in the 1990's without Dave

Photo by Mike Goldstein, VE3GFN

This is Dave VE1ADH and the Shack in 2002

ELMER VE1OD SHACK

Elmer Naugler's, VE1OD, Shack in the 1980's

VE1OD 20-foot dish

VE1OD antenna in the 1970's

SCOTT VE1QD AND HIS STATION 2013

This is HARC President, Scott Wood, VE1QD, with his station in 2013.

THE HIGHLANDS CAR RALLY

The Highlands Rally was a car rally held from various points to various points in the province of Nova Scotia. HARC provided communications for its first car rally during July, 1975. This was a rally run from Truro to New Glasgow and on to Antigonish. This took place on all dirt roads and some were nothing but trails through the woods. A lot of fun was had by all and there were amateur stations from all over Nova Scotia providing the communications.

The rally run from 12 noon on Saturday July 29th to 12 noon Sunday July 30th, 1978, was run from Halifax to Truro and back to Halifax. This was run on various back roads and involved 15 car teams. The Gore VHF 2-meter repeater provided most of the communications with this rally. According to Aaron Solomon, VE1OC, in his Maritime News, page 95 of the November, 1978, issue of QST there were 22 Nova Scotia and New Brunswick amateur radio operators providing the communications for this rally. The race officials were so pleased with the HARC communications they had them relay the official times.

The 1979 Highland Rally was on July 28th starting at 1500 and finished at 0630. This rally was all the back roads and dirt trails from Dartmouth to Truro, Earlton and back to Dartmouth. HARC provided the communications for this rally.

Lynn Bowser VE1ENT

Brit Fader, VE1FQ, and George Dunfee, VE1AGT, at Keddy's Motel, Truro, while they were net control for the Highlands Rally believed to be sometime in the 1980's.

THE LOBSTER TRAP CAR RALLY

The Lobster Trap Rally in Moncton was another car rally that had amateur radio for communications and often required help from the members of HARC. They required 50 amateur radio operators for the 1979 rally. A couple of friends of mine, Dan, VE1AUT, and Jeanette, VE1AZT, a husband-and-wife team often entered this one. They enjoyed it so much they "took off", him driving and she navigator and went from Moncton to Montreal and back just for the heck of it. They were using a small Volkswagen car with a rear air-cooled engine commonly referred to as a bug or beetle. They often practiced for this rally in a parking lot before heading off into the bush on the real thing. Aaron Solomon, VE1OC, recorded their transfer west in his maritime news for March, 1980. Dan has been VE6YB and Jan VE6BRX since the move.

1986 HIGHLAND CAR RALLY

ASM Aaron Solomon, VE1OC, stated in his Maritime News for October, 1986, that 30 Halifax, Dartmouth and Truro amateur radio operators provided the communications for the 1986 Highland Car Rally. He stated that Herb Bradley, VE1ADA, and Brit Fader, VE1FQ, used the VE1QQ call sign. This would indicate that they were the net control station. Herb swapped his VE1ADA call sign for VE1HX a few years after this event.

HX was the call sign of the first marine radio station that Mr. Marconi opened in 1905 in the approaches to Halifax harbour. That bit of detail is for anyone besides me who would be interested.

FLEA MARKETS

The Halifax Amateur Radio Club has held an annual flea market for many years. HARC has also handled flea markets at special times for various reasons, but mostly to make money in one way or another. It was stated at the monthly meeting of HARC on September 19th, 1979, that a flea market would be held at the club on October 20th. There were various tables for this flea market and there would be no charge by HARC. One of the main reasons for this flea market was to sell the estates of Doug Johnson, VE1OM, Ian MacLeod, VE1GC and West Street, VE1EK. All three had become recent silent keys.

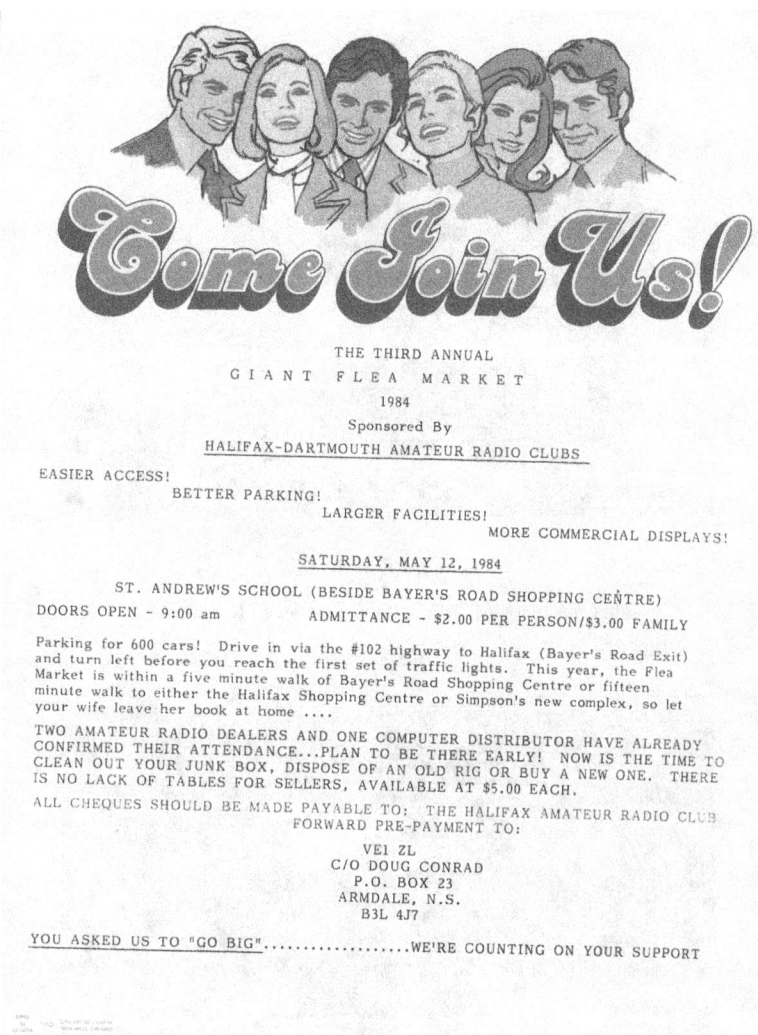

THE THIRD ANNUAL
GIANT FLEA MARKET
1984
Sponsored By
HALIFAX-DARTMOUTH AMATEUR RADIO CLUBS

EASIER ACCESS!

BETTER PARKING!

LARGER FACILITIES!

MORE COMMERCIAL DISPLAYS!

SATURDAY, MAY 12, 1984

ST. ANDREW'S SCHOOL (BESIDE BAYER'S ROAD SHOPPING CENTRE)

DOORS OPEN - 9:00 am   ADMITTANCE - $2.00 PER PERSON/$3.00 FAMILY

Parking for 600 cars! Drive in via the #102 highway to Halifax (Bayer's Road Exit) and turn left before you reach the first set of traffic lights. This year, the Flea Market is within a five minute walk of Bayer's Road Shopping Centre or fifteen minute walk to either the Halifax Shopping Centre or Simpson's new complex, so let your wife leave her book at home ....

TWO AMATEUR RADIO DEALERS AND ONE COMPUTER DISTRIBUTOR HAVE ALREADY CONFIRMED THEIR ATTENDANCE...PLAN TO BE THERE EARLY! NOW IS THE TIME TO CLEAN OUT YOUR JUNK BOX, DISPOSE OF AN OLD RIG OR BUY A NEW ONE. THERE IS NO LACK OF TABLES FOR SELLERS, AVAILABLE AT $5.00 EACH.

ALL CHEQUES SHOULD BE MADE PAYABLE TO: THE HALIFAX AMATEUR RADIO CLUB
FORWARD PRE-PAYMENT TO:

VE1 ZL
C/O DOUG CONRAD
P.O. BOX 23
ARMDALE, N.S.
B3L 4J7

YOU ASKED US TO "GO BIG"..................WE'RE COUNTING ON YOUR SUPPORT

HARC Files

Mel Lever, VE1VX, and Wayne King, VE1CBK, organized a successful flea market for HARC in 1985 according to Maritime News for September, 1985.

These flea markets became an annual affair and were a combined effort on the part of HARC and DARC (Dartmouth). The one in 1986 was advertised in the March Maritime News as the annual Halifax and Dartmouth flea market/dinner dance and was held on May 30th and 31st, 1986, a two-day affair. The September Maritime News states that approximately 350 attended this flea market and it was enjoyed by all – a great success. The HARC and DARC flea market was held on May 23rd, 1987.

John Brady VE1WZ

This was held at Dartmouth and believed to be 1969.

John Brady VE1WZ

This was held at Dartmouth and believed to be 1969.

This is John Brady, VE1WZ.

HARC REPEATERS

HARC operates four repeaters:

VE1HNS on 146.9 MHz -

VE1PSR on 147.270 MHz +

VE1PSR 444.350 MHz +

VE1PSR on 53.550 MHz –

Packet:

VE1NSD on 145.050 LAN NODES

HARC Files

This is the construction of the VE1PSR Repeater Station on Cowie Hill in Halifax about 1992.

VHF NET

There has been a Take 15 Net run by HARC and in 1980 it was meeting each Sunday evening on the CBC Repeater at 8:45 PM. The Take 15 Net moved from the CBC Repeater to this VE1PSR Repeater when it became operational. The purpose of this net is to provide news regarding the Club's operation, notices of general interest to all radio amateurs and a means of allowing members and non-members alike to meet on the air on a regular basis and to exchange news of mutual interest. The Net is sponsored by HARC.

CALL LETTER MV PLATES

The Halifax Amateur Radio Club and the Nova Scotia amateur radio operators were first issued call letter license plates for their motor vehicles in 1959. Prior to that Canadian Assemblies Limited, Amherst, Nova Scotia, the company that produced the Nova Scotia license plates, would make one for any amateur operator on receipt of one of their QSL cards. They did this free of charge and amateurs were informed of this in the Maritime News for May, 1956.

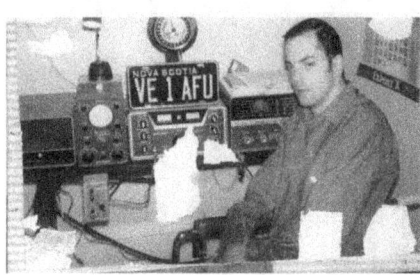

HARC Files

Dave E. LeCain, VE1AFU, with his pre 1959 license plate.

This is a very poor copy of this photo but the best I could find.

One of the first projects of the Nova Scotia Amateur Radio Association, when it was formed, was to obtain Amateur Radio Call Letter Plates for all amateurs in Nova Scotia. This was a great challenge and required a lot of hard work. A committee of six members was formed to compile statistics and material for a brief to be submitted to the Minister of Highways for the province of Nova Scotia. This work commenced as soon as NSARA was formed in 1957 and on December 3rd, 1958, the Minister of Highways informed them that it appeared feasible. A total of 160 amateurs made application for call sign plates for the year 1959. Hugh Corkum, VE1VN, was president of NSARA in 1959 when these plates were first issued and he received the first set. Each amateur had to pay a three-dollar fee for the call letter plates over and above the cost of the vehicle's regular plates.

The state of Florida was the first state issued with call letter plates. They received their plates in late 1949. Every issue of QST seems to list another two or three states that approved these plates throughout the 1950's. By 1958 all but nine states had been issued call letter plates. New Brunswick was issued call letter plates in 1957 but for some reason this had been terminated and was reinstated in 1969. Therefore, Nova Scotia was not that far behind the majority. Newfoundland received their call letter license plates in 1962; the same year New York State received theirs. This left Kentucky, Massachusetts and New Jersey as the only states that did not issue call letter plates by 1962.

Massachusetts managed to get their plates four years later in January, 1966, leaving Kentucky and New Jersey the last two states. When Massachusetts stated they would issue these plates they requested applications from those interested and felt they would receive between 700 and 800. They received 2,018 so their quesstimate was a bit off and will give one an idea of the interest shown in these plates. The CB crowd in Florida was trying to get call letter license plates in 1970. If approved amateur plates would go up from $1.00 to $5.00 and ARRL was after those involved to do what they could to veto the government bill making this possible. The Florida CB crowd received their call letter plates in 1972 for the $5.00 fee, but the amateur plate remained the same at $1.00.

Kentucky finally received their amateur radio call letter license plates in 1973. Ontario was the very last state, district, province or territory to receive their call letter license plates. Ontario received theirs on March 18, 1976. Mr. R. G. Gammon, VE3ACL, was one of the amateurs working hard to obtain the Ontario plates and he said the battle for them lasted 17 years.

Alaska certainly gave their amateur radio operators a deal on call letter license plates in the 1970's. They took 97% off the $35.00 fee if one had an 80 through 10-meter station in their vehicle available 24 hours a day. In other words, they received their call letter license plates for $1.00 providing the radio was permanently wired in the vehicle.

This is a collection of Canadian motor vehicle license plates that Ron Allen, W3OR, has collected and appears on page 21 of the January, 2001, issue of QST.

When first issued in Nova Scotia one had to bolt the call letter plate over the vehicles actual plate. This was a bit out of the ordinary and was not that bad if the actual plate was the same colour as the call letter plate. When the call letter plate was bolted over the yellow plate of a small commercial vehicle it could be quite noticeable. At least it was better than the state of Wisconsin. One had to have mobile equipment mounted in the vehicle with the call letter plates in that state. A few amateur radio operators with call letter plates and mobile radio equipment in their vehicle have had to show their radio license to a police officer that stopped them.

Nova Scotia was more or less part and parcel with all the others when it came to these motor vehicle plates. Before the silhouette of the schooner, the plate had a dot in the centre for some unknown reason, and with the silhouette the call letters are not evenly spaced across the plate. At least it is a call letter plate but one is only allowed the one vehicle assigned these plates.

In some areas one can have all the vehicles they own with the same call letters on each vehicle. I have seen photographs of as many as four vehicles, all belonging to the same amateur operator, with his call letter plates on each and every vehicle. One can see a photograph of four vehicles with the same call letter plates on page 13 of the May, 1988, issue of QST. When Nova Scotia went from a series of numbers, to three letters and three digits, I brought this to the attention of the Motor Vehicle Branch to see if they would agree with this feature. The reply was a routine letter sent in reply to any of a number of requests, and my letter probably reached the first waste basket only.

This is the front cover of The Canadian Amateur for April 1959:

HARC Files

This is the Honourable G. I. Smith minister of highways presenting the first set of call letter plates to Hugh H. Corkum, VE1VN Lunenburg. This same photograph appears on page 71 of the June, 1959, issue of QST. Hugh Corkum was president of the Nova Scotia Amateur Radio Association at the time. He was a radio operator in rum-running vessels and retired chief of police at Lunenburg. He was the author of the book "On Both Sides of the Law".

Also present was E. S. Campbell, centre, register of motor vehicles for Nova Scotia. Mr. Campbell was also an amateur and held the call VE1QQ. One wonders if it was, he who was 1DJ in 1926. Mr. Campbell replaced this VE1QQ call with VE1AHQ. The VE1QQ call has been assigned to what is now the Emergency Measures Organization shortly after this photo was taken.

Page 58, QST August, 1962

The May, 1959, Maritime News stated that about 150 Nova Scotia amateur radio operators were issued these call letter license plates.

248

THE CANADIAN AMATEUR 1959

Page 86, QST, August, 1972

There is a lot of HARC history in this old issue of The Canadian Amateur. This was a new publication at the time. This April, 1959, issue of this magazine was the fourth and their headquarters were in British Columbia. A good portion had been written by HARC member Cyril Boudreau, VE1AFB, who held call VE1RJ at the time. Cyril was the Nova Scotia correspondent for this magazine.

This is in the IARU News Section and the caption reads: "Major W. C. "Bill" Borrett, VE1DD, was one of the founders of the International Amateur Radio Union. Bill Represented Canada at those important first meetings in 1925. Still enjoying good health at the age of 80, he is a founder and vice-president of CHNS Radio. Pictured with VE1DD is VO1FX, president of the Society of Newfoundland Radio Amateurs." Bill became a silent key in 1983 and is recorded in silent keys for December, 1983.

Friday, June 29, 1979

VE1DD

LIFE FELLOWSHIP — Colonel William C. Borrett, noted Halifax historian and author of such popular Nova Scotian works as Tales Told Under the Old Town Clock, was awarded a life fellowship in the Royal Nova Scotia Historical Society by RNSH society president, Dr. A. E. Marble. Making the presentation at Camp Hill Hospital are, from left, Dr. A. E. Marble; Janet Dauphinee, RN, head nurse; and Colonel Borrett, who also started the Halifax Corps of Commissionaires. (Wamboldt-Waterfield)

HARC Files

PRESIDENT IARU

The president of ARRL is also the president of IARU but he can decline these duties if it is felt that the two positions were too demanding. Harry Daniels, W2TUK, was president of ARRL and declined the position of IARU. Noel B. Eaton, VE3CJ, vice president of ARRL was selected and approved by the other members of the ARRL board to act as president of IARU. Noel Eaton took over as president of IARU in January, 1974, and the June issue of QST lists some of the work he managed in his first four months in the position. A lot of it was travelling around the world visiting the various amateur radio organizations that were members and that made up the International Amateur Radio Union.

Amateur Radio World is the title of the IARU news in QST on page 94 of the April, 2009, issue of QST. It appears as though there are some changes in appointing the executive of the IARU and that the president of ARRL is no longer automatically the president of IARU. The opening paragraph of this article explains that as follows: "Every five years, the International Amateur Radio Union (IARU) engages in a consultative process to nominate a President and Vice President. The consultation is initiated by the ARRL, as the IARU International Secretariat and is conducted with members of the IARU Administrative Council (AC) representing the three regional organizations. The result of the process is the nomination of a single candidate for each of the two positions. The IARU Member-Societies then decide whether to ratify the nominations."

League Lines in QST for:

April, 1983, stated 117 countries
November, 1985, stated 124 countries
December, 1987, stated 129 countries
January, 1992, stated more than 125 countries

These are the member countries of the IARU and the members of the IARU vote on whether or not to allow a new member to join.

The RAC web site states there are160 countries in the IARU in 2008. This International Amateur Radio Union is as good as one can get. It would be so much better if each country that signed agreements within this union would abide by the rules and regulations. This is not the case and one can check with Radio Amateurs of Canada for the most recent up-dates on violations. If one hears or suspects a violation, they can also check with RAC for the proper reporting procedure. Only one amateur radio association or society in each nation can be a member and represent that nation's amateur radio operators. The Canadian Radio Relay League represented Canada as soon as it separated from the American Radio Relay League. As soon as the Canadian Amateur Radio Federation and the Canadian Radio Relay League became Radio Amateurs of Canada it represented the Canadian amateur radio operators in the IARU.

BRIT FADER VE1FQ AMAZING RECORD

Brit Fader, VE1FQ, left a record that every Amateur Radio Operator would like to leave. His record of achievement is nothing short of amazing. They claim if you want something done ask a busy person to do it. The record that Brit left makes one wonder where he found the time to tuck his shirt tail in.

There is an excellent photograph of Brit operating his station in the April 1959 issue of The Canadian Amateur, but unfortunately it will not reproduce well enough to use here. It appeared with the story titled "The Northern Messenger" by Aaron Solomon, VE1OC. This was the name given Brit in 1951 for all the messages he handled with those in isolation at our northern communities. He held the highest number of messages handled consistently in the VE1 call area. He was up around 400 messages per month at times. I doubt the local marine radio station in the approaches to Halifax Harbour with call sign VCS could equal that record. While doing the research for this history project I would simply list Brit's traffic total at the end of each monthly Maritime News to indicate the end of that item. When he went below one hundred for the month, I began to wonder what was wrong.

When *HMCS MAGNIFICENT* returned to Halifax in 1957 Brit was invited aboard for a dinner. He was presented a plaque that included the ship's crest at this dinner for all his work in handling traffic from the "Maggie" while she had been to Egypt and the British Isles. The "Maggie" was using amateur radio call sign VE0ND. She had been fitted with bunks here in Halifax in order to transport the army with their vehicles and equipment from Halifax to Egypt during the Suez crisis at the time. My father was a Chief Petty Officer at the Mechanical Training Establishment at *HMCS STADACONA* and was instrumental in fitting these bunks.

There are a couple of photographs and a description of the Canadian troops at the Gaza Strip in Egypt on page 86 of the October, 1957, issue of QST. But Brit of course had already been working them. VE3AHU/SU was one of the main operators at that time but there were two others; VE1ACK and VE6QK.

1958 was a big year for Brit. His traffic totals were way down on the monthly reports and I was wondering what was wrong. Brit managed to get DXCC that year and was also married at about the same time. His traffic totals started back up shortly after that.

The records indicate that Brit was one of the amateur operators to supply emergency radio communications equipment for the Springhill Mine disaster of 1958. Prince Philip was in the area at the time and landed at the Moncton airport and drove down to visit that disaster. I remember watching his motorcade leave Moncton.

In 1959 Brit had the best signal in the north from this area via a homemade 3 element beam antenna. He handled a lot of messages for the D.O.T. personnel, members of the RCMP, DND, and others posted to Canada's north. He started this service in 1946 when he returned from overseas with the Army. In 1948 a lot of this communication was between some Halifax area men working up north. Mickey McWilliams, VE8NB, at Resolution Island and A. W. Wilson, VE8OE, at Nottingham Island in the Hudson Strait were two of the northern stations in 1948.

Brit Fader, VE1FQ, worked for and retired from the Halifax Central Post Office in 1973. His retirement was announced in the Maritime News of QST in May, 1974. He is first mentioned in the Maritime Division news section of QST in 1935. He completed school that year and went to work with an equipment company. He joined Canada Post in 1940 and in 1942 was a mail car postal clerk on the Maritime Express running from Halifax to Campbellton, New Brunswick. He was filing a report with all the news he could find on any amateur radio operators for QST and asked anyone living along the Maritime Express railway line to drop down and visit with him.

Brit started his "The Month in Canada, Nova Scotia – VE1" news article in January 1943 as follows: "Well, the Army finally got hold of me. I tried to get in the Air Force, but without success. So here I am with the Royal Canadian Corps of Signals, down here in Sydney on a course. After this course is finished, I don't know just where I will be, but as long as I am around this part of the Maritimes I will try to gather up as much dope as possible on the gang." This was shortly after the mail car experience and he served in the army until 1946.

Brit, VE1FQ, went overseas to England in November, 1944, and although he sailed from an American port his ship stopped at Montreal enroute. There he went ashore and met Alex Reid, VE2BE, the Canadian General Manager with ARRL. By this time the world of amateur radio was making plans on the restructuring of the hobby on the termination of the war. Alex gave Brit a copy of the plans and proposals for amateur radio, as presented before the FCC in the United States. Brit also visited Fred George, VE2BV, who was a big gun in the RCA plant in Montreal and was given a tour of this plant. He met more hams at the plant. One was W1III who was visiting from the plant at Camden, New Jersey. This is the RCA plant Frank Follweiler, K2SQM and VE1SQM would retire from many years later. Frank was a machinist in a U.S. Navy Landing Ship Tank (LST) off the coast of Italy at the time Brit did the tour of the Montreal plant.

Brit, VE1FQ, was a very active member of the amateur radio community and HARC had a lot of faith in him. At the HARC meeting on April 12th, 1946, it was suggested that a letter be sent to ARRL and XTAL magazine advising them that Brit was still the QSL manager. Don Bain, VE1LZ, agreed to fill in as QSL manager until Brit returned from overseas with the Army. Within the lists of QSL managers in QST they simply listed VE1 as VE1FQ will resume service soon. One wonders if ARRL realized he was still serving in England. Brit returned to Canada Post on arrival back from overseas.

Brit was controller for the Maritime Net that meets daily at 7-PM on 3750 kilohertz for 12 years and would often fill in when needed after relinquishing this position. He did this with 8 watts of phone power up until he started running a pair of 813's late in 1956. He often ran phone patches from traffic received on this net and from his service into the north. This managed to grow until he was handling phone patch traffic from around the world. There were no phone plans back then and long-distance telephone service was expensive.

In 1961 Brit was recruited by a team of scientists from the Geophysics Division of E.M.R. Ottawa engaged in a seaborne magnetic data acquisition project. His job was operating as well as maintaining the shore-based monitoring station which continuously recorded the magnitude of the Earth's magnetic field. This station was located in his basement. 1956 was the International Geophysical Year and there was a detailed description of this in an article printed in the July, 1956, issue of QST. This was five years before Brit had this experience but was no doubt connected in some way.

Brit attended the Civil Defense College at Arnprior, Ontario, during the winter of 1960 – 1961.

At the Moncton Amateur Radio Convention in 1967 Brit received two awards. He received the VE1GR Trophy. HARC created the VE1GR 75-meter DX trophy in memory of Hal Ward, VE1GR. A second presentation was made to Brit to recognize his 30 years as the VE1 QSL manager. He received an NCX3 and Power Supply to recognize this 30 years' service.

John Perkins, VE1FH, and Clarey Pelley, VE1AVP, organized a successful roast for Brit in 1985.

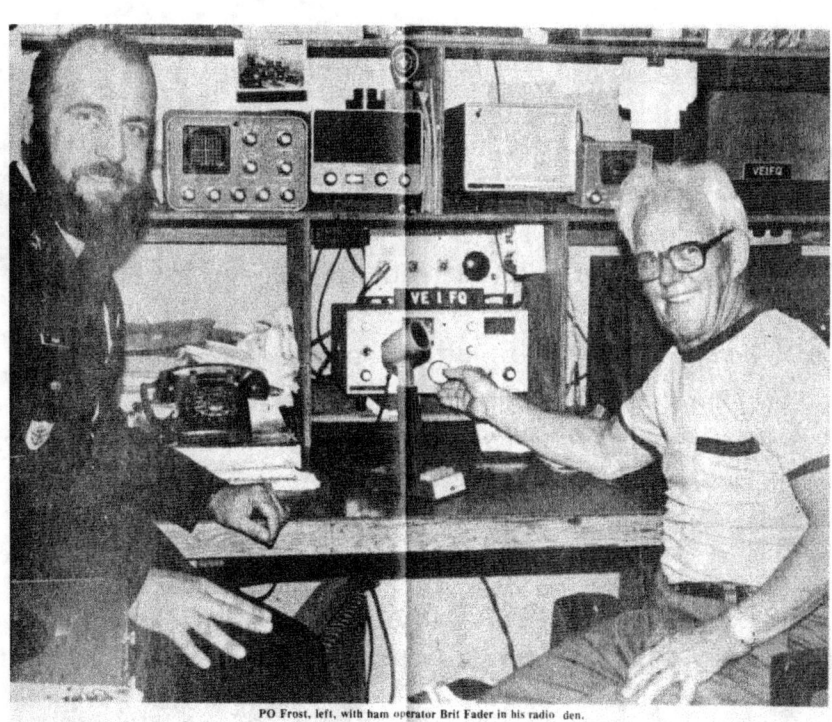

PO Frost, left, with ham operator Brit Fader in his radio den.

Ham Operator Brit Gets A Thank You

Crew members of HMCS Skeena organized a social evening Friday to thank Sackville ham radio veteran Brit Fader for his voluntary communications work between the ship and the men's families.

Skeena was recently away for two months on Northern Wedding, a North Atlantic exercise. France, Spain and the Mediterranean were also visited.

"Mr. Fader was just fantastic patching in radio messages from the ship so that we could talk to our families by telephone," says

Petty Officer Dennis Frost.

PO Frost, radio operator for Skeena and a ten-year resident of Beaver Bank Villa, says Mr. Fader sometimes spent as much as four or five hours a night relaying messages to and from the ship.

TRUSTEE

A ham operator since 1935, Mr. Fader lives on Main Highway, just above the Beaverbank lights, and is well-known for his work as a school trustee.

Before satellites made their appearance a great deal of his communications were

with the armed forces, government and Hudsons Bay personnel in the Canadian North.

His calls on behalf of the armed forces have made life happier for the families of men aboard navy vessels, on fishery missions and on UN duty in such places as Egypt and the Golan.

PO Frost says Mr. Fader is the sort of person who goes out of his way to be helpful.

"If flowers were to be sent for an anniversary or a new baby or something, he would call the florist and arrange for them to go."

Mr. Fader, a retired post office employee and father of two sons and a daughter, is modest about his help to the servicemen.

However, he and fellow Sackville ham operator Bertus Backer are very well known for their voluntary communications efforts.

Mr. Fader says the main difference between ham radio and CB radio equipment is the fact that the ham equipment is far more powerful.

To make his equipment more effective Mr. Fader has a rotatable antenna.

The Skeena ceremony took place aboard ship. The ship's captain, Commander B.E. Derible, presented Mr. Fader with a plaque.

HARC Files

The Canadian Division Maritime News by Aaron Solomon, VE1OC, stated that Brit, VE1FQ, had received a plaque from *HMCS SKEENA* for phone patching in his news for April, 1979.

Brit announced at the HARC monthly general meeting held on June 20th, 1979, that he had just been awarded the ARRL 40-year membership plaque. He received over forty certificates of appreciation from Canadian military units and ships away from home for his phone patch activities to dependents in the Halifax area at all hours of the day. Brit was given at least one letter from the Admiral, and several mess dinners in appreciation for the many phone patches he had handled for the navy.

In 1979 DND expressed its appreciation for his efforts in a unique way when he was invited to join a Canadian Forces destroyer for a cruise. Brit described his recent trip to the Great Lakes in a naval destroyer at the September 19th, 1979, HARC monthly meeting. He also recorded the trip in the September issue of his monthly Bulletin. This was a 21-day holiday excursion aboard *HMCS SAGUENAY* with international call sign CZFX. They departed Halifax July 30th, 1979, proceeded up the St. Lawrence River to Montreal; from there to Hamilton; to Welland; to Collingwood, back to Toronto and then home via Air Canada. Brit stated it was a very enjoyable trip thanks to the navy. He had the pleasure of meeting several old friends

HARC Files

Left to right: Brit Fader, VE1FQ,
Angus MacDonald, VE1YO
and Helen Logan, VP9JD.

in the various ports of call. This was a very rare event for a member of the general public. Aaron Solomon, VE1OC, mentioned this trip in his Maritime News for October, 1979.

In 1984 Brit, VE1FQ was citizen of The Year in his home town of Sackville, Nova Scotia. This was in recognition of Brit's efforts as a parent volunteer and school trustee for the local school system.

Brit's term as manager of the Maritimes QSL bureau ended after 48 years of dedicated service on January 1st, 1987. Brit had acted as editor of the HARC news bulletin for a period of 34 years. He let these positions go because of failing eyesight. In addition to all this and everything else, Brit was the communications manager for the Waverly Ground Search and Rescue that became part of Halifax Regional Search and Rescue.

The Halifax Amateur Radio Club has retained Brit's VE1FQ call sign and has the QSL Bureau named in his honour as The Brit Fader Memorial QSL Bureau. Brit's biography would simply be a history of HARC during his life time. How he accomplished so much in only one lifetime is truly amazing.

VE1 QSL Bureau manager since 1938, the most senior in time of all the ARRL QSL Bureau managers, L. J. Brit Fader, VE1FQ. Brit is active in the Halifax, NS, Amateur Radio Club and is currently editor of their paper, the *Bulletin*. He is also past president, vice president and activities manager. His past positions have also included tours of duty as EC, SEC, OPS and NCS.

There is also the Brit Fader Scholarship Fund held by HARC. The first to receive a scholarship from this fund was awarded at the HARC 75th anniversary celebrations in 2008. Joe Koechl, VA3KOE, was the first to be awarded a scholarship from this fund.

Page 76, QST December, 1976

VE1BXC HANDLED 501 MESSAGES

The most traffic I found recorded for a month was VE1BXC recorded in Maritime News for April, 1979, as having handled 501 messages.

1943 CANADIAN NICKEL

The 1943 Canadian nickel was the one that had "We win when we work willingly" engraved in Morse code around the outside edge.

ANGUS MACDONALD VE1YO SILENT KEY

DARC Web Site

I will never forget the time Angus, VE1YO, became a silent key on September 17th, 1980. The Japanese tuna fleet had visited Halifax as it often does each year. There were several Japanese amateur radio operators in this fleet and some had been over to visit Angus. Angus became a silent key between the time of their visit and when they sailed. They called in on the Maritime Net at 1900 local time and wanted to speak to Angus. I forget who the net control was at the time but whoever it was had a heck of a time trying to explain this to this Japanese operator. I'm not sure the Japanese operator really understood the net control.

The Dartmouth Amateur Radio Club (DARC) holds the VE1YO call sign in memory of Angus.

NO HISTORY OF THE CANADIAN AMATEUR RADIO ORGANIZATIONS

There has been no actual record kept on much of the history of the Canadian Amateur Radio organizations. At least if there has, I have been unable to find it. One record I found states that the Canadian Amateur Radio Federation (CARF) was formed on September 2nd, 1967. CARF was formed in Winnipeg with temporary officers on that date. Jim Roik, VE4UX, was acting president. Jim Couprie, VE4CS, was acting secretary-treasurer. CARF was formed "to promote the welfare of the Canadian amateur in the national field". VE1XG and Wimpy Mills, VE1NZ, attended the Radio Society of Ontario and the CARF meeting at Brantford, Ontario. This is recorded in the Maritime News by Bill Gillis, VE1NR, in the January, 1969, issue. I am unable to identify the one who held the VE1XG call at that time. CARF tried to merge with the Canadian Division of ARRL in 1970 but ARRL did not agree and turned them down.

This is an entry in the minutes of the monthly HARC meeting on June 18th, 1975, "Reported that the ARRL Canadian Division from now on will be known as the Canadian ARRL". That is the creation of CRRL and as close as I have come to the exact date. CRRL remained the Canadian Division in print in QST but I feel confident those in ARRL felt it would eventually break from their ranks.

CRRL had a section in QST called Canadian Newsfronts and in it in the August, 1978, issue it stated that CARF and CRRL were often submitting proposals to D.O.C. that were not the same. Therefore, they were attempting to form The Canadian Amateur Radio Council, a combined membership from both CARF and CRRL to form a joint voice to D.O.C. Bud Punchard, VE3UD, was at the monthly meeting of HARC held on June 21st, 1978, and he explained this difference between CARF and CRRL. He seemed to lean more in favour of CARF claiming they were closer and could see more of D.O.C.

Ron Hesler, VE1SH, stated in Canadian Newsfronts on page 53 of the September, 1978, issue of QST that CRRL was the better organization to represent Canadian Amateurs. In other words, a lot of time and space was wasted between the two organizations CRRL and CARF in arguing that they were the better organization to represent we the Canadian Amateur. This was the main topic with everything one read from those two organizations and it became very boring.

In the February, 1979, issue of QST CRRL announced that it had produced its first training manual with the title "Traffic Training Manual". In the next issue of QST the March, 1979, issue CRRL stated it had created public relations assistants across Canada with Camille Maillette, VE1RO, the public relations assistant for this the VE1 area.

The Canadian Newsfronts index was not included in the main index of the September and October, 1979, issues of QST for some unknown reason. It was included in the November, 1979, issue. I found the Canadian Newsfronts section without the index in October, 1979. The big news in that section is the fact that CRRL Incorporation had been approved with the following executive:

President, Ron Hesler, VE1SH
Vice President, Bill Loucks, VE3AR
Secretary, Gordon Steane, VE3BMG
Eastern Director, Albert Daemen, VE2IJ
Central Director, Tom Atkins, VE3CDM
Western Director, George Spencer, VE4IM became VE6XN
General Counsel, Bob Benson, VE2VW

This was to be the executive until 1200 hours on January 1st, 1983, but it soon changed according to the March, 1980, issue of QST.

President, Ron Hesler, VE1SH
Executive Vice President, Mitch Powell, VE3OT
Honorary Vice President, Noel Eaton, VE3CJ
Secretary Fred Towner, VE6XX
Central Director, Tom Atkins, VE3CDM
Western Director, George Spencer, VE6XN
Eastern Director, Albert Daemen, VE2IJ
General Counsel, Bob Benson, VE2VW

This new executive created a large staff with various titles but there was no VE1 within the lot.

CRRL received a new Director and Vice Director in January, 1980. Mitch Powell, VE3OT, became the New Director beating Ron Hesler, VE1SH. Mitch Powell was first licensed as a 13-year-old kid living in Yarmouth, Nova Scotia, with call sign VE1WN back in 1948. He started off as a radio operator on the northern stations and then went back to university. He was teaching at a university in London, Ontario, when he became Director of CRRL. Fred Towner, VE6XX, from Calgary became Vice Director. He beat out HARC member Aaron Solomon, VE1OC. Fred received 1,423 votes to Aaron's 850 votes.

The Canadian Newsfronts section had been removed in the front index of QST but there was a lot of Canadian detail in the October, 1979, issue. This issue contained the record of the 1979 field day activity. The front cover had a photograph of the tent and tower that made up the field day site of VE2XS. Inside of this October issue one of the strays contained a photograph and description of the complete homebrew station of VE3CTP. It was a nice-looking station.

Another attempt of a merger with CRRL by CARF took place in 1980 but was rejected. In Maritime News by Don Welling, VE1WF, for May, 1981, he stated that he, Andy McClellan, VE1ASJ, and HARC member Aaron Solomon, VE1OC, had been made assistant directors of CRRL. According to CRRL in Canadian Newsfronts, page 62, QST, January, 1982, the Canadian Division of ARRL became the Canadian Radio Relay League Incorporated in 1979. Mitch Powell, VE3OT, resigned from CRRL on September, 20th, 1982, and Tom Atkins, VE3CDM, took his place. Harry MacLean, VE3GRO, moved up into Tom's position. There were nearly 5,000 members of CRRL in 1983. In May, 1985, CRRL stated there were 5,120 members in CRRL at the end of 1984. CRRL and CARF still had their ongoing discussions on a merger and forming the one Canadian organization in 1987 and it was as boring as ever. The last paragraph in Canadian Newsfronts for November, 1987, stated that if CRRL and CARF were to merge into the one organization and take over the administration of the amateur radio service it would be more cost effective than if DOC continued to do it.

CRRL went off on its own on January 1st, 1988, after 67 years that we Canadians had been a section of ARRL. They actually removed the Canadian flag from in front of ARRL headquarters at Newington, Connecticut, on that date. Carl Anderson, VE1BQO, became the Maritimes – Newfoundland SCM in the May, 1988, issue of QST. The Maritimes – Newfoundland section or division became the Atlantic division in the CRRL organization in the May, 1988, issue by Carl, VE1BQO.

The CRRL list of President and Vice President was removed in the February, 1988, issue of QST. The last Maritime News in the Canadian Division Section was the March, 1988, issue. The last Canadian Division News in Section Activities was the May, 1988, issue of QST. There were no VE call signs in the June, 1988, issue of Silent Keys but a VE call does show up in that section now and then.

Industry Canada would not deal with two organizations representing Canadian Amateurs so CRRL and CARF were combined and formed Radio Amateurs of Canada (RAC). CRRL and CARF had their final individual meetings on May 1st, 1993. They merged the next day May 2nd, 1993, to become RAC. Their respective publications, *QST Canada* and the *Canadian Amateur*, published their final editions in May and June, followed by their new joint magazine, *The Canadian Amateur*, in July.

The executive of RAC when it was formed on May 2nd, 1993:

President, J. Farrell Hopwood, VE7RD
First Vice President, Dana Shtun, VE3DSS
VP International Affairs, George Spencer, VE3AGS
VP DOC Affairs, Earle Smith, VE6NM

VP Administration, Clayton Bannister, VE3LYN
Treasurer, William Loucks, VE3AR
Secretary, Eric Hott, VE3XE
Pacific Director, David Fancy, VE7EWI
Alberta/NWT/Yukon Director, Ken Oelke, VE6AFO
Mid-West Director, Bob Sheyn, VE5FY
Ontario North Director, Bob Bishop, VE3JAB
Ontario South Director, George Gursline, VE3YV
Quebec Director, Jean-Guy Riveria, VE2JGR
Atlantic Director, Carl Anderson, VE1UU
Honourary Legal Counsel, Timothy S. Ellam, VE6SH
Honourary Legal Counsel, Timothy D. Ray, VE3XV

Pat Doherty, VE3PD, replaced "Hoppy" Hopwood, VE7RD, as president in 1998.

BRIT FADER VE1FQ CRRL AMATEUR OF THE YEAR

In 1976 the Canadian Radio Relay League (CRRL) created an award titled Radio Amateur of the Year. To qualify an individual should have made an outstanding contribution to Amateur Radio in the last year, or have contributed consistently to the welfare of Amateur Radio over several years. The first recipient of this award was Brit Fader, VE1FQ. It took a bit of work to get Brit to Toronto and surprise him with this award. Bob Eschauzier, VE1ST, and Dave Oldridge, VE1BFV, were involved in this, with the help of Mrs. Fader. They talked Brit into going to Moncton to visit

CRRL AMATEUR OF THE YEAR
At the Toronto National Convention last June, Brit Fader, VE1FQ, accepted the CRRL Amateur of the Year award, the *first* such presentation. In his long-time career in amateur radio, VE1FQ has amassed an impressive list of contributions, including duty as EC, SEC, OPS and NCS. Most senior in time of all the ARRL QSL bureau managers, he is active in the Halifax (NS) ARC.
Send your nominations for the 1977 award to your assistant director or SCM by April 30, 1978. Recipients of the Division Certificate of Merit become eligible for ballot listing automatically. The division vice director and all assistant directors select the final candidate.

Page 66, January, 1978, QST

Bob. Dave was already in Moncton and he and Bob met Brit at the bus. When they had him in Bob's Volvo they headed straight to Toronto. Brit figured it out before they got to Fredericton that something was amiss. He grumbled for about half the trip, but once at the CRRL annual convention, he was fine, because there were so many people there that he knew from working them on the air. Dave said you could have knocked him over with a feather when they announced he was "Ham of the Year". It was a complete surprise.

Brit was inducted into The Canadian Amateur Radio Hall of Fame in 1990. Unfortunately, Brit passed away the day before this plaque marking this honour arrived at the door of Geoffrey Smith, VE3KCE, who was chairman of the board of trustees. This plaque was presented to his family.

BILL ELLIOTT VE1MR 2ND AMATEUR OF YEAR

The second VE1 to be awarded this prestigious award was another member of HARC, a president, Bill Elliott, VE1MR. Bill received this award for the year 2003. Howard Dickson, VE1DHD, made the submission, Dave Nimmo, VE1NN, was the Director and Bill Gillis, VE1WG, was RAC President at the time. Lynn managed to "spruce up" Bill to the point he actually wore a tie to the HARC Christmas dinner. They made this presentation at that dinner. One of the few times anyone has seen Bill wearing a tie. Bill knew he had been nominated but the presentation was a complete surprise. Bill is an excellent operator, technician and a very good president. He certainly deserves any recognition amateur radio has to offer.

Courtesy Bill Elliott VE1MR

This is Dave Nimmo, VE1NN, on the left, Bill Elliott, VE1MR and Bill Gillis, VE1WG, on the right presenting the RAC – Radio Amateur of the year award for the year 2003.

SCOTT WOOD VE1QD 3RD AMATEUR OF THE YEAR

A third HARC member received the Canadian Amateur Radio of the year award. HARC president Scott Wood, VE1QD, was presented this award for 2012 at the Dayton, Ohio, Hamvention on May 18th, 2013, by RAC president Geoff Bawden, VE4BAW.

Standing with Scott from left to right are: John Sluymer, VE3EJ, George Nicholson, VE1GRN and N4GRN, Gary Bartlett, VE1RGB, Scott Wood, VE1QD, Geoff Bawden, VE4BAW RAC President, and John Scott, VE1JS. John Sluymer and George wrote letters of support and Gary and John Scott together with Howard Dickson (not pictured) prepared the submission.

This is what Geoff, VE4BAW, stated in his report for June, 2013. "It was a pleasure to be able to take advantage of the fact that Scott Wood, VE1QD and I were both in Dayton and I was able to present Scott with the 2012 Canadian Radio Amateur of the Year Award for his contribution to Amateur Radio in Canada.

Scott has demonstrated outstanding leadership locally, regionally and internationally in his 60 years as an Amateur. Scott's championing and organizing of the Maritime DX Forum created a Forum for eight years that was world class."

HARC must be the only club to have three members that have received this prestigious award.

FRANK FOLLWEILER K2SQM AND VE1SQM

HARC Files

Frank Follweiler
K2SQM VE1SQM

Frank worked as a machinist for the RCA Corporation at Camden, New Jersey. He purchased a cottage at Peggy's Cove, Nova Scotia in 1950 and spent as much time there as possible. This was back when his annual taxes were fifty-three cents for the cottage per annum. He joined the Halifax Amateur Radio Club shortly after he purchased the cottage. Frank's wife Marie was a school teacher and amateur radio operator, WA2HSN. Peggy's Cove is one of the big tourist attractions in Nova Scotia. The antenna on Frank's cottage produced some interesting entries in their Guest Book. Marie passed away several years before Frank. She had been quite ill the last couple of years she and Frank spent at the cottage. After Marie died Frank and his old alley cat, as he called his old orange tom cat, spent their summers at the cottage. Frank was a machinist in the U.S. Navy during World War II. He sailed in LST's around Italy during the war. The coffee in the U.S. Navy must have been pretty bad because every time Frank was offered a cup he declined, stating he had a cup in Italy during the war. Frank and Marie did not have a family and when Frank died on

HARC Files

261

March 16th, 1994, he left the cottage and all of the furnishings, including his radio gear, as well as some money to the Halifax Amateur Radio Club. The cottage was sold and all this money has been invested by the club making HARC one of the better off financially clubs in the country. To sum Frank up in one sentence it would simply be that he was one fine guy and a great amateur. Frank did not own a broadcast receiver, TV, or have a telephone and ordered his cars new so he could get them without a broadcast receiver. He died at his home in Camden doing what he liked best. He had just signed off from a QSO with his best amateur friend.

This is Marie Follweiler, WA2HSN, on the left with the mike and Evelyn Bligh, VE1BC, on the right. This was taken about four months after Bill Bligh became a silent key. Evelyn exchanged her VE1OW call sign for Bill's VE1BC. She kept the VE1BC call sign for nearly ten years when I picked it up in January, 1975. Evelyn was first licensed in 1937 so had been operating for 28 years when this photo was taken in 1965.

APRS

I kept running into something called APRS and thanks to Stan Horzepa, WA1LOU, and his Digital Dimension Department in the March, 1998, issue of QST I learned a simple description. It had to be simple because simple is all I can understand. Stan states: "APRS (Automatic Packet Reporting System) is DOS/Macintosh/Windows Software that uses unconnected packets to indicate the position of objects on computer displayed maps. It's useful for tracking station locations in all sorts of networks and public service activities." I have seen HARC members tracking a vehicle from the club station with that equipment and found it rather neat.

HAM RADIO ACRONYM'S

One should have a dictionary of Ham Radio Acronym's and I am sure it would be a fairly large document if one included everything they found. Visual Call Sign Database – VCD, Commonwealth of Independent States – CIS, Low Power Crummy Antenna – LPCA, Capitol Region Malicious Interference Tracking – CRMIT (pronounced Kermit), Space Amateur Radio Experiment – SAREX, and the list goes on and on.

SWISSAIR 111

On September 2ⁿᵈ, 1998, Swissair Flight 111 departed New York at 2118 or 18 minutes past 9-PM Atlantic daylight savings time. This was a scheduled flight to Geneva, Switzerland, with 215 passengers and 14 crew members on board. This aircraft was a McDonnell Douglas MD-11 and was registered in Switzerland with registration HB-IWF. About 53 minutes after departure, while cruising at flight level 330, the flight crew smelled an abnormal odour in the cockpit. Their attention was then drawn to an unspecified area behind and above them and they began to investigate the source. Whatever they saw initially was shortly thereafter no longer perceived to be visible. They agreed that the origin of the anomaly was the air conditioning system. When they assessed that what they had seen or were now seeing was definitely smoke, they decided to divert.

They initially began a turn toward Boston; however, when air traffic services mentioned Halifax, Nova Scotia, as an alternative airport, they changed the destination to the Halifax International Airport. While the flight crew was preparing for the landing in Halifax, they were unaware that a fire was spreading above the ceiling in the front area of the aircraft. About 13 minutes after the abnormal odour was detected, the aircraft's flight data recorder began to record a rapid succession of aircraft systems-related failures. The flight crew declared an emergency and indicated a need to land immediately. About one minute later, radio communications and secondary radar contact with the aircraft were lost, and the flight recorders stopped functioning. About five and one-half minutes later, the aircraft crashed into the ocean about five nautical miles southwest of Peggy's Cove, Nova Scotia. The aircraft was destroyed and there were no survivors.

One can bring up several photographs of this aircraft on the internet simply by inserting the registration HB-IWF in Google. Swissair aircraft were all white and the tails were like the flag of Switzerland. They were all red with the Greek white cross identical to the flag. Swissair went bankrupt on October 2ⁿᵈ, 2001, and was reformed as Swiss International Air Lines. This new company does not use that colour scheme.

Transport Safety Board Photograph

These are some of the vessels at the crash site. CCGS HUDSON with call sign CGDG in the foreground, HMCS ANTICOSTI with call sign CGAA in the middle and HMCS HALIFAX with call sign CGAP. This photo was taken on September 14ᵗʰ, 1998. CCGS HUDSON is the only ship that has circumnavigated both the North and South American continents. She did this as CSS HUDSON when painted all white. She made this voyage under the command of Captain David Butler from November 19ᵗʰ, 1969, to October 16ᵗʰ, 1970. I sailed with Captain Butler shortly after this voyage and the late Philip Rafuse from Annapolis, Nova Scotia, made this memorable voyage as radio officer. Phil held call VE1ASM and operated as VE0MX on this voyage.

101 amateur radio operators from HARC provided communications for this disaster at various sites throughout Halifax, Shearwater and the Peggy's Cove areas. Al Penny, VO1NO, received the Cover Plaque Award for his article on this disaster in the February, 1999, issue of QST.

> • **February** *QST* **Cover Plaque Award:** The winner of the February *QST* Cover Plaque Award is Al Penny, VO1NO/VE1, for his article "The Crash of SwissAir Flight 111." Congratulations, Al!
> • More b...

Page 73, May, 1999 QST

Congratulations Al. It was near impossible to scan this without breaking the back of QST. Al was a Lieutenant Commander in the Canadian Navy and was president of HARC in 1999. He states in this article that the amateur radio community was alerted to this disaster at 11:45 PM local time. He goes on to describe the involvement in detail and that amateur radio equipment had to be modified in order to work on military, coast guard and police frequencies. This is something that would not have been permitted a few years before this incident. HARC provided communications until the morning of September 13th. Al goes on to credit the excellent work from the training received via EMO. This training is still provided by EMO and many HARC members have taken this training. Al also makes a list of the lessons learned at the end of his article. One will probably always learn something from these disasters and one should always be prepared for the next one.

Those involved with the actual search and recovery on the water will carry the memories of this for the rest of their lives. One could not describe what they saw and went through at the site of the actual impact of the aircraft with the water. They could not believe the human mutilation and material destruction this accident caused.

HARC member Neil Hughes, VE1YZ, was over the crash site in his yacht, *JENNIFER AND JEAN* with call sign VE0MBM but did not make contact with HARC. Neil is a retired Air Canada Airline Captain and involved with Winlink. One can find him listed with those involved with this communications system on the Winlink web site.

This Swissair disaster is one of many incidents one can use to convince one and all to keep their emergency equipment and training up to date. Another good example was the train wreck just west of Toronto at Mississauga, Ontario. This occurred on November 10th, 1979, and involved a chemical train loaded with tank cars containing a very deadly brew of various kinds and types. 200,000 people had to be evacuated to a safe area. Amateur radio played a big part in this incident and a full report can be found on page 50 of the March, 1980, issue of QST.

The ice storm that destroyed the power lines in Quebec over a four-day period in January, 1998, is another good example. A description of the service provided by Radio Amateurs du Quebec (RAQI) can be found starting on page 52 of the June 1999 issue of QST. The Y2K scare over December 31st, 1999, and January 1st, 2000, probably involved more amateur radio operators than any other incident. Everyone predicted a major crash of all computers passing that date and all kinds of precaution took place. Over ninety million dollars in cash was brought into Halifax to make wage payments to a large group of employees. A whole fleet of armoured trucks were used to transport this money from the Halifax International Airport to downtown Halifax. Two railroad tank cars of diesel fuel were parked next to some banks in Montreal to supply emergency power if needed. Amateur radio operators everywhere were to be used to provide communications in the event of various communication facilities collapsing, especially the 911 emergency service.

RESEARCH REFRESHED MEMORIES

The research for this project certainly refreshed a lot of memories. It was a lot like going home after being away for several years. It was an emotional experience, especially with the entire American made equipment from so many well-known manufacturers who are no longer in business. My first receiver was a National NC125 that I purchased at Eaton's department store Moncton, New Brunswick, in 1958. The first transmitter I operated was a Heathkit DX40. We all spent a lot of time dreaming of the ultimate station from the advertisements for all that equipment back then.

Anyone who reads this should realize that I have spent years of research in order to record some of this detail. I have learned a lot trying to record this history of HARC. While looking at every page of QST that was ever printed I kept thinking that it is a shame that QST is the ARRL only. It would be nice if it were a product of the IARU but that is not possible for many reasons.

We are a family of the world and not any one nation because our signals know no political boundary. Because of this we make many friends outside our own little national pasture. There have been very few Canadian amateurs recorded in the Silent Keys column of QST since CRRL removed us from the ARRL on January 1st, 1988. Mind you I know several American amateurs that did not make the Silent Keys column. It would be nice to have the entire world recorded in the Silent Keys column and there should be a master record of this so that one could look up any previous amateur operator. People like Keith Russell, Alex Reid, Noel Eaton, Brit Fader, Doug Smith, should be remembered and there must be an endless list of such people. Every nation must have an endless list of them and we should be able to look them up especially when at my age one can remember working some of them years ago.

Carl Zelich, AA4MI, mentions this on page 25 of the May, 2006, issue of QST. Carl is suggesting that a tax-deductible donation to the ARRL Foundation of $20.00 or more would ensure that your personal history be remembered. Carl, has it labelled "Silent Key Remembrance" and claims it would be a small project, computer programming-wise, to implement? He is talking of having your own record recorded and sums it up "Wouldn't it be great to know that you will not be forgotten?" I feel it should be a worldwide record and that a club should be able to have former members recorded.

One accumulates many memories after many years of operating radio. I have a lot of them and a familiar call sign will bring back many. Fortunately, I do not have many of those who died shortly after our communication. But one was bound to collect a few. It was amazing how calm they were. They not likely realized they would be dead in a few minutes. I monitored the silent periods and the distress frequencies of 500, 2182 and 8364 kilohertz faithfully for many years but did not hear a lifeboat radio. I do not know anyone who has heard a lifeboat radio in use as designed, although we did test these radios now and then when at sea.

I also monitored the aircraft distress frequencies of 121.5 and 243.0 megahertz for the many years I was with aeradio. I broke out in a cold sore after a few close ones with aeradio for some unknown reason. I blamed that on stress, tension, whatever the terminology one wants to give it. After any real serious incident, I would lay my head on the table, close my eyes and replay each step of the recent incident by memory, to make sure each and everything I did was the best I could do, and to see if there was anything to learn that would help with the next one.

On the other side of this experience, we often heard something that gave us a good laugh. One of the best that I heard was on the 80-meter amateur radio band. There were four amateur operators who are now a silent key that most of us miss dearly. Dave LeCain, VE1AFU, Yarmouth, Tommy Doucette, VE1TJ, Belliveau Cove, Captain Desire Doucette, VE1ADU, New Edinburgh and Captain Adrien Blinn, VE1SZ, Church Point.

This is Tommy, 20 years old and when he first received his amateur license and call sign VE1ALV. Tommy taught himself both the theory and code and managed to earn a bit of money repairing radios in his local area. His call was changed to his initials, VE1TJ, a few years before he became a silent key. One can see the devil in his eyes in this photograph.

Page 81, QST, December, 1963

Dave LeCain, VE1AFU, was studying to be a medical doctor, was involved in a car accident while attending college and was a paraplegic from then on and Tommy Doucette, VE1TJ, developed muscular dystrophy when 8 years old. I feel certain amateur radio extended their lives by a decade and possibly decades. I saw Tommy a few weeks before he became a silent key.

Tommy is the only amateur radio operator that I knew who has his call sign engraved on his tombstone. Tommy reminded me of my uncle who died from the same disease when he was 16 years old. They both were very quick with a reply. Tommy and Doug Johnson,

VE1OM, were recorded in the same Silent Keys Section of the May, 1979, issue of QST. Doug was a long-standing member of HARC and a former president.

Dave LeCain and Tommy were probably the two best phone operators one could hear. They both were very quick and you never knew what they would come back with. They started meeting on 3760 kilohertz each evening and this soon became the "Nut Net" with anyone and everyone joining in. It eventually broke off into two sections with another section on 3740 kilohertz. Both began before and ended after the Maritime Net on 3750 kilohertz each evening. Many HARC members will remember it well.

Captain Desire Doucette, VE1ADU, was known as "Dupont" and he was a master of the fishery patrol vessels operating around Western Nova Scotia. Captain Adrien Blinn, VE1SZ, was a master of the Gypsum ships running from Nova Scotia to Mexico and the West Indies with numerous visits along the Gulf Coast and Eastern Seaboard of the United States. I had many memorable experiences as his radio officer and was with him when he received his amateur radio license. I then bought his CB radios for my father-in-law.

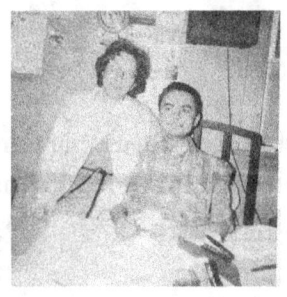

Atlantic Amateur Radio Newsletter

I was tuning, around 80-meters and heard Adrien, VE1SZ, talking with Dupont, VE1ADU, and naturally stopped to hear what they had to say. While they were talking Tommy, VE1TJ, swung down out of nowhere transmitting "hello test 1 2 3 4 5 hello test VE1TJ" on top of them. Dupont had a bit of a speech impediment especially when he got excited and he came back to Tommy with "tee tee Tommy what are you doing?". Tommy as quick as possible said "I wanted to test my equipment and needed a frequency with nothing on it".

Tommy looks very happy in this photo. This was taken in 1965 when Tommy was 23 years old and had been bedridden for 8 years. That is Miss Barb Robinson, VE1AQD, with Tommy. This was taken on the evening of May 6th when 22 members of the Annapolis Basin Amateur Radio Club held their monthly meeting at Tommy's home. This was a great thrill for one and all, I am sure. Tommy's room was on the north side of their home at St. Bernard. This was on the edge of St. Mary's Bay. The room was like a large sun porch with lots of windows and a beautiful view of the bay.

HARC File

Most of the area amateur radio clubs have their club crest or logo.

Lunenburg County ARC Web Site

Through this research I found a few biblical passages that may interest some members of HARC. If you plan on some antenna or tower work you may want to consult Luke 14:28. If you are operating break-in, try Isaiah 65:24.

Have you seen The Tonight Show with Jay Leno where CW beat the kids with their text messaging via cell phone in passing the same message? This was recorded on Friday May 13th, 2005. Chip Margelli, K7JA, did the transmitting and Ken Miller, K6CTW, did the receiving. The show failed to mention the fact CW beat the text messaging in three practice sessions before the actual show. They were actually beaten four times. ARRL was not that happy with the amount of coverage amateur radio received from this episode. This event is described in the July, 2005, issue of QST.

"Amateurs on 600 meters" is the title of a brief description of a proposal for the use of 500 kilohertz, or 600 meters, at the World Administrative Radio Conference to be held in 2011 (WARC 2011). This is on page six of the January/February 2009 issue of TCA. There will not likely be another radio frequency that has done more to save lives than 500. I am a member of a large world-wide group of former ship radio officers. The German members of this group are trying hard to have this frequency held as a museum or memorial to the history of this frequency. They have quite a proposal they have submitted to their amateur radio group that represents them in the IARU.

I made the suggestion to this group that it would be nice to see very low powered, solar powered beacons that came up with "de" and the call sign of the former coastal station in that area on that frequency. These beacons could be spaced out to key every 20 minutes or so and this would make for interesting copy. They should be spaced out so that one would not transmit on top of another as much as possible. I would lay awake at night and enjoy listening to each one. The local amateur radio community could operate and maintain these beacons although once constructed there would be little maintenance. They could be mounted on the roof of a building or the home of some member in the area. There were a lot of members in this group that agreed with me. We all have spent many hours listening to that frequency over the years. I spent all but six weeks of 1975 listening to it from two different ships.

There are three reasons this will not work that make the suggestion a bit ridiculous. The first reason is where one would find the amateur radio interest to operate one at VCN Grindstone on the Magdalene Islands, VAX Canso, VOJ Corner Brook, VCT Sable Island, 8RB Guyana, C6N Bahamas, 8PO Barbados, TXU St. Pierre, and such small places. The second reason is that some of those three-character call signs, like VCS are still assigned to present stations, although those operating these stations not likely know it because the call sign is not used. The third reason is the one that really puts an end to it. I am one of the youngest that have a fond memory of these stations. I often wonder why I am still here because so many that I have been with over the years are now a silent key. Once those remaining with this memory are gone what difference will it make? There probably will be some future interest in this simply from the way CW is "hanging on" and so many are finding CW fun. If the amateur radio world is assigned a band of frequencies around 600 meters, it will be just another amateur radio band much the same as all the other amateur radio bands.

The International Maritime Organization (IMO) is one of several international organizations that make the rules and regulations for marine safety at sea. Whether or not any nation follows these rules and regulations is another story. This is their ruling on 500 kHz or 600 meters from a recent IMO COMSAR meeting in all its technical glory. COMSAR is the acronym for Radio Communications and Search and Rescue. Each of these meetings is numbered and this was meeting 12. IFSMA is the acronym for International Federation Ship Masters Association. WP probably means Working Paper at least that would be my first guess. No doubt WRC-11 is what we know as WARC-2011.

4.75 The Sub-Committee recalled that, at COMSAR 12, it had considered the proposal by IFSMA to preserve the heritage of the important frequency 500 kHz, and that it was considered that this frequency could be better used in future. The Sub-Committee had also considered it necessary to be very careful not to lose access to this very important frequency band, currently controlled in the maritime environment.

4.76 The Sub-Committee noted that WP 5B had sent a liaison statement to WP

5A on studies related to WRC-11 Agenda item 1.23 stating that, prior to identification of preferred frequency bands for secondary amateur allocations in the 415-526.5 kHz bands, the maritime service must first consider existing and future requirements for ship and port safety spectrum in existing maritime spectrum to solve Agenda item 1.10. It was also noted that the band was also under study for the provision of future systems for enhancing of safety of navigation at sea (e-navigation applications).

Whether or not the Amateur Radio World gets 600-meters in our life time will remain to be seen. 490 kHz and 518 kHz have been the only frequencies in use on this band by the maritime world since CW was removed in the 1990's. This has been used by NAVTEX, an international narrow band direct printing radio telex system that broadcasted navigational information. This was part of the Global Maritime Distress and Safety System (GMDSS). There is still some GMDSS equipment in use but as I understand it that is now obsolete. NAVTEX is the acronym for the words navigational and telex. The GMDSS has been replaced with VHF and Satellite. 518 kHz did the English transmissions and 490 kHz the non-English.

DICK GRANTHAM VE1AI

There have been a number of very active amateur radio operators that have made HARC a very efficient and interesting club. Dick Grantham, VE1AI, is but one of them. He was the youngest club president way back in 1966 and is still one of the clubs most active members.

Dick Grantham, VE1AI

The picture above is Dick holding his IOTA 100, 200 ,300, and 400 certificates all neatly framed, which he received from the RSGB (Radio Society of Great Briton) in 2011. IOTA is the "Islands on the Air" award program, sponsored by the RSGB and is for working 100 or more of the Islands on the Official IOTA List (www.rsgbiota.org). Reg, VE7IG, is the Canadian QSL checker for IOTA and kindly checked Dick's 400 plus QSLs while he was here at the DX Forum during the summer of 2011.

GARY BARTLETT VE1RGB

The 2010 winner of the British Commonwealth Contest Medal was Gary Bartlett, VE1RGB, a very active member of HARC. Congratulations Gary. One can find so many and so much via the internet today. If one enters Gary's call sign in their search engine on their home computer, they will find a lot to read on Gary. It is members like Gary that make and have made HARC what it is. Unfortunately, and sadly, Gary became a silent key on July 22, 2017 at 73-years of age. This was way too young.

Photo by Mike Goldstein VE3GFN

Photo by Mike Goldstein VE3GFN

Gary Bartlett, VE1RGB

1100ᵀᴴ ISSUE OF QST

The March, 2009, issue of QST is Volume 93 Number 3 and on page 94 it states that the issue you are holding in your hands is the 1100th issue of QST. Volume 1 Number 1 was the December, 1915, issue. I have looked at every page and I trust I chose what was of interest to this project.

THE LOSS OF HMS BOUNTY

Metro Goldwyn Mayer or MGM is the movie company that has the lion growl just before the movie begins. In 1960 MGM had the Smith and Rhuland Shipyard at Lunenburg, Nova Scotia build them a replica of the *HMS BOUNTY* that experienced a mutiny in the South Pacific in 1789. This replica was the first time in history that a ship was built from the keel up as a prop for a movie. This replica was built for the movie "Mutiny on the Bounty" starring Trevor Howard and Marlon Brando. When this replica was completed, it was registered in Lunenburg to MGM's Canadian office at Toronto, Ontario. This replica was registered as a yacht. There was no legal requirement for this vessel to carry a radio officer. While the vessel was owned, operated and sailed by MGM it carried a radio officer. I was the second and last radio officer to sail in this vessel.

This is a post card produced by MGM with a photograph taken in the South Pacific. The black line from the top of the main mast to the top of the mizzen is the ship's main antenna and the one I used for amateur radio. This is the card used as the QSL card for the VE0MO amateur radio station.

Fifty years have passed and it would still be a pleasure to thump a few characters I encountered back then. I walked into a mess. Those characters are the reason Canada has no merchant navy. For this reason, when one asked me if I was radio officer when the movie was made, I simply said yes. This was true. They did a few scenes while I was this ships radio officer. Unfortunately, this was included in Bruce Nunn's book "History with a Twist" ISB 1-55109-255-7 page 22.

When Captain Robin Walbridge, KD4OHZ, sailed the *BOUNTY* on October 25th, 2012 from New Haven, Connecticut for St. Petersburg, Florida I would have loved to have been with them. Especially if I could have had my old radio room and radio station that I had in 1962. The main radio station was an MF transmitter (250 watts), an HF transmitter (300 watts) and two receivers from the RCA 5U merchant ship station. There was also a small marine radiotelephone for small vessels (about 20 watts). My amateur radio station was a Collins KWM-1 final amplifier plate input power was 175 watts PEP. The amateur station was assigned radio call sign VE0MO.

When we left *BOUNTY* we left her at Jacobson's Shipyard in Oyster Bay, New York. There they gutted her below deck and made her more like the original. She was placed in the New York World's Fair of 1963. Apparently, this did not work out very well. MGM made a Tahitian Village at St. Petersburg Florida and I was told they towed *BOUNTY* to this Florida location. Ted Turner bought MGM and eventually removed *BOUNTY* from the MGM organization and sold her. *BOUNTY* began sailing again sometime after this move as an American vessel. Anyone who visited *BOUNTY* after 1962 did not see our living accommodation. That accommodation was pretty good actually with a lot of varnished wood, a good galley, refrigerators and all the comforts of home. It was quite comfortable.

Doug Faunt was part of the last crew as an Able Seaman of a Watch in *BOUNTY*. Doug was also an amateur radio operator with station N6TQS. Doug found me on the internet and had made contact with me several times. He invited Joan and I aboard the last time *BOUNTY* visited Halifax in July 2012. We had a great visit and met the crew. *BOUNTY* appeared to be in great shape and no reason for her not to sail on October 25th. Yes, hurricane Sandy was out there but when you have a schedule and sail the North Atlantic you are bound to run into weather. They often sail a ship when a hurricane is expected because it is felt that a ship at sea is better than in harbor during such weather.

I have been told that the U.S. Coast Guard said *BOUNTY* was having steering problems just before she sank. When we sailed in *BOUNTY*, we carried two large timbers ten inch by ten inch with canvas sewed between them. We planned to use this as a sea anchor if needed. Had Captain Walbridge still had this and tossed it over the bow he would have solved any steering problem. The replica had the same steering system as that fitted in the *BETHIA* (the vessel that became the original *HMS BOUNTY*) in 1783. The wheel was mounted on the end of a wooden drum. A rope was secured to the rudder and went to this drum via pulleys and wound around this drum making a simple, effective and reliable steering system.

At least thirty years ago I did a lot of research on *BETHIA* and now believe she visited Halifax, Nova Scotia. The admiralty bought a vessel of this exact name and description, renamed it *BOUNTY*, put Captain William Bligh in command and sent it to Tahiti for breadfruit plants. The breadfruit plants were to be delivered and transplanted in Jamaica for food in that area. I was never able to prove the *BETHIA* that visited Halifax was one and the same that became *BOUNTY*. There have been so many versions of the *BOUNTY* mutiny in 1789 that one will never know the real story.

The pumps in *BOUNTY* failed and caused her to sink on Monday October 29th, 2012. At the inquiry on this sinking held in Portsmouth, Virginia commencing February 12th, 2013 we learned the truth. *BOUNTY* was not fit to sail and her Captain knew it. Some of the planking in the ship's bottom was rotten. She was taking on so much water she had to be pumped out every four hours. The fuel line to the engines became defective shutting down her generators and main engines. Her pumps operated on electricity and were therefore unable to work. It really is sad, a shock and hard to believe. 15 of the 16 on board were rescued by the U.S. Coast Guard. All 16 crewmembers were washed into the sea as they went to get into two life rafts. They all managed to swim to the life rafts and helped each other into the rafts. Doug managed to swim to a life raft but was so tired those in the life raft had to haul him on board. Crewmember Claudine Christian was found, rescued, but died in a hospital on shore. The search for Captain Walbridge terminated on November 1st, 2012.

Each person on board left in a survival suit. This would have been a requirement by the U.S. Coast Guard. These survival suits are very expensive. No ship owner would buy them unless forced to do so by law. There has been no mention of Captain Walbridge's qualifications. The U.S. Coast Guard would have made certain he had a proper certificate because BOUNTY was carrying paying passengers. The U.S. Coast Guard has a Sailing Masters Certificate. Apparently, Captain Walbridge and the two mates had certificates of some type but they were the only certificates held by the crew. Our crew in 1962 consisted of three Master Mariner Certificates, one a Sailing Master, three Engineer Certificates and my Radio Officers Certificate. Our captain was a sailing master and had been the sailing instructor for all Canadian naval officers.

Captain Walbridge made it known he enjoyed sailing close to hurricanes and may have gone too close to this one. One wonders if Captain Walbridge wanted to be rescued. Captain Walbridge held amateur radio call sign KD4OHZ. Claudene Christian probably wore herself out trying to swim to a life raft in her survival suit. She would have probably survived had she just relaxed and lay in the water in her suit. These suits are equipped with a lighted beacon and the reason the Coast Guard found her.

I was a twenty-year-old naval radio operator in 1959 when James Gallagher published his book "Fire at Sea". I have read this book many times and more or less used it as a text book. It is a description of Chief Radio Officer George Rogers, how he became Chief, how he set fire to the American Passenger Liner MORO CASTLE and gave a good description of his life up until he died in penitentiary in 1958. George simply sat and listened to the radio stations asking each other if they knew the ship that was on fire off Atlantic City, New Jersey. George was chastised for not stating he was on fire and waiting orders from the captain. George killed around one hundred and fifty people but managed to get away with this MORO CASTLE incident. He went to the penitentiary for a separate murder. It is a shame society has not found a means of identifying these animals and eliminating them before they create so much grief.

When I went aboard BOUNTY and had a good look, I was convinced that if she was ever lost at sea, it would be a fire and not sinking. Let's face it; she was a real fire hazard. All wood, canvas, rope, varnish, paint, and you name it. It is amazing she lasted fifty years. There was no very high frequency radio back when I sailed in her. It was strictly radiotelegraph with a small audio modulated radiotelephone that was not much better than a good megaphone. Her international call sign was VYFM. I was half scared to death of her because she had no emergency radio equipment. In other words, if anything went wrong, I would have to alert the world while the electric generators were still operating. I intended to be chatting with someone, preferably a U.S. Coast Guard radioman if there was a serious problem and I had no word from the captain. For this reason, I made the habit of leaving the radio station already for a quick distress call if needed. A habit I kept until my last trip at sea in September 1978. *BOUNTY* was one of twelve ships I sailed in from 1957 until 1978. She was the last one still in service and now they are all gone. The others were broken up in various shipyards around the world.

There was a big difference between what it was and what it could or should have been. What it should have been is what I dream about today.

BOUNTY's international call sign was VYFM when registered in Canada as stated, and was listed as such for several years while she belonged to Metro Goldwyn Mayer and was alongside their Tahitian village in Florida. When Ted Turner sold her and she was registered in the United States her international call sign was changed to WCP4944. The United States ship call signs change quite often for some unknown reason. *BOUNTYs* was changed to WDD9114 and this was her call sign when she sank.

I find it hard to believe that anything has been constructed that has created more proverbial crap than this *BOUNTY* replica. Captain Walbridge stated on television that Marlon Brando would not star in the movie if they burnt this replica at the end of the movie. They burnt a beautiful 36-inch model at the end of the movie and did it at the MGM studios in Culver City, California. This is typical of the stories around this vessel and this is one that surfaced in the last few years. The crew running around in those fire scenes in the movie was on a hull built in a tank at Culver City. The same hull they used for the storm scenes. I do not believe this story and if it were true, we would

Pitcairn Islanders earn a living by making and selling hand crafts. Here Len Brown carves a wall fish, Millie Christian weaves a basket, Norma Clark displays a basket and sun hat, while Anderson Warren shapes a walking cane. All items are hand-made; carvings are usually made from ''Miro wood'' (Thespesia populnea), and weavings from the leaves of the Pandanus palm. As of 1970, Pitcairn's population was approximately 90.

have known about it when they completed the movie. Marlon Brando was interested in one thing only; Marlon Brando. He had no interest in the *BOUNTY* replica. The best description I have heard of Marlon Brando was right after he died his former wife stated publicly that "there must be a special place in hell for that bastard". I'm sure anyone who knew Marlon Brando would agree with her whether they will admit it or not.

The one thing that is true is that Marlon Brando wanted his character Fletcher Christian to die at the end of the movie when *BOUNTY* was burning. We all know that was not fact because there would be no Tom Christian, VR6TC, Betty Christian, VR6YL, Irma Christian, VR6ID, and the rest of the Christian's on Pitcairn Island. This includes Claudine Christian lost in this incident. She was a descendant of Fletcher like the others. The VR6 call sign prefix was changed to VP6 a few years ago.

I received some interesting photographs on April 15, 2015, from Bill Parker, VE1VP, formerly of Bedford, Nova Scotia, now living in Port Alberni, British Columbia. I will include them here for possible interest.

This is a model of BOUNTY by Len Brown, Pitcairn Island.

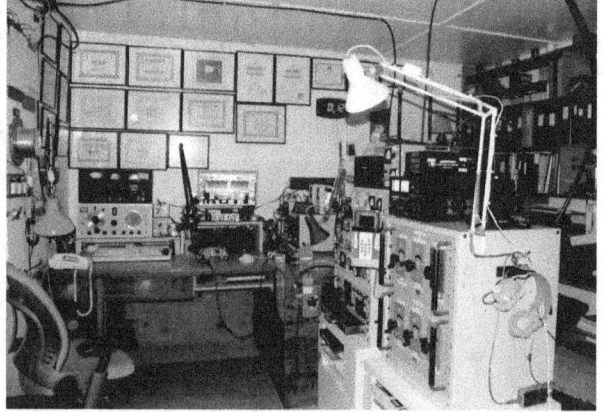

Bill stated this is what is left of his VE1VP radio station.

Per visited Halifax in the cruise ship *ROYAL VIKING SEA* about 1980 and more photos of him back then can be seen in my section on satellite communications in my marine radio history. Per was Chief Radio Officer in *CRYSTAL SYMPHONY* laying off Pitcairn Island when this photo was taken.

Tom, VP6TC, was diagnosed with possible Parkinson's and early signs of Alzheimer's/ dementia in December, 2009, while on a family visit in New Zealand. His wife Betty Christian, VP6YL/VR6YL, said his health "deteriorated all too quickly," and the last few months were "cruel ones to watch such a strong, vibrant man reduced to where he was not really aware of his surroundings and then was unable to walk and swallow food or liquid." Tom passed away on July 7th, 2013, and was buried on July 8th in the cemetery on Pitcairn Island. Tom was 77 years old. Lack of available transportation prevented most of Tom and Betty's children from making it back for the funeral. Tom was known as the "Voice of Pitcairn." He held the Member of the British Empire medal and served on the

Pitcairn Island Council as the Governor's Representative for 40 years and Tom was also the Officer in Charge of the Pitcairn Island Marine Radio Station with call sign ZBP.

There are a multitude of stories surrounding this mutiny and one will never know the facts. It is believed that Fletcher Christian was killed by one of the mutineers on Pitcairn Island. When the mutineers were found in 1808, they were all dead but one. They had killed each other fighting over the women they had taken to Pitcairn Island from Tahiti. They did not take enough women and the mutineers wound up fighting over the few they had. It is believed that one of the women killed one of the mutineers because he killed her husband. John Adams was one of the mutineers and the only sur-

Per Mikalsen, LA3FL, from Alta, Norway and Tom, VP6TC

vivor when found in 1808. Adamstown on Pitcairn Island is named for him. John had saved the bible carried in *BOUNTY*. I have seen photographs of it. After the other mutineers were dead John held Sunday church services with this bible. This apparently helped stabilize things on the island.

This movie "Mutiny on the Bounty" has been twisted around a number of times as well. The last version I saw had Marlon Brando listed before Trevor Howard. Trevor Howard was listed first on the earlier versions. The premier showing of the movie on November 6th, 1962, started and ended with those who found the mutineers in 1808. Several of the other versions start with *BOUNTY* getting ready to sail from England and some do not mention the finding of the mutineers.

This is the BOUNTY on July 21ˢᵗ, 2012 during her last visit to Halifax, Nova Scotia. There was a difference than when we sailed in her in 1962. Her hull and spars are black, the colour the Americans painted their ships whereas the spars were varnished and the hull royal blue when we sailed in her. Her decks below were completely different. Where our cabins and living accommodation had been was wide open. The living accommodation was now down on the next deck and the cargo hold was gone along with the two water tight doors either side of where the hatch to the cargo hold had been. I found a mark on her deck where our ten-inch radar had sat behind the chart table when she visited Lunenburg years ago. I could not find that mark this trip. Her engine room hatch appeared to be the same hatch but was in the open. She was fitted with John Deere generators and main engines on this visit. The Caterpillars we had were removed long ago. Her lower mizzen mast is grey and black. That was one of her engine exhausts but we did not use it. We used one on the starboard side in the hull. I did not ask them but it appears they were using the mizzen exhaust.

I did not see any radio or electronic equipment while on board. The captain told me they had two radars but I did not see them or the antennas for them.

BOUNTY had Greenport, New York on her stern where the Lunenburg N.S. used to be. While we were alongside the Southwind Marina at Long Beach, California in 1962 a tour boat loaded with sight seers used to go past. It had a public address system and the operator used to state the ship had been built in New Zealand for the movie. Close enough for tourist work. That boat had the name *SHEARWATER*, the same as the naval air station at Dartmouth, Nova Scotia and the reason I remember it so well.

This is "Spud" Roscoe, VE1BC and Doug Faunt, N6TQS on July 21ˢᵗ, 2012.

Doug was sailing as a volunteer and was not paid for sailing in *BOUNTY*.

Note the hatch cover, the hatches had been replaced with these shed like coverings.

Robin Walbridge, KD4OHZ, Missing at Sea after Sinking of Tall Ship *Bounty*;
Ship's Electrician Doug Faunt, N6TQS, Rescued

Every DXer knows the story of the HMS *Bounty* and Pitcairn Island. VP6: In 1789, the HMS Bounty — a small three-masted sailing vessel sent by Britain's Royal Navy to the Pacific on a supply expedition — was roiled by tension between its crew and its captain, William Bligh. After landing in Tahiti and taking on a cargo of breadfruit, the *Bounty* set sail for the West Indies; it never reached that destination. Instead, Master's Mate Fletcher Christian led the men in a mutiny, eventually allowing Bligh and his loyalists to sail off in a longboat. After an arduous journey, they reached

Doug Faunt, N6TQS, on the deck of the tall ship *Bounty*.
[Photo courtesy of Doug Faunt, N6TQS]

safety at the Dutch-owned port of Kupang. Christian and his followers ended up on Pitcairn Island where they burned the *Bounty* and settled on the island. Passing ships did not discover the enclave until after the turn of the 20th century.

On October 29, a replica of the *Bounty* — built in 1960 for a remake of the 1962 film *Mutiny on the Bounty* — sank off the coast of North Carolina as Hurricane Sandy made its way toward New Jersey. Of its 16 crew members, 14 were rescued by the US Coast Guard. *Bounty* Captain Robin Walbridge, KD4OHZ, never made it

to one of the two deployed life rafts and is presumed dead. Claudene Christian, who claimed to be a direct descendent of Fletcher Christian, was found unresponsive and passed away at a North Carolina hospital that evening.

Doug Faunt, N6TQS, of Oakland, California, was one of the 14 who was rescued by the Coast Guard: Faunt served as a deckhand and was also the ship's electrician. According to Spud Roscoe, VE1BC, Faunt had satellite communications equipment and *Winlink* capabilities on board the *Bounty*, but he was not the ship's radio officer. "Sailing on replica ships was a hobby of Doug's," Roscoe told the ARRL. "He had previously sailed across the Great Australian Bight on a replica of the HMB *Endeavour*, Captain Cook's ship. He was an able seaman of the watch." Roscoe was the radio officer on the replica *Bounty* for its original voyage to France in 1962.

Faunt told the ARRL that the *Bounty* crew tried various methods, including a satellite phone, to call for help, "but we got nothing when tried calling out on HF. We tried calling the Maritime Mobile Net, but nothing was out there. We had *Winlink* on the ship that we used for e-mail and accessing the Internet to post to blogs and to Facebook, and we finally found an e-mail address for the Coast Guard. As

a last-ditch effort, we used *Winlink* to e-mail the Coast Guard for help. Within an hour, we heard a C-130 plane, and later, a helicopter overhead." According to Faunt, it was Walbridge, as master of the ship, who sent out the distress messages.

"I don't know how I made it off the ship," Faunt recalled. "I had finished serving a long watch, and then we started going down. I was exhausted. I had to swim to get to the life raft. The water was full of rigging, and here I am, in my Gumby suit, trying to swim. It was so difficult. While swimming to the raft, I came up for air and a spar was coming at me. I finally found a raft and tried to climb into it, but I almost didn't make it, tired as I was. Through the help of my shipmates who were already aboard the raft, I got on." The two life rafts were out about 100 miles from shore when they were rescued.

The vessel left Connecticut on Thursday, October 25 with a crew of 11 men and five women, ranging in age from 20-66. After being treated at a hospital in Elizabeth City, North Carolina, Faunt arrived back home in California on October 31. "I'm looking for a new boat to sail and a DXpedition to go on," Faunt told the ARRL. "Ham radio got me into my position on the *Bounty*, and ham radio got me out alive!"

January 2013 issue of QST

HARC member Doug Conrad, VE1UY, told me he had the Maritime Mobile Net tuned in during this incident but heard nothing.

The following appeared in the Halifax Chronicle Herald on Wednesday, July 24, 2013.

BRIDGEWATER — The United Nations' International Maritime Organization has named two United States Coast Guard members recipients of its awards for exceptional bravery at sea for rescuing nearly the entire crew of HMS Bounty after it capsized in a hurricane last year.

The organization says Randy J. Haba and Daniel J. Todd overcame the effects of cold, fatigue and ingesting sea water to rescue 14 of the 16 crew members who had been aboard the stricken tall ship during hurricane Sandy on Oct. 29.

Capt. Robin Walbridge and crew member Claudene Christian died. Walbridge's body was not recovered.

The replica ship was built in Lunenburg over 50 years ago.

The two rescue swimmers were nominated by the U.S. government.

The organization said Haba spent an hour battling strong currents and 10-metre waves to get Bounty survivors from life-rafts into the helicopter's rescue basket. It says a huge wave knocked off his mask, but Haba "demonstrated the utmost determination and perseverance" by rescuing two more people without the equipment.

Todd is said to have displayed strength and ingenuity by saving nine crew members. He was helping a survivor into the rescue basket when a large wave knocked over the life-raft that held four more survivors. Todd grabbed the life-raft's sea anchor and was able to save those inside, then made it to a second life-raft and rescued the three who were inside.

Haba and Todd will be invited to attend the Nov. 25 awards ceremony at IMO headquarters in London, England.

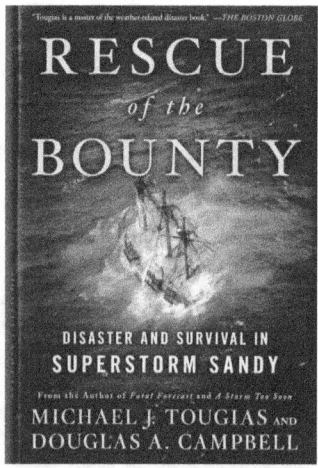

This book is rather interesting. We do not learn if Captain Walbridge had a marine certificate and if so what type of certificate. This book states nearly all in the crew when she sank had 100-ton certificates including the engineer. This book does not mention our crew. The *BOUNTY* had been modified so many times since 1962 that she was not the same ship when she sank.

VE1CPA

Chris Hadfield was a former Royal Canadian Air Force pilot who became an astronaut. He was the first Canadian to walk in space and flew in two space shuttle missions. He served as commander of the International Space Station.

The local students in various schools in this area made radio contact with Chris who held amateur radio call sign KC5RNJ and VA3OOG. Probably the best way for me to record this experience is to simply reproduce some of what has already been published on the subject. The local Halifax contacts were made at Bedford, Nova Scotia in the Charles P. Allen High School using call sign VE1CPA.

The Halifax Amateur Radio Club *Reflector*   Page 9

Charles. P. Allen High School ARISS contact January 5th 2013

Friday January 4th was the start of set up at the school about noon. We assembled the rotators, mast, crossboom, mounted the antennas and ran cables - both the 2 RF and 2 rotator cables. We were using a KLM 22C (a 22 element circularly polarized beam) for the primary antenna and a Diamond 500 vertical for the secondary antenna. Both had receive pre amplifiers installed at the antenna. The rotator system was a Yaesu G-5600B azimuth rotator and elevation rotator. The radios were both Yaesu FT-736R for the primary and back up. The set up was done in freezing temperatures with windy conditions - it was cold out there. Helping with set up was Doug, VE1LDL, Wayne, VE1WPH, Eric Thibodeau (our sound technician) and several students including Will Harris, VA1INN a recent graduate of the HARC basic course. I know there were several others I have missed. Tired out we quite about 8 PM with the plan to finish set up on Saturday morning.

On Saturday morning we finished setting up the gear and then Murphy struck. The azimuth rotator seemed to have a problem. I went home and picked up my rotator control box and when trying it found the same problem at which point we were too close to the contact to trouble shoot the problem. We made a decision to use the back up equipment with the original primary equipment to be used as a back up if we could hear the space station on it. Helping with set up on Saturday were most of the same people with the welcome addition of Terry, VE1TRB.

As the time of the contact drew closer the tension was mounting. We initiated the call to the space station and heard Chris Hadfield reply "VE1CPA this is NA1SS how do you copy" we immediately went back but did not hear any reply. We tried calling several times and sometimes heard Chris calling us - it was obvious we were not being heard and we could not always hear him. By this time everyone was very tense and then we heard Chis say he was switching to the secondary frequency. We immediately called on the secondary frequency and contact was established We had lost about half of the 10 minute pass. The students were very quick in their questions and Chris was doing a fine job of responding quickly, the contact was progressing very smoothly but unfortunately time ran out. After the seventh student (there were 9) we heard Chris come back but noticed the signal was beginning to drop down. At this point Wayne stepped up to the microphone and bid goodby and had the crowd send a cheer to Chris

Then it was time to tear down and we were helped by most of the people who helped set up with the addition of Ian Harris, VE1INN. We are still trying to determine what went wrong with the equipment and will try to report our findings later once we have any idea what happened. Overall it was a successful contact and the students and audience were quite thrilled to be able to hear directly from the space station.

This contact generated a huge amount of publicity for Amateur Radio and the space station on radio, television and the internet. I am a little hazy on all the details here since Friday and Saturday seem to be a blur. I know I missed naming several people who helped with set up and tear down. We did a lot in one and a half days to make this contact happen. A lot of effort was put into the preparation, set up and tear down but it was worth the effort. The only down side was that 2 students were not able to ask their questions.

Bill Elliott, VE1MR

ARISS contacts

Saturday, January 5, 2013, marked the beginning of a historical series of ARISS contacts between Canadian students from coast to coast to coast and Canadian Space Agency Astronaut Chris Hadfield, VA3OOG. This first contact included students from Charles P. Allen High School, Bedford and began at approximately 1:30PM.
The event was open to the public.

A special guest speaker was Canada's first Astronaut Marc Garneau.

Coverage from media was extensive with national coverage provided by CBC. Check out the video on these web sites:

.http://www.cbc.ca/player/News/
Canada/NS/ID/2323450116/

http://www.youtube.com/watch?
v=gg4qLUYTrug

http://www.n2yo.com/?s=25544

http://www.hrsb.ns.ca/

The C.P.Allen contact was the first of 5 ARISS contacts planned for the 4 Atlantic Canada in January. Currently scheduled are:

January 12 - Royal Canadian Air Cadets, Newfoundland Detachment, St. John's NFLD (Direct)

Week of January 14 - 20, Saint Rose Elementary School, Saint John, NB (Direct)

Week of January 14 - 20, Stonepark Intermediate School, Charlottetown, PEI (Direct)

Week of January 28 - February 3, Fountain Academy of the Sacred Heart, Halifax, NS (Telebridge)

Chris Hadfield has recently joined 5 other crewmates as part of Expedition 34. In March of 2013, as the crew increment changes to Excpedition 35, Chris will assume command of the orbiting platform. In doing so he will become the first Canadian to command a spacecraft.

73, Wayne, VE1WPH,
ARISS (Canada) Mentor"

Photo by Ron, VE1AIC

Hams & friends who contributed to the success of the contact
Back row, left to right: David Cosh, VY2DAC; Ron MacKay, VE1AIC; Ron, Huybers, VY2HR; Brent Taylor, VY2HF; George Meggison, VY2GM;
Front row left to right – Eric Thibodeau, Bill Elliott, VE1MR; Wayne Harasimovitch, VE1WPH (ARISS mentor); Ken McCormick, VY2RU.

The ARISS / C.P.Allen High School contact did not go smoothly but, as the tension mounted, communication was finally achieved using the backup frequency about half way through the pass. The students wasted no time asking their questions.

What you don't see on the TV news coverage or the internet videos is all the prep time including the planning, the negotiating, the preparing, packing up & transporting of equipment, the "-C" temperature outdoor setup of the club trailer with telescoping mast, adding azimuth & elevation rotator, the running of cables, indoor equipment deployment, etc.

Present were the students & teachers & amateur radio techs & ops then there were the CBC camera crew & interviewer, and the public who responded to the previous evening news announcement (surprising to me) that the event was open to the public.

Good fortune to Ken and the other St. John's hams with your ARRISS contact next weekend   - ed.

As reported by Kenneth, Stonepark Intermediate School, Charlottetown, PEI, Canada was a success. The students completed all of the 18 planned questions with enough time remaining for closing remarks from the school followed by audience applause. Chris was able to respond and offer a farewell to PEI. Contact was concluded with 30 seconds remaining in the pass.

Comm was established at the scheduled AOS time. Initial signal was noisy but improved as elevation increased (as expected). D/L Signal strength and audio quality were good during the first half of the pass but did begin to drop off at approximately TCA. This condition continued to degrade gradually as the pass progressed through to LOS. While signal strength and audio quality fluctuated (estimated 3 cycles) above and below the mean. Generally speaking, the comm was still readable above background "white" noise through the TCA to LOS portion of the pass.

Audience estimated to be *450 to 500* with School Board, local government and municipal representatives in attendance. The event was recorded by undergraduate media students from Holland College. A video release is expected shortly. Radio, television and newspaper were represented. I invite you to view the following CBC Television link. Allow the piece to run for a moment then advance (slide); the run time goes to 20:44

http://www.cbc.ca/player/News/Canada/PEI/Compass/ID/2341381530/

73 - Wayne, VE1WPH

Also see video of the contact at

http://www.cbc.ca/player/News/Canada/PEI/ID/2341239360/

VE1ENT AND VE1MR LIFE MEMBERS HARC

Lynn Bowser VE1ENT and Bill Elliott VE1MR were made life members of HARC at the monthly meeting held on November 19th, 2014. Both have put in many hours of hard work for HARC over the years. The list of HARC presidents will indicate that Bill has been the longest serving president of HARC.

NIGEL SERVICE VE1NPS SILENT KEY

The funeral for Nigel Patrick Service, VE1NPS, was held on February 4, 2015 in the Calvin Presbyterian Church in Halifax. Nigel's son Peter is VE9NPS. Tom, VE1GTC, and Jeremy, VE1JHF, participated in this service and did an excellent job of describing the meaning of 73. Tom was at the Dias with a hand held and Jeremy was at the back of the church with another. Jeremy called Nigel while Tom held the hand held next to the microphone. I have Hearing aids and a heck of time hearing. I understood very little of the service but this is one demonstration I heard loud and clear.

Brian, VA1CC, Murray, VE1BB, and his XYL Heather sat behind me during this service. Murray asked me if I had seen the piece of anchor from the Halifax Explosion on December 6th, 1917, just outside the church. I describe this explosion on page 10 of this exercise and Murray gave me a brochure on this piece of anchor that landed in the area of the present church. No wonder it was the biggest man-made explosion up to the dropping of the atomic bomb.

VE1TTT VE1WC

Wayne Catchpaw
December 4, 1945 – November 1, 2016

Wayne and his buddy Sam Semple VE1YVN were retired Navy Chief Divers. Wayne was the only coxswain in the navy that told his captain that he was a disgrace to the naval uniform because of the way he treated the crew on HMCS CORMORANT. He was drafted (transferred) the next day, but he did his job as Sherriff of that vessel. HMCS CORMORANT was the navy's diving tender.

VA1MMA

The Halifax Amateur Radio Club set-up a Wireless Room in the Maritime Museum of the Atlantic and assigned it the call sign VA1MMA. HARC member Dick Grantham VE1AI is in charge of this project. Dick states: This has been the greatest undertaking for the Halifax Amateur Radio Club in the club's almost 8 decades history. You have all proven that it could be done and done WELL! We have concluded our second full season in 2017. The success continues to grow in popularity by bringing history to life in a museum steeped in our maritime history!

Just a few statistics:

Certificates issued just this year 2017: 6376

This is the Wireless Room at the Maritime Museum of the Atlantic. The modern amateur station is as follows: Transceiver – ICOM IC-7600 matching power supply and speaker. Vibroplex paddles for CW. Software for logging and computer control of the ICOM radio is "Logger 32". Antennas: tower mounted multiband vertical (model MA-5V") on the roof for 10 through 20 meters with a dual band inverted "V" for 40/80 meters. MFJ switch to enable the selection of which radio and which antenna you wish to use on the air. For VHF/UHF we have a Yaesu dual Band mobile radio, model FT-8800, and a Ten Tec power supply. The antenna is a KLM dual band antenna on the tower. The large unit on the right under the clock is one of the Canadian Marconi CM11 Transmitter Receivers that first entered service in 1943. Each and every Royal Canadian Naval unit had at least one of those stations from 1943 until the late 1970's. HARC member Fred Archibald VE1FA received this one from a museum in Quebec and restored it to working order. It is now capable of working the world. The unit on the bottom of this transmitter receiver with the dark face is the receiver, a CSR5. If one removes the receiver this unit is then known as a TM11.

This is the VA1MMA Icom tag.

There are a number of various volunteers from the Halifax Club that open the station at various times to entertain any visitors that visit the Wireless Room while visiting the museum. They are shown a list of the continental Morse code characters and then they are invited to transmit their name on an oscillator. The club has issued thousands of these certificates that have gone around the world. This is very popular especially when a cruise ship is visiting Halifax. The cruise ships dock near where the museum is located on the water front.

Dick went on to state the certificates this year, 2017 went to such countries as: Japan, China, Taiwan, Australia, New Zealand, England, France, Germany, Czech Republic, Hungary, Ukraine, Brazil, Argentina, Mexico, Spain, Costa Rica, Norway, Sweden, Switzerland, Nigeria, Several of the West Indies Islands, Colombia, Venezuela, Poland, almost all of the states of the United States including Alaska, and all of the provinces of Canada, and on and on.

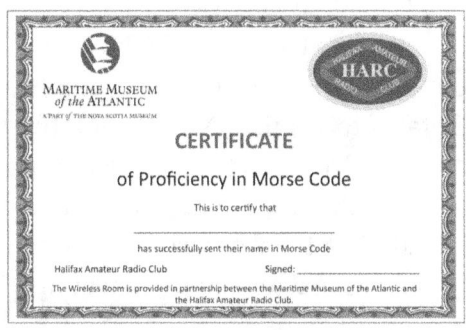

One would have to agree that it is a very interesting and exciting project for the Halifax Amateur Radio Club,

THE WIN AWARD

Winfield Alexander Hartlin, VE1WIN became a silent key on March 20th, 2017 and right after the executive of HARC decided to create the "Win" Award. This award was to be presented annually to the one who had worked the hardest for HARC during the past year. Win was a very hard-working member of HARC and the welcome inspiration for this award.

The first to receive this award was Barry Manuel, VE1JRG who was presented this award at the annual HARC Christmas Dinner held at Boston Pizza on December 20th, 2017.

Barry Manuel, VE1JRG with the first "Win" Award, December 20, 2017

THE CANADA C3 EXPEDITION

The Canadian Coast Guard sold the old icebreaker SIR HUMPHREY GILBERT and she was renamed POLAR PRINCE. In 2017 this vessel became part of the C3 Expedition. Canada celebrated its 150th anniversary as a nation and this ship sailed from Toronto, Ontario to Halifax, Nova Scotia then up past Labrador and across the Arctic Ocean and down to British Columbia celebrating this 150th anniversary. The name C3 was for the three oceans, Atlantic, Arctic and Pacific. The POLAR PRINCE arrived in Victoria on October 22, 2017. It then departed the West Coast, sailing down the Pacific Ocean, through the Panama Canal and up to Lunenburg, Nova Scotia arriving on January 9, 2018.

HARC member Alexey "Alex" Tikhomirov, VE1RUS met the ship when it arrived in

Halifax and was in contact with Barrie Crampton, VE3BSB via telephone while he tested

the equipment. Barrie Crampton, VE3BSB was in charge of this project.

While making the run from Toronto to Victoria this station used call sign CG3EXP and while making the run from the West Coast to Lunenburg it used call sign VE0EXP. This was with a WSPR station; Weak Signal Propagation Reporter computer program used for weak-signal radio communication between amateur radio operators.

Alex makes two trips per year to Eureka in the far north and operates the club station at 80N 86W with call sign VY0ERC while on these excursions. The following is from the VY0ERC log when the ship was approaching the Panama Canal on its way back to Nova Scotia:

Timestamp	Call	MHz	SNR	Drift	Grid	Pwr	Reporter	RGrid	km	az
2017-12-22 00:28	VE0EXP	7.040116	-23	0	EK31	0.2	VY0ERC	ER60tb	7634	

The VE0EXP WSPR station disappeared off the east coast of the United States on its way to Lunenburg. POLAR PRINCE had hit some weather at the point where the WSPR station had disappeared and they found that the station had slid off the rack and landed on the deck behind the rack. It had not been secured properly but received no damage and when connected back up it worked well. In the navy our equipment was bolted down with spring-loaded connections to protect the equipment from vibration. In the merchant and Coast Guard ships the equipment was simply bolted down. This WSPR station should have been bolted to the rack it was sitting on; it was lucky it managed to go that far before sliding off.

CLUB STATION FEBRUARY 2018

This is Brandon Fowler VE1BMF operating the HARC Club station VE1FQ on February 21, 2018.

VE1JMB

Congratulations to Jim Bignell VE1JMB for making the front page of the Halifax Chronicle Herald newspaper on January 4th, 2021 with this photograph and a detailed description of the Get on the Air winter event.

HMCS SACKVILLE VE0CNM

Sam, VE1YVN has donated a Yaesu transceiver for an amateur radio station in HMCS SACKVILLE. This is not history but we hope when you read this it will be. Sam, VE1YVN, Dick, VE1AI and Bruce, VE1NB hope to set-up this station in SACKVILLE in her original radio room. They want to install a wire antenna as the vessel was fitted during World War II. This antenna was known as an Inverted L and they want to use it as the amateur radio station antenna. They hope to get the sea cadets interested in this project, have a few obtain amateur radio licenses and have them explain things to those who visit the ship.

Canadian shipyards built many of these Corvettes during the war and Canada managed to have 111 as part of the Royal Canadian Navy. The other Corvettes Canada built were commissioned in the other navies, such as Britain and the United States. SACKVILLE is the last of these Corvettes. The Canadian Corvettes were named after Canadian towns and this one is named for Sackville, New Brunswick. During the war a leading telegraphist was in charge of her radio room with two or three telegraphists to do the radio operating. There was also a coder or two to code the out going messages and decode the incoming messages. There was a couple of signalmen to operate the signal light in Morse code, communicate with flags and this made for a good-sized communications staff.

The VE0CNM amateur call sign helps describe the vessel as the **C**anadian **N**aval **M**useum.

HARC EXECUTIVE MEMBERS

HARC has been fortunate to have had outstanding leadership through the years. Unfortunately, the records of the entire executive are not available in order to record them in this exercise. The following Club Executive Members are to be recognized and congratulated:

1933

Maritime Amateur Radio Association

President – C. S. Taylor, VE1BV

John MacKasey, VE1DE, replaced VE1BV as president on February 16th, 1934

1934

The Maritime Amateur Radio Association was renamed

The Halifax Amateur Radio Club on March 16th, 1934

President – John MacKasey, VE1DE

Secretary Treasurer – Wes Street, VE1EK

Assistant Secretary Treasurer – John Roue, VE1FB

John MacKasey, VE1DE, moved to Ontario and Len Foster, VE1EF, replaced John on June 15th, 1934, and Art Grant, VE1EP, replaced Len on September 28th, 1934.

1935

President – Art Grant, VE1EP

Secretary Treasurer – John Doull, VE1FN

Wes Street, VE1EK, was elected president at the September 27th, 1935, meeting.

1936

President – Wes Street, VE1EK

A. A. "Steve" Stevens, VE1EC, replaced Wes Street, VE1EK, on September 11th, 1936

1937

President – Steve Stevens, VE1EC

1938

President – Cliff Short, VE1AW

Vice President – Wes Street, VE1EK

Secretary Treasurer – John Roue, VE1FB

Assistant Secretary Treasurer – Doug Smith, VE1FO

1939

President – Bill Bligh, VE1BC

Vice President – Ron Hart, VE1MZ

Secretary – Barclay Dowden, VE1HK

Treasurer – Doug Smith, VE1FO

Chairman of the Transmitter Committee – Ralph Fraser, VE1HJ

1940

All amateur radio operating ceased in Canada on September 5th, 1939. The HARC Fiscal year was set as September to September at the club meeting August 18th, 1939 and the election of officers was to take place at the September meetings.

President – Ed MacLaughlin, VE1JH

Vice President – Brit Fader, VE1FQ

Secretary Treasurer – Barclay Dowden, VE1HK

Assistant Secretary Treasurer – Ainsley Croft, VE1EU

Chairman of the Transmitter Committee – Ralph Fraser, VE1HJ

The executive was reorganized at the monthly meeting held on January 26th, 1940 as follows:

President – Brit Fader, VE1FQ

Vice President – Ed MacLaughlin, VE1JH

Secretary Treasurer – Barclay Dowden, VE1HK

Assistant Secretary Treasurer – Gordon Phalen, VE1KG

Chairman of the Transmitter Committee – Ralph Fraser, VE1HJ

Those present at the February 16th, 1940 meeting were: VE1KG, VE1EP, VE1MZ, VE1HP, VE1FQ, VE1JH, VE1HJ, VE1HK, and as their guests from the services, VE5AJV, VE1OK, VE4ABA, VE3HV, VE5AJU, VE1HH, VE3HG, and Mr. Moore.

1944

The Halifax Amateur Radio Club was reorganized February 28th, 1944, having been inactive since the last recorded meeting on February 16th, 1940.

President – Gordon Phalen, VE1KG

Vice President – Don Sutherland, VE4FK

Secretary Treasurer – Ed MacLaughlin, VE1JH

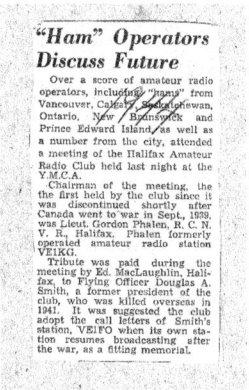

"Ham" Operators Discuss Future

Over a score of amateur radio operators, including "hams" from Vancouver, Calgary, Saskatchewan, Ontario, New Brunswick and Prince Edward Island, as well as a number from the city, attended a meeting of the Halifax Amateur Radio Club held last night at the Y.M.C.A.

Chairman of the meeting, the the first held by the club since it was discontinued shortly after Canada went to war in Sept., 1939, was Lieut. Gordon Phalen, R. C. N. V. R., Halifax. Phalen formerly operated amateur radio station VE1KG.

Tribute was paid during the meeting by Ed. MacLaughlin, Halifax, to Flying Officer Douglas A. Smith, a former president of the club, who was killed overseas in 1941. It was suggested the club adopt the call letters of Smith's station, VE1FO when its own station resumes broadcasting after the war, as a fitting memorial.

1945

President – Fritz Webb, VE1DB

Vice President – Harold Bishop, VE1OB

Secretary Treasurer – Ed MacLaughlin, VE1JH

Membership and Program Chairman – Wes Street, VE1EK

1946

President – Wes Street, VE1EK

Vice President – Walt Wooding, VE1ET

Secretary – Ed MacLaughlin, VE1JH

Treasurer – Harold Bishop, VE1OB

Harold Bishop, VE1OB, resigned as Treasurer at the March, 1946, meeting and Don Bain, VE1LZ, replaced him. Wes Street had to resign at the May 17th, 1946, meeting and vice president Walt Wooding finished the year as president.

This is Walt Wooding the former VE1ET.

Walt moved to Ottawa in 1958 and became VE3CLJ.

Walt did not have the VE3WW call long before he became a silent key in 1990.

1947

President – John Roue, VE1FB

Vice President – Ron Hart, VE1MZ

Secretary – Holland H. "Shep" Shepherd, VE1RR

Assistant Secretary – Tommy Baker, VE1SF

Treasurer – Art Grant, VE1EP

Ron Hart, VE1MZ, was very active with HARC for many years. This ended in 1948 when he moved to the United States. The Section Communications Manager, Art Crowell, VE1DQ, stated "The HARC lost one of its most ardent and energetic members when VE1MZ moved to Chicago". Ron became W9IVP in 1955.

1948

President – C. E. Wigle, VE1PR

Secretary – Tommy Baker, VE1SF

Secretary – Ralph R. Pattison, VE1RP

Ralph, VE1RP, signed a number of letters as Secretary regarding the Bicentennial Hamfest for 1949. These letters were dated in 1948. According to Art Crowell, VE1DQ, in April, 1948, VE1SF was reelected Secretary so Holland Shepherd, VE1RR, must have been replaced in 1947.

1949

President – Doug Johnson, VE1PQ

This Doug Johnson moved to the U.S. and became WA7MGK.

1950

President – Brit Fader, VE1FQ

1951

President – Doug Johnson, VE1PQ

This Doug Johnson moved to the U.S. and became WA7MGK.

Chairman of Membership Committee – Don Bain, VE1LZ

1952

President – Ralph R. Pattison, VE1RP

Vice President – C.H. Baker, VE1HD

Secretary – Ray Wilson, VE1WL

Treasurer – D.M. Copp, VE1NO

Bulletin Editor – Doug Johnson, VE1OM

1953

President – Doug Johnson, VE1OM

Vice President – unknown, VE1DG

Secretary – Trevor Burton, VE1CP

Treasurer – George Sandoz, VE1PT

Bulletin Editor – Brit Fader, VE1FQ

George, VE1PT, resigned shortly after his appointment because of commitments at work. An HARC activities committee was formed in 1953 with VE1RY as chairmen and the members were Fritz Webb, VE1DB, Ralph Beach, VE1WD, Jack Mather, VE1LY and Don Bain, VE1LZ.

1954

President – Shep Shepherd, VE1RR

Vice President – Harry McLean, VE1ED

Secretary – Ralph Beach, VE1WD

Treasurer – Binks Fisher, VE1AFN ex VE3EA

Bulletin Editor – Brit Fader, VE1FQ

1955

Bulletin – Brit Fader, VE1FQ

1956

President – Jack Mather, VE1LY

Vice President – Fritz Webb, VE1DB

Secretary – Binks Fisher, VE1AFN

Treasurer – Wes Street, VE1EK

Bulletin Editor – Brit Fader, VE1FQ

1957

President – Wes Street, VE1EK

Vice President – Harold Bishop, VE1OB

Secretary – Binks Fisher, VE1AFN

Treasurer – Ian Macleod, VE1GC

Bulletin Editor – Brit Fader, VE1FQ

1958

Bulletin – Brit Fader, VE1FQ

1959

President – Ray Wilson, VE1WL

First V.P. – Father Henry Boudreau, VE1HY

2nd V.P. – Tommy Baker, VE1SF

Secretary – Binks Fisher, VE1AFN

Treasurer – Norm Weedmark, VE1QV

Bulletin – Brit Fader, VE1FQ

1960

President – Tom Clahane, VE1SP

Vice President – Cyril Boudreau, VE1RJ

Secretary – Binks Fisher, VE1AFN

Treasurer – Perry Bauchman, VE1YQ

Bulletin – Brit Fader, VE1FQ

1961

President – Norm Weedmark, VE1QV

First V.P. – Cyril Boudreau, VE1RJ

2n^d V.P. – H. E. "Danny" Danielson, VE1MM

Secretary – Binks Fisher, VE1AFN

Treasurer – Perry Bauchman, VE1YQ

Bulletin Editor – Brit Fader, VE1FQ

1962

President – Perry Bauchman, VE1YQ

Vice President – Danny Danielson, VE1MM

Secretary – Harold Bishop, VE1OB

Treasurer – Art Wentzell, VE1YE

Bulletin Editor – Binks Fisher, VE1AFN

Ham-of-the-Month

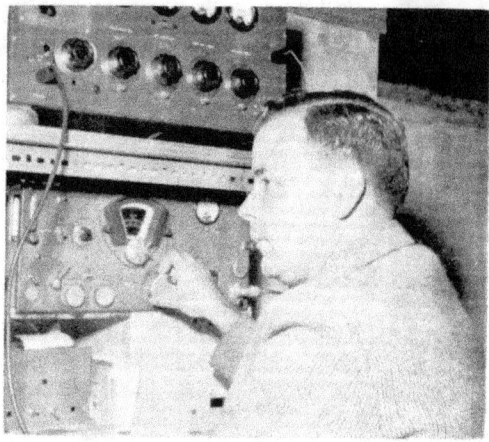

is H. E. "Danny" Danielson, VE1MM, of Halifax, Nova Scotia.

Danny was born at Strasbourg, Saskatchewan, in 1927 and got his senior matriculation there. In 1946, he graduated from the Provincial Normal College in Moose Jaw and taught school in Saskatchewan for the next three years.

Prior to this period, however, in 1944, Danny enlisted in the RCAF as a "Boy Airman" and this early association with the Air Force led later to six years of flying with the RCAF as an aircrew Radio Officer During 1950-51 he served with the Communications and Rescue Flight, Edmonton, and as Staff Instructor No. 2 Maritime Operational Training Unit at Greenwood, N.S. from 1951-53. In 1953 he was transferred to 404 Buffalo Squadron, also at Greenwood, and has fond memories of trips to Ireland, England, Greenland, Bermuda, the Azores and the Northwest Territories.

Then came June 1955 when Danny joined CBHT at video operator. He received his amateur radio license in 1960 and his phone ticket in January of 1961.

On the technical side, his equipment consists of a Collins 32G rig, running 65 watts input. Though his operation is on 80 - 40 - 20 metres, he enjoys 80 metre DX — his best DX on 80 being Switzerland and Czechoslovakia. He also has WANS No. 6 for working all Nova Scotia counties.

The local boys report Danny has a receiver which only receives signals that are 579... Could this be right?

— Don Bain

OCTOBER 1961 ● PAGE 19

HARC Files

This is from a publication titled CQ – CBC HAMS!

This was a Canada wide publication.

Thank you Bill, VE1MR, for the fine job of scanning this.

1963

President – Danny Danielson, VE1MM

First V.P. – Orest Chaban, VE1AFQ

2n^d V.P. – Wylie Barrett, VE1YN

Secretary – Binks Fisher, VE1AFN

Treasurer – Brian Pass, VE1AHR

Bulletin – Brit Fader, VE1FQ

There were 53 licensed amateur radio operators and 5 short wave listeners as members of HARC.

1964

President – Doug Johnson, VE1OM

First V.P. – Ian McLeod, VE1GC

2n^d V.P. – Joe Kennedy, VE1AJE

Secretary – Binks Fisher, VE1AFN

Treasurer – Harold Bishop, VE1OB

Technical Manager – Orest Chaban, VE1AFQ

Bulletin – Brit Fader, VE1FQ

There were 65 licensed amateur radio operators and 5 short wave listeners as members of HARC according to one report on file. This is a list of the HARC members dated February, 1964.

1.	VE1AAC	Gerald Harris	29.	VE1 IZ	Don Bain
2.	" ADR	Bob Grantham	30.	" MM	Danny Danielson
3.	" AFN	Charles Fisher	31.	" NW	George Cooke
4.	" AFQ	Orest Chaban	32.	" NX	Doug Shaffner
5.	" AFZ	A. G. Giffin	33.	" OB	Hal Bishop
6.	" AHD	Walt Jakeman	34.	" OC	Aaron Solomon
7.	" AHG	Bugs Grundy	35.	" OD	Elmer Naugler
8.	" AHQ	Frank Hulsman	36.	" OG	Scott Mosher
9.	" AI	Dick Grantham	37.	" OI	M. F. Helpard
10.	" AIC	Elvin Veale	38.	" OK	Martain Raine
11.	" AIH	Art Gunn	39.	" OM	Doug Johnson
12.	" AIX	Gordon Sanborn	40.	" PT	George Sandoz
13.	" AJE	Jim Gray	41.	" QI	F. Hunt
14.	" AMI	Mike Clarke	42.	" RJ	Cyril Boudreau
15.	" AML	Eileen Jakeman	43.	" RW	E. S. Campbell
16.	" DB	Fritz Webb	44.	" SD	John MacEwen
17.	" DD	W. Borrett	45.	" SP	Tom Baker
18.	" DR	Frank Johnson	46.	" UB	Tom Wilcox
19.	" EH	Walt Huskins	47.	" WD	Ralph Beach
20.	" EK	Wes Street	48.	" WL	Ray Wilson
21.	" EO	Francqis DeLisle	49.	" YE	Art Wentzell
22.	" FQ	Brit Fader	50.	" YN	Wylie Barrett
23.	" GC	Ian MacLeod	51.	" YQ	Jean Bilodeau
24.	" GI	Maurice Rickard	52.	" ZP	Joe Wild
25.	" HL	Tom Logan	53.	" ZQ	Earle Smith
26.	" IF	Bob Schultz	54.	" SWL	Irving Balcom
27.	" JX	Jan Klyn	55.	" "	Bill Nickerson
28.	" KG	Surge Szpilfogel	56.	" "	Graham Powell
			57.	" "	Eric Simms
			58.	" "	Bernie Flynn

HARC Files

There is a wealth of memories looking back through the pages of time over this list. My first was when I was six years old and remember the great time SWL Bill Nickerson and I had on his mother's kitchen floor. That was a fine-looking set of homemade toy bobsleds. Bill was from Grafton and Sylvia was from Barrington Passage. Bill was simply too busy with university to get a VE1 call from hanging out with this group. He managed to participate in three field days with them from 1955 to 1958. He went to Ontario later in 1964 and became VE3IBC. After a few years in Ontario, he went on to Winnipeg, Manitoba. His first VE4 call was VE4AKJ but he found the phonetics difficult with Parkinson disease and asked for VE4DL. He found that call much easier to pronounce. He was a member of the Winnipeg Amateur Radio Club (WARC) but had to more or less terminate his activity with ARES because of the Parkinson. He participated with ARES in the 1997 flood. Medical Doctor Peter Nickerson (internal medicine) who has specialized in kidney transplants and lectures around the world as professor/researcher/lecturer with the University of Manitoba is their son. William Nickerson, VE4DL became a silent key in May 2012. Jean Bilodeau,

VE1YQ, must have been away and returned to Nova Scotia. He was VE1IJ four years before in 1960. All Canadian amateur radio operators should have a two letter call permanently as long as they are Canadian. Anything less is discrimination. They should have the call of their choice if available. If they like dits, CH5HH, or dahs, XO0OO, Etc. One assumes E. S. Campbell, VE1RW, is the one and the same as 1DJ, VE1QQ and VE1AHQ.

1965

President – Ian Macleod, VE1GC

First V.P. – Frank Johnson, VE1DR

2nd V.P. – Walter Jakeman, VE1AHD

Secretary – Binks Fisher, VE1AFN

Treasurer – Art Gunn, VE1AIH

Membership – Graham Powell, VE1ANI

Bulletin – Brit Fader, VE1FQ

President Ian, VE1GC, stated in correspondence that HARC had 66 members.

OFFICERS OF CITY RADIO CLUB — The Halifax Amateur Radio Club has elected a new slate of officers. Carrying out the business of the local ham operators for 1964-1965 will be, left to right, Binks Fisher, secretary; Ian MacLeod, new president; and Walter Jakeman, 2nd vice-president. (DeBaie, Halifax Photo Service).

1966

President – Dick Grantham, VE1AI

First V.P. – Art Gunn, VE1AIH

2nd V.P. – Mike Goldstein, VE1ADH

Secretary – Binks Fisher, VE1AFN

Treasurer – Gerry Harris, VE1AAC

Bulletin – Brit Fader, VE1FQ

Dick, VE1AI, was the youngest president HARC ever had and Dick, VE1AI, claimed it was the youngest executive that HARC had ever had. Art Gunn, VE1AIH, moved to Montreal in May, 1966, and the first vice president position was left open.

NEW CLUB OFFICIALS — Members of Halifax Amateur Radio Club have elected a new slate of officers. Looking over some of their "rare" DX cards are Mike Goldstein, second vice-president; Dick Grantham, new and youngest, president of the club, and C. "Binks" Fisher, secretary. (Bryden, Halifax Photo Service)

1967

President – Gerry Harris, VE1AAC

First V.P. – Don Bower, VE1AMC

2nd V.P. – Graham Powell, VE1ANI

Secretary – Binks Fisher, VE1AFN

Treasurer – Art Smith, VE1AQS

Bulletin – Brit Fader, VE1FQ

The CBC Halifax Club Executive for 1967 was:

President – Tom Wilcox, VE1UB

Vice President – Doug Shaffner, VE1NX

Secretary Treasurer – Fred Hunt, VE1QI

1968

President – George Cousins, VE1TG

First V.P. – Jim Shand, VE1ASN

2nd V.P. – David Carstairs, VE1ATC

Secretary – Binks Fisher, VE1AFN

Treasurer – Art Smith, VE1AQS

Bulletin – Brit Fader, VE1FQ

Membership – Don Bower, VE1AMC

Entertainment – Dave Carstairs, VE1ATC

Technical – Jim Shand, VE1ASN

1969

President – Jim Shand, VE1ASN

First V.P. – Ian MacLeod, VE1GC

2nd V.P. – Murray Alary, VE1ALS

Secretary – Binks Fisher, VE1AFN

Treasurer – Ralph Anderson, VE1ADK

Bulletin – Brit Fader, VE1FQ

Auditor – Doug Johnson, VE1OM

There were 36 members in 1969 and all were licensed amateur radio operators.

1970

President – Bob Shultz, VE1IF

First V.P. – Camille Maillette, VE1RO

2nd V.P. – Dick Grantham, VE1AI

Secretary – Binks Fisher, VE1AFN

Treasurer – Ralph Anderson, VE1ADK

Bulletin – Brit Fader, VE1FQ

Past President – Jim Shand, VE1ASN

Auditor – Doug Johnson, VE1OM

Membership – Peter Bradford, VE1ASX

Technical Chairman – Sauli Arosankari, VE1AIH ex OH5SG

Dick Grantham, VE1AI, had to resign all positions with the club due to work commitments at the April 17th, 1970, monthly meeting. Dick recommended Jim Christian, VE1ASR, who was appointed 2nd vice president and Program Chairman at the same time. Jim had recently received his advanced license permitting phone operation.

1971

President – Harley Grimmer, VE1MX

First V.P. – Barry Hyndman, VE1XW

2nd V.P. – Cam Maillette, VE1RO

Secretary – Binks Fisher, VE1AFN

Treasurer – Murray Alary, VE1ALS

Bulletin – Brit Fader, VE1FQ

Technical Chairman – Barry Hyndman, VE1XW

Membership – Clary Pelley, VE1AVP

Auditor – Doug Johnson, VE1OM

1972

President – Barry Hyndman, VE1XW

First V.P. – Dick West, VE1AGX

2nd V.P. – Clarey Pelley, VE1AVP

Secretary – Binks Fisher, VE1AFN

Treasurer – Murray Alary, VE1ALS

Bulletin – Brit Fader, VE1FQ

Membership – Tony Nichols, VE1MN

Total membership in 1972 was 32 members and all licensed amateur radio operators.

1973

President – Dick Grantham, VE1AI

First V.P. – Don Bower, VE1AMC

2nd V.P. – Jim Shand, VE1ASN

Secretary – Binx Fisher, VE1AFN

Treasurer – Murray Alary, VE1ALS

Bulletin – Brit Fader, VE1FQ

Chairman of Membership Committee – Morris Levy, VE1ZF

1974

President – Sauli Arosankari, VE1AIH ex OH5SG

First V.P. – Gerry Von Klein, VE1AYJ

2nd V.P. – Jim Shand, VE1ASN

Secretary – Binks Fisher, VE1AFN

Treasurer – Harley Grimmer, VE1MX

Bulletin – Brit Fader, VE1FQ

Membership – Morris Levy, VE1ZF

Gerry Von Klein, VE1AYJ, moved back to the United States and became WA9GYS.

1975

President – Brit Fader, VE1FQ

First V.P. – Bill Delehay, VE1BAD

2nd V.P. – Barry Mouzar, VE1AZX

Secretary – Binks Fisher, VE1AFN

Treasurer – Harley Grimmer, VE1MX

Bulletin – Brit Fader, VE1FQ

1976

President – Brit Fader, VE1FQ

First V.P. – Perry Bauchman, VE1AYZ

2n^d V.P. – Bill Delehay, VE1BAD

Secretary – Binks Fisher, VE1AFN

Treasurer – Harley Grimmer, VE1MX

Bulletin – Brit Fader, VE1FQ

Bill, VE1BAD, worked for Imperial Oil and was transferred to Toronto in July.

1977

President – Don Bower, VE1AMC

First V.P. – Mike Pothier, VE1AJP

2n^d V.P. – Clive Bagley, VE1AMZ

Treasurer – Bill Leithead, VE1ABR

Secretary – Binks Fisher, VE1AFN

Bulletin – Brit Fader, VE1FQ

Harley Grimmer, VE1MX, replaced Bill Leithead, VE1ABR, as Treasurer during the January 19th, 1977, HARC meeting but it does not state the reason for this change.

CBC Halifax Amateur Radio Club:

President – Tom Wilcox, VE1UB

Vice President – Baz Miller, VE1APP

Secretary Treasurer – Fred Hunt, VE1QI

1978

President – Don Bower, VE1AMC

First V.P. – Mike Pothier, VE1AJP

2n^d V.P. – Ralph Campbell, VE1QU

Secretary – Binks Fisher, VE1AFN

Treasurer – Harley Grimmer, VE1MX

Bulletin – Brit Fader, VE1FQ

The only change in the 1978 executive from the 1977 executive is that Ralph, VE1QU, replaced Harley, VE1MX, as Treasurer.

1979

President – Mike Pothier, VE1AJP – VE1UG

First V.P. – Bob Brown, VE1BFX

2nd V.P. – Ross Kaye, VE1ARE

Secretary – Binks Fisher, VE1AFN

Treasurer – Harley Grimmer, VE1MX

Bulletin – Brit Fader, VE1FQ

Mike Pothier was using his new VE1UG call sign at the September 19th, 1979, monthly meeting of HARC.

1980

President – Clive Bagley, VE1AMZ

First V.P. – Bob Brown, VE1BFX

2nd V.P. – Mel Lever, VE1BSH became VE1VX in 1983

Secretary – Binks Fisher, VE1AFN

Treasurer – Harley Grimmer, VE1MX

Bulletin – Brit Fader, VE1FQ

Public Relations – Scott Wood, VE1BLA became VE1QD

1981

President – Cliff McMullen, VE1BNK

First V.P. – Tom Nepjuk, VE1BSM

2nd V.P – John Turney, VE1BNL

Secretary – Clarey Pelley, VE1AVP

Treasurer – Terry Gillespie, VE1BMF

Bulletin – Brit Fader, VE1FQ

Public Relations – Scott Wood, VE1BLA

When I first met Cliff, VE1 **"Bad Navy Kid"**, he was a Chief Petty Officer in the Navy. This was back in 1960 when he was a Chief Radio Technician in *HMCS SWANSEA*. One of the Canadian Marconi CSR5 receivers quit working. I asked Chief McMullen to take a look at it. He came up to the radio room and the first thing he did was light his pipe. He unbolted the receiver and slid it out of its cabinet. Then he looked around to see who was looking and gave it a great slam on the table. He then bolted it back in saying "that will do more for it than I can". I'll be darned. The receiver worked "the finest kind" and I have had a good laugh for fifty years. Believe it or not I have seen that procedure stated in electronic manuals for those old tube receivers, claiming it often worked.

1982

President – Cliff McMullen, VE1BNK

First V.P. – Tom Nepjuk, VE1BSM

2nd V.P. – John Turney, VE1BNL

Secretary – Clarey Pelley, VE1AVP

Treasurer – Terry Gillespie, VE1BMF

Bulletin – Brit Fader, VE1FQ

One will note there is no change in the 1982 executive.

1983

President – Don Bower, VE1AMC

Bulletin – Brit Fader, VE1FQ

1984

President – John Perkins, VE1FH

Bulletin – Brit Fader, VE1FQ

1985

President – John Perkins, VE1FH

Bulletin – Brit Fader, VE1FQ

1986

President – John Perkins, VE1FH

Secretary – Bob Swinwood, VE1PQ

Bulletin – Brit Fader, VE1FQ

Aaron Solomon, VE1OC, records Bob, VE1PQ, as president in his Maritime News in December, 1986.

1987

President – Dave Nimmo, VE1NN

Bulletin – Brit Fader, VE1FQ

HARC membership was 87 and was up 50% from 1986 according to President Bob Swinwood, VE1PQ, in a letter he had written to all HARC members.

1988

President – Bob Swinwood, VE1PQ

First Vice President – Bill Elliott, VE1MR

2nd Vice President – Jim Cleveland, VE1CHI

Secretary – Bernie Conrad, VE1BLM – became VE1PT

Treasurer – Mel Lever, VE1VX

Bulletin Editor – Tom Fullerton, VE1CES

Tom stated he had mailed 110 Bulletins at the January meeting. The call sign VE0MMA was received for *ACADIA*. Bob, VE1PQ, informed the November meeting that 1988 HARC membership was 78 members just 9 members less than 1987.

1989

President – Bob Swinwood, VE1PQ

First Vice President – Bill Elliott, VE1MR

2n^d Vice President – Jim Cleveland, VE1CHI

Secretary – Bernie Conrad, VE1BLM

Treasurer – Mel Lever, VE1VX

Bulletin Editor – Tom Fullerton, VE1CES

1990

President – Terry Galaugher, VE1KWG

1991

President – Jack Kiuru, VE1ZK

1992

President – Martha DeVanney, VE1LE

1993

President – Rick Gardiner, VE1RGG

1994

President – Bob Swinwood, VE1PQ

First Vice President – Rob Ewart, VE5BE – became VE1KS

2n^d Vice President – Basil Coady, VE1JEB

Treasurer – Andy Hodder, VE1BV

Secretary – Pearson Friars, VE1SWL

1995

President – Gerry Burnett, VE1GUN

1996

President – Bill Elliott, VE1MR

1997

President – Bill Elliott, VE1MR

1998

President – Bill Elliott, VE1MR

1999

President – Al Penny, VO1NO

2000

President – Al Penny, VO1NO

First V.P. – S. Marsden, VE1YB

2nd V.P. – Helen MacRae, VE1HMR

Secretary – Rick Gardiner, VE1RGG

Treasurer – J. Fowler, VE1JHF

Member at large – L. Riley, VE1LFR

Club Station Manager – Pierce Friars, VE1SWL and L. Riley, VE1LFR

2001

President – Al Penny, VO1NO

First V.P. – S. Marsden, VE1YB

2nd V.P. – Trevor Bast, KA8ZUO/VE1

Secretary – Nigel Service, VE1NPS

Treasurer –Jeremy Fowler, VE1JHF

Member at large – Dave Nimmo, VE1NN

Club Station Manager – T. Hemming, VE1RX

2002

President – Bill Elliott, VE1MR

First V.P. – Dave Nimmo, VE1NN

2nd V.P. – Murray MacDonald, VE1MMD

Secretary – Wayne Ernst, VE1GPK

Treasurer – Jeremy Fowler, VE1JHF

Member-at-Large – Terry Gillespie, VE1RQ

Club Station Manager – John Goodwin, VE1CDD

Past President – Al Penny, VO1NO

2003

President – Dick Grantham, VE1AI

First V.P. – Murray MacDonald, VE1MMD

2nd V.P. – Trevor Bast, KA8ZUO/VE1

Secretary – Wayne Ernst, VE1GPK

Treasurer – Jeremy Fowler, VE1JHF

Member-at-Large – Howard Dickson, VE1DHD

Club Station Manager – John Goodwin, VE1CDD

Past President – Bill Elliott, VE1MR

2004

President – Bill Elliott, VE1MR

First V.P. – Fraser MacDougall, VE1WO

2nd V.P. – Rick Gardiner, VE1RGG

Secretary – Howard Dickson, VE1DHD

Treasurer – John Goodwin, VE1CDD

Member-at-Large – Tom Caithness, VE1GTC

Club Station Manager – Pat Kavanaugh, VE1PHK

Past President – Dick Grantham, VE1AI

2005

President – Bill Elliott, VE1MR

First V.P. – Fraser MacDougall, VE1WO

2nd V.P. – D. Perrin, VE1HUP

Secretary – Howard Dickson, VE1DHD

Treasurer – John Goodwin, VE1CDD

Member-at-Large – Tom Caithness, VE1GTC

Club Station Manager – Pat Kavanaugh, VE1PHK

2006

President – Bill Elliott, VE1MR

First V.P. – Rob Ewert, VE1KS

2nd V.P. – Howard Dickson, VE1DHD

Secretary – Murray MacDonald, VE1MMD

Treasurer – Fraser MacDougall, VE1WO

Member-at-Large - Vacant

Club Station Manager – Brian Allen, VE1AZV

Past President – Dick Grantham, VE1AI

2007

President – Bill Elliott, VE1MR

First V.P. – Dick Grantham, VE1AI

2nd V.P. – Peter Whalen, VE1PJW

Secretary – Helen MacRae, VE1HMR

Treasurer – Fraser MacDougall, VE1WO

Director-at-Large – Doug LeBlanc, VE1LDL

Club Station Manager – Brian Allen, VE1AZV

Past President – Bob Swinwood, VE1PQ

2008

President – Bill Elliott, VE1MR

First V.P. – Dave Nimmo, VE1NN

2nd V.P. – Tom Gaum, VE1BMJ

Secretary – Helen MacRae, VE1HMR

Treasurer – Fraser MacDougall, VE1WO

Director-at-Large – Doug LeBlanc, VE1LDL

Club Station Manager – Brian Allen, VE1AZV

Past President – Bob Swinwood, VE1PQ

Bill Elliott, VE1MR, has been the longest serving president of HARC. He felt it time he had a break and did not re-offer to serve another term as president.

2009

President – Murray MacDonald, VE1MMD

First V.P. – Brian Allan, VE1AZV

2nd V.P. – Tom Gaum, VE1BMJ

Secretary – Helen MacRae, VE1HMR

Treasurer – Rod Padmore, VE1BSK

Director-at-Large – Don Trotter, VE1DTR

Club Station Mgr. – John Goodwin, VE1CDD

Past President – Bill Elliott, VE1MR

It is a shame Helen MacRae, VE1HMR, became a silent key on May 27, 2017 and only 73-years old.

2010

President – Murray MacDonald, VE1MMD

First V.P. – Bill Elliott, VE1MR

2nd V.P. – Sheldon Page, VE1SJP

Secretary – Howard Dickson, VE1DHD

Treasurer – Keith Landra, VE1STN

Director-at-Large – Don Trotter, VE1DTR

Club Station Manager – Wayne Harasimovitch, VE1WPH

Past President – Dick Grantham, VE1AI

2011

President – Fraser MacDougal, VE1WO

First V.P. – Bill Elliott, VE1MR

2nd V.P. – Rod Padmore, VE1BSK

Secretary – Betty Caithness, VE1BSW

Treasurer – John Goodwin, VE1CDD

Director-at-Large – Sheldon Harling, VE1GPY

Club Station Manager – Wayne Harasimovitch, VE1WPH

Past President – Bob Swinwood, VE1PQ

2012

President – Fraser MacDougal, VE1WO

First V.P. – Bill Elliott, VE1MR

2nd V.P. – Betty Caithness, VE1BSW

Secretary – Vacant and done by the President

Treasurer – John Goodwin, VE1CDD

Director-at-Large – Doug LeBlanc, VE1LDL

Club Station Manager – Jim Flowers, VE1JIM

2013

The following is the header of the January issue of the Reflector. This lists those elected at the November 2012 club meeting when the annual election of club executive is made:

Executive

President - Scott Wood, VE1QD, 823-2761 ve1qd@rac.ca
First V.P. - Bill Elliott, VE1MR 865-8567 bowser.elliott@ns.sympatico.ca
2nd V.P. - **VACANT**
Secretary - Jeremy Fowler, VE1JHF VE1JHF@Gmail.com
Treasurer - John Goddwin, VE1CDD 865-5731 VE1CDD@eastlink.ca
Co-Treasurer – Jim Guilford, VE1JG 466-2124 jimguilford@gmail.com
Director-at-Large: - Doug LeBlanc, VE1LDL 465-4665 leblanc@accesscable.net
Club Station Mgr. - Jim Flowers, VE1JIM 443-8657 ve1jim@ns.sympatico.ca
Past President - Fraser MacDougal, VE1WO 865-4198 fmac@ns.sympatico.ca

Committees/Offices/Prime Contacts

Government liaison – VACANT
QSL Bureau Mgr - Tom Caithness, VE1GTC 477-7081 tom.caithness@ns.sympatico.ca
EMO Coordinator - Dave George, VE1AJP 466-8723 dgeorge@dal.ca
EMO Trailer coordinator – David Musgrave, VE1EDA 435-4333 ve1eda@rac.ca
Reflector editor - Lynn Bowser, VE1ENT 865-8567 bowser.elliott@ns.sympatico.ca
Reflector Dist. - Carol Wood, VE1HAZ carolwood@accesswave.ca
Membership - Carol Wood, VE1HAZ carolwood@accesswave.ca
Web page – Brandon, Detmers VA1BSD brandon.detmers@dal.ca
Basic ham course - Erik Hein, VE1JEH 826-7145 bravojeh@gmail.com
2012 Flea market Chair – Murray MacDonald, VE1BB
Callbook Editor – Howard Dickson, VE1DHD, 823-2024 dhdickson@eastlink.ca
Field Day coordinator – **VACANT**
Safety Officer - Terry Bigelow, VE1TRB ve1trb@eastlink.ca
NSARA Director - Barry Diggins, VE1TRI 861-3719 ve1tri@rac.ca
Honorary Legal Counsel – Paul Radford, VE1ARH

Non-Club Contacts

RAC Atlantic Director - Everett Price, VO1DK vo1dk@rac.ca
RAC Section Manager – Craig Seaboyer, VE1DSS cseaboye@stfx.ca
RAC Assistant Director for HRM Scott Wood, VE1QD, 823-2761 ve1qd@rac.ca

2013 will be the last time I record the HARC Executive. This is recorded in every issue of the Reflector and copies of the reflector are held at the club and archives.

This is HARC President, Scott Wood, VE1QD, in 2013.

HARC MEETINGS WERE HELD AT THE FOLLOWING LOCATIONS

1933

The 1933 meetings were all of the Maritime Amateur Radio Association and held in Room B at the **Nova Scotia Technical College**

1934

The MARA was renamed HARC on **March 16th, 1934** and this meeting was held at the **Nova Scotia Technical College**.

April 20th so few members showed up that the meeting was cancelled

The meetings for the rest of the year were held at the **N.S. Technical College** until:

October 26th after a supper at the **Green Lantern**

November 23rd meeting was held at the **N.S. Technical College**

December 16th meeting was held at the home of **Trevor Burton, VE1CP**

1935

January 22nd meeting was held at the home of **Wes Street VE1EK**

February no meeting

March meeting held at the home of **John Doull, VE1FN**

April meeting was held at the home of **Bill Horne, VE1GL**

May 3rd at the home of **Bill Bligh, VE1BC**

June, July, August no meeting

September 27th meeting at the home of **Bill Horne, VE1GL**

October 25th at the home of **Trevor Burton, VE1CP**

November meeting was held at the home of **Cliff Short, VE1AW**

December no meeting

1936

January meeting was held at the home of **Trevor Burton, VE1CP**

February meeting was held at the home of **Doug Smith, VE1FO**

March no meeting

April meeting was held at the home of **Hal Ward, VE1GR**

May 20th at the home of **Bill Bligh, VE1BC**

June 17th at the home of **Art Grant, VE1EP**

July 25th at the home of **A. A. "Steve" Stevens, VE1EC,** in Dartmouth

August 11th at the home of **Doug Smith, VE1FO**

September 23rd at the home of **Wes Street, VE1EK**

October 30th at the home of **Ralph Fraser, VE1HJ**

November 27th at the home of **Bill Bligh, VE1BC**

December 16th at the **YMCA**

After supper they went up to the fourth floor and had a look at the new HARC club room.

1937 – 1940

YMCA

This was the old YMCA building on Barrington Street later known as the CPR building.

1941 – 1945

Moir's Cafeteria on Market Street

According to Brit, VE1FQ, meetings were held sporadic during the war but all at this location.

Brit also stated the building was on Grafton Street.

Halifax-HRM West Community Herald Volume 2 Number 35 Monday May 25, 2009, page 2

Moir's chocolate factory was at this site from when it founded in 1830 until it moved to Woodside in 1975.

This building was constructed in 1927. This is the site of the metro center in 2009.

1946 – 1948

Moir's old chocolate factory on Argyle Street

The factory must have covered Market, Grafton and Argyle Streets

On the 3rd Friday each month

Thanks to Don Watters VE1BN former VE1QG

This is a photograph of an HARC meeting in the Moir's Chocolate Factory taken in August 1946.

Left to right in the front row:

Oscar Sandoz, VE1QZ, Ron Hart, VE1MZ, Art Crowell, VE1DQ, Mr. Nolan or Jean Paul Roy a Department of Transport Radio Inspector, Walt Wooding, VE1ET, unknown, Ralph Fraser, VE1HJ, unknown and Fritz Webb, VE1DB.

Left to right second row:

Unknown, Don Watters, VE1QG, unknown, unknown, unknown, unknown, unknown and unknown

Left to right third row:

Unknown, Scott Mosher, VE1OG, unknown, unknown, unknown, unknown, Aaron Solomon, VE1OC, unknown, unknown, Bob Langford, VE1QR

Left to right the fourth and fifth row or the remainder in the back:

Unknown, Don Topple, VE1RU, Pete Payzant, VE1QF, unknown, unknown, unknown, Reg Hamm, VE1RH, unknown, Steve Malcom, VE1SW, unknown, unknown, Jack Whitley, VE1OK, unknown, unknown, unknown and unknown

January to October 1949

Old Wren Barracks

This was the geology lab. Of Dalhousie University

This was at the corner of Coberg Road and Oxford Street

October 1949 to October 1951

RCAF Gorsebrook Barracks

At the North side of the old Gorsebrook Golf Course

South Street

November 1951 to October 1958

Cathedral Barracks

Morris Street

This is the site of the Rehab Center on University Avenue in 2009

November 1958 to January 1968

The Old Halifax Police Station on Brunswick Street

One had to show their HARC membership card to the duty officer in order to gain admittance to the meetings. This was a great way to make certain one and all were paid up members.

January 1968 to February 1977

Nova Scotia Technical College

Spring Garden Road

The first meeting at this site was January 19th, 1968.

The VE1FO Club Station was set-up at this location in 1968 and has operated in all the club meeting locations since then. This has included an outdoor tower at each location.

The last meeting at the Nova Scotia Technical College was held on February 19th, 1977.

N.S.I.T., Leeds Street

The first meeting at N.S.I.T. was held on March 16th, 1977.

1977 – 1985

N.S.I.T., Leeds Street

1986

T.U.N.S., Leeds Street

1987 – 1988

Teacher's Resource Center, 6225 Chebucto Road

The last meeting at this location was on September 21st, 1988

1988

Queen Elizabeth High School

There were two meetings at this location, the one on October 19th, and the other on November 16th, 1988.

1989 – 1991

Public Archives, 6016 University Avenue

T.U.N.S. (Technical University of Nova Scotia) November 20th, 1991

1992 – 2007

Bloomfield Community Centre, corner of Almond and Robie streets

The Bloomfield Community Centre, the former Bloomfield High School

The home of HARC from 1992 until 2007

The VE1FO radio station and meeting rooms were on the second floor.

2007 – 2018

Saint Andrew's Community Centre, Bayer's Road

The first meeting at Saint Andrews Community Centre was on September 19th, 2007.

The Community Centre did not move but from construction around the area it was given a new address as: 3380 Barnstead Lane, Halifax in 2010.

Saint Andrew's Community Centre, the former Saint Andrew's High School

The home of HARC from 2007 until 2018

The top window on the right corner was the home of radio station VE1FO. The window just to the left of the radio station was the club library. The window just to the left of the library was the home of the VE1FQ QSL Bureau. There was a lecture room just behind these three rooms with no window. The club meeting room was on the lower floor and the windows of this room can be seen just through and behind the shrubs in the fence. The base of the club antenna tower was inside that fence.

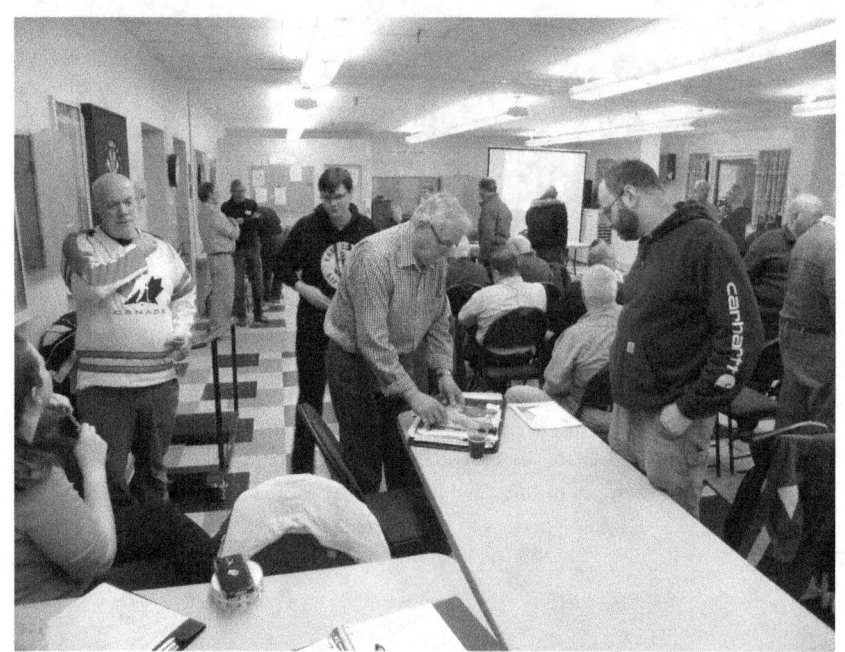

These three photographs are of those in attendance at the February 21st, 2018 club meeting at 3380 Barnstead Lane, Halifax during the ten-minute break. The club station was on the second floor, the floor above this room.

2018 – Present

The Halifax Amateur Radio Club moved again in 2018 to Fire Station 50 in Hammonds Plains. The antennas and towers were moved on June 9, 2018. The first meeting in this club room was held on June 20th, 2018.

The HARC meetings were held in the large room on the second floor on the left of the building with the three windows as shown in this photo. The VE1FO club radio station, the VE1FQ QSL bureau and the club library were located in small rooms/offices down the left side of the building on the second floor. The Halifax Regional Municipality is involved with this club because the club is involved with the Emergency Measures Organization and provides communications in any emergency.

This is the first meeting of HARC at Hammonds Plains on June 20, 2018.

HARC club meetings were cancelled because of the COVID 19 pandemic and the club station was closed. On Wednesday, January 20th, 2021 the club meeting was held via ZOOM on the computer. I felt one would be able to watch this meeting although I do not have a camera or microphone for my computer. I managed to tune it in but the sound was very poor so I gave up and did not watch it. My tinnitus consists of a 6,000 hertz steady sizzle at 68db and makes it very difficult to hear especially something like this. Hopefully those who could hear enjoyed the meeting.

HARC FIELD DAYS

Field Day is the largest and most popular event in the Canadian and American amateur radio world. Field day is mentioned in nearly every monthly meeting in one way or another and every issue of QST has one thing or another that mentions field day in one way or another. Therefore, I contacted ARRL thinking they would have a record of all field days. The only record ARRL has kept is that recorded in the various issues of QST. They told me the **first Field Day was 1933** and that there were none during World War II.

The three main or better-known events in amateur radio are; the DX competition, the Sweepstakes competition and the Field Day competition. The DX competition was first started in 1927 and the sweepstakes in 1930. Here is another case where the records do not agree. One early record state the above dates and in the September, 1986, issue of QST it states the DX contest was created in 1937. In other words, are both correct and it is my interpretation that is in error. The DX competition became the DX contest in 1937.

Some of these events were held for over a week when first created but have gradually been cut back so that each is held over one weekend only. HARC members have participated in all three over the years but Field Day is the only one that involves the whole club. And therefore, is the only one I have recorded in detail.

The rules for each Field Day are found in the May or June issue of QST for each year.

The 1933 rules state that one should call CQ FD in CW and call CQ FIELD DAY on phone. The point system for 1933 worked as follows:

> A. up to and including 20 watts multiply by 3
> B. over 20 watts up to and including 60 watts multiply by 2
> C. over 60 watts multiply by 1

These first field days were quite different from the field days held today and those that have been held for several decades. Most of the equipment back then was "home brew" and was made by each operator including any generators that powered the equipment. There were no transistors and each piece of radio equipment had tubes. Most of the stations put out less than 50 watts. It was hard to tell whether one was within the assigned amateur band with this equipment. The lucky ones were the ones with crystal control. One did not make 3 contacts a minute. They were lucky to make a contact every 15 minutes. One had to call CQ and then listen over the whole band for a reply.

The most popular bands were the 7 and 3.5 mc bands that we call 40 and 80 meters today. About a third of that number was found on the 14 and the 56 mc bands. That is 20 meters to we today and the 56 mc band was known as 5 meters. There was a bit of activity on the 1.7 mc band or 160 meters in today's terminology. Most of the activity was CW and most of the little phone activity was found on 5 meters.

1933 – This first field day was held on June 10th and 11th, 1933. The Field Day record for 1933 is recorded in the September 1933 issue of QST. Club Scores: VE3KC – The London Amateur Radio Club was listed 2nd with a score of 770 points. Note the VE3KC call sign on Field Day 1934. VE5EZ – The Victoria Short Wave Club was listed 6th with a total score of 24 points. VE5 was the British Columbia call sign prefix at that time. Other Leading Scores: VE3GT – was listed 16th with a score of 663 points. These three were the only VE stations recorded for that year. There were records for a total of fifty stations that reported their Field Day results for 1933 the first year of Field Day.

1934 – The detail on the 1934 Field Day was recorded in the September 1934 issue of QST. VE3KC – Western Ontario Amateur Radio Association was 7th overall with a score of 210 points and the only VE recorded. This same VE3KC call sign was in use in 1933 as the London ARC.

There was a special note that the VE stations required Special Permission to work portable on frequencies other than the 56-mc band. This band was known as 5-meters.

A record in the history of the Hamilton Amateur Radio Club, VE3DC, in the book from spark to space on page 94 states that their club participated in this 1934 Field Day. They state that they were 3rd overall and the only VE that participated in this Field Day. QST does not mention them. Note that the Hamilton Club is using call sign VE3KM in 1935, 1936 and 1937 but nothing matches for their claim for 1934 in the book from spark to space.

1935 – Wes Street, VE1EK, stated the 1st HARC Field Day was 1935. The 1935 Field Day Scores are recorded in the September 1935 issue of QST. VE3KM – the Hamilton Amateur Radio Club was 5th overall with 594 points and the 1st Canadian station. There were six Canadian stations that participated in the 1935 Field Day according to QST. VE3KM, VE3GT, VE3TM, VE3SG, VE3GI and VE2CO are the six stations and there is no record of a VE1 station.

1936 – There is no record of HARC participating in this Field Day. The January 1937 issue of QST describes the 1936 field day. VE3KM was 3rd overall with a score of 612 points. The Moncton Amateur Radio Club, VE1DC, was the only VE1 entry and they were 12th overall with a score of 188 points.

1937 – The record for this year is a bit of a mess. There is none for VE1MK, the HARC station that received that call sign in February. The only VE1 call listed is VE1DC, the Moncton Amateur Radio Club. They are listed as the 4th Canadian station, 56th overall with a score of 188 points. The same score they had the year before. In the text describing this field day there is a photograph of the VE3GT field day station listing all the operators. They state this station had a score of 936 points with 59 QSO's and was the 2nd highest Canadian station, but the station does not appear in the list of stations and no station has 936 points on this list. The 1st Canadian station was VE3KM, the Hamilton Amateur Radio Club with 1,341 points, 113 QSO's and 9th overall.

We found a list of the Halifax Amateur Radio Club Field Day Sites in the station files. This states the first HARC Field Day was held in 1938 and gives the site of each Field Day up to and including 1988. I will include the site with each yearly entry as follows:

1938 – Boy Scout Camp, Millers Lake – HARC with call sign VE1MK scored 477 points and was the 4th Canadian station and 62nd overall. This Field Day was recorded in the December 1938 issue of QST as simply Field Day Participation. The best HARC DX was a G call sign worked in the United Kingdom. Bill Bligh, VE1BC, was the high scorer. There were nine operators operating VE1MK according to the September issue of QST and they were: Bill Bligh, VE1BC, Doug Smith, VE1FO, Brit Fader, VE1FQ, John Roue, VE1FB, Harry Scott, VE1KB, Ralph Fraser, VE1HJ, Ron Hart, VE1MZ, John Doull, VE1FN and Ainsley Croft, VE1EU.

1939 – Dr. Morton's camp, Upper Sackville – The HARC station, VE1MK, made 77 contacts. They had a power rating of up to 20 watts and had 3 complete stations in use. The operators are listed as Barclay Dowden, VE1HK, Gordon Phelan, VE1KG, Ralph Fraser, VE1HJ, Ainsley Croft, VE1EU, Ron Hart, VE1MZ, Bill Bligh, VE1BC, Jim Mathews, VE4AFT, Harry Scott, VE1KB, Dr. John Morton, VE1JM/VE3ALK and Doug Smith, VE1FO. They earned a total of 1,071 points. Dr. John Morton's call sign was VE1JM and he was first licensed in 1936. Dr. Morton transferred to Ottawa and held the two call signs for several years. Jim Mathews, VE4AFT, was stationed here in Halifax with the Royal Canadian Navy.

QST December 1939 page 41: "We operated from a camp at Upper Sackville, Nova Scotia, about 20 miles from Halifax. Three complete units were operated simultaneously, one each on 3.5, 7 and 14 MCS. The equipment was all within 100 feet of one point, the 3.5 and 14 MCS rigs being approximately 20 feet apart and the 7 MC rig about 70 feet away from the other two. All operation was from battery power. *VE1MK – the Halifax Amateur Radio Club*"

VE1MK was 58th overall and the 4th Canadian station among the amateur radio clubs.

The Hamilton ARC, VE3KM, was 1st Canadian station with 2,907 points. The Montreal ARC, VE2CO, was the 2nd Canadian station with 1,539 points. The West Side Radio Club, VE3JJ, was third with 1,224 points and they had the same 77 contacts as HARC but were running the one station only so this brought the points up. HARC had 1,071 points as recorded above. I would say HARC put in a fine performance for 1939.

QST also lists a score of 360 points for Cliff Short's station, VE1AW, and states the operators were Cliff Short, VE1AW, Fritz Webb, VE1DB, Wes Street, VE1EK and Eddie MacLaughlin, VE1JH.

1940 – 1945 – No Field Day was held because of World War II.

1946 – HARC earned 900 points on Field Day according to the minutes of one of the HARC monthly meetings and according to the Maritime news section of QST. The operators were: Ralph Fraser, VE1HJ, Fritz Webb, VE1DB, Wes Street, VE1EK, and Harry Scott, VE1KB. The station used call sign VE1FO/1 and was powered by a gas-driven "chore horse". The power output rating was A – that translated to up to 30 watts. The antenna was a half-wave Zepp on 3.5 Mc and a three-element beam on 28 Mc. Fifty-one contacts were made, twenty-four being Field Day stations.

A Zepp antenna was a copy of the antenna used on the zeppelin air ships that were quite popular before World War II and carried a radio officer/operator.

1947 – HARC was the 1st Canadian and the 9th station overall in the three transmitters operating simultaneously category. HARC with call sign VE1FO/1 had three transmitters operating simultaneously in class A, up to 30 watts, and earned a score of 4,320 points.

1948 – Sambro Creek – June 12th and 13th, 1948 – HARC was the 1st Canadian station and 28th overall with 2,538 points in the three transmitters operating simultaneously category. This involved 253 contacts in class A – up to and including 30 watts. Twenty members of HARC operated three stations; 3.5, 7 and 14 Mc using the VE1FO/1 call sign. The Dartmouth Club with call sign VE1DN/1, the Loyalist City Amateur Radio Club, Saint John, the Annapolis Valley Amateur Radio Club and the Yarmouth Club also had stations participating in this field day. Yarmouth was using call sign VE1HN/1 and ran a war surplus army 19 set powered by a gas driven generator. The station list states that Don Bain, VE1LZ, had a station at ST. Margaret's Bay. Wes Street, VE1EK, operated at home for this field day per this QSL card.

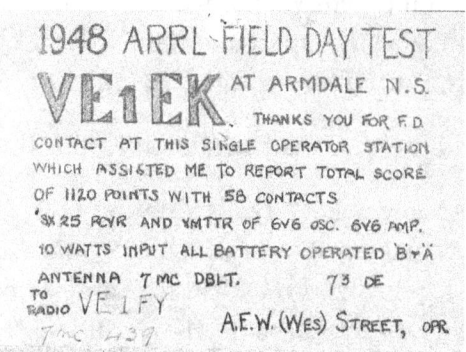

Thanks to George Crowell VE1LB ex VE1FY

Someone went to a lot of trouble in typing up the HARC Field Day Log for 1948. This is the first and last page of five lengthy pages.

1745

FIELD DAY REPORT OF VE1FO - HALIFAX AMATEUR RADIO CLUB.

NO. OF TRANSMITTERS IN USE AT ONE TIME - - - 3   80,40 AND 20 METERS CW

POWER INPUT TO ALL TRANSMITTERS 29 WATTS

RECEIVERS IN USE ON ALL BANDS - ---- BC 348'S

SOURCE OF SUPPLY FOR RECEIVERS - - - 6 VOLT STORAGE BATTERIES

           45 VOLT B BATTERIES

SOURCE OF SUPPLY FOR TRANSMITTERS   KATOLIGHT 500 WATT POWER PLANT.

NO. OF OPERATORS - - - - - 7   VE1HJ,VE1FQ,VE1BK,VE1RP,VE1CP,VE1HT,VE1AS

LOCATION - - - - - SAMBRO CREEK   20 MILES SOUTH WEST OF HALIFAX N.S.

ANTENNAS USED - 80 METERS   80 METER ZEPP

        40 METERS   VEE BEAM 275' PER LEG

        20 METERS   TWIN TRI-PLEX BEAM

- -

80 METER LOG

DATE	TIME	STATION	HIS REPT.	OUR REPT.	OPERATOR
JUNE 12	1618 A.S.T.	VE1WU	559	569	BK
	1635	VE1ME	569	599	BK
	1645	VE1OL	559	579	BK
	1650	VE1JK	579	569	BK
	1800	VE1EA	589	599	FQ
	1808	VE1QN	599	589	FQ
	1833	VE1HN	57 FONE	589	FQ
	1840	W1HSW	569	579	HT
	1900	VE1KN	599	599	HT
	1905	VE1CW	599	599	HT
	1909	VE1UU	579	579	HT
	1925	VE1LZ	579	459	HT
	1950	VE1WA	559	579	AS
	2014	K1NRE-1	569	559	BK
	2030	W1KKS-1	579	579	BK
	2103	VE1VJ	589	589	BK
	2108	W2AR-2	559	569	BK
	2118	W1LXT-1	559	459	BK
	2141	W1BJP-1	569	569	BK
	2155	VE1ET	589	599	BK
	2159	W1OSA-1	569	569	BK
	2240	W1ACT-1	569	569	BK
	2307	W1HSW	579	579	FQ
	2311	VE2XR	569	479	FQ
JUNE 13	0001	W3QV	579	579	CP
	0015	W2UBU-2	589	579	CP
	0030	W1RBG-3	559	569	CP
	0038	VE1JK	579	589	CP
	0108	W2SV-2	569	559	HT
	0115	W1KAE-1	569	579	HT
	0120	VE1EV	569	589	HT
	~~0125~~	~~W1RRB~~	NO CONTACT ALREADY WORKED		
	0131	W4KEK	449	--- QRM	BK
	0150	W2AK	579	379	BK
	0207	W1DEO	469	579	BK

```
VE1FO                            20 METER LOG   CONTINUED.
      DATE       TIME      STATION    RST REPT.    RST REPT.   OPERATOR
JUNE 13      0940 A.S.T.   W2AD-2       589          589         RP
             1005          W3AFU-3      559          579         RP
             1015          W2WIW        559          579         RP
             1030          W2RTL        559          579         AS
             1115          W2SGK        579          599         AS
             1120          W3WOX        569          599         AS
             1150          GI5UR        449          469         AS
             1215          W6AMT-6      579          579         RP
             1220          VE2CB        559          569         RP
             1233          W5FGE-5      449          579         RP
             1250          W2ALO-2      559          589         RP
             1300          W2ANW-2      559          469         RP
             1305          W1KKS-1      589          579         RP
             1340          W2GSN        589          579         RP
             1346          VE3BNG-S     579          579         RP
             1410          W0EMS      4-559        4-559         AS
             1416          W8OCA        559          569         AS
             1429          W2AGM        579          559         AS
             1425          W2AW         559          565         AS
             1432          W3OMX        559          569         AS
             1444          W1HLK        559          579         AS
             1459          W5SCQ        449          579         AS
             1507          W3NEW        559          569         AS
             1515          W2QYV        559          559         AS
             1522          W9EHS        559          579         AS
             1532          G3CSL        449          579         AS
             1545          W9AIV        559          579         AS


             NO CONTACTS.   97


TOTAL NO. CONTACTS   65
                     91
                     97
                  --------
                     253
      TOTAL

MESSAGES ORIGINATED   1

MESSAGES RECEIVED     4

MESSAGES RELAYED      0
```

HARC Files

I found a couple of old logs within the HARC Files. My favourite is the one of Frank Higgins, VE1KY. His was an old RCN Cypher Log book. His first entry was in 1937 and his last is in 1960. Brit, VE1FQ, is the second entry in 1937 on June 6th, 1937. Frank had worked Murray, VE1KW, on June 3rd, 1937. All three stations were in the city of Halifax.

One of the many things I found interesting in Frank's old log is that he has recorded his frequencies simply as 80 meters, 20 meters, and so on in meters and not frequency in either kilocycles or megacycles. In the various other records, I found back then everything is in megacycles such as 3.5, 7.0 and so on. This old VE1FO log records meters so they must have used meters in the logs and frequency in megacycles elsewhere. You will note there is no actual frequency recorded in these logs. The majority of these old transmitters were "Rock Bound" or crystal tuned and on the one frequency only. This may be the reason the frequency was not that important.

1949 – Dr. Sieniewicz's camp, Hubbards – June 18th and 19th – HARC using the VE1FO/1 call sign and managed 299 contacts in the three transmitters operating simultaneously category with a score of 2,916 points. HARC was the first Canadian station and 26th overall. There were twelve operators operating in the class "A" category – up to and including 30 watts.

1950 – Fritz Webb's cottage, Queensland – June 24th and 25th – HARC was the 1st Canadian station and 29th overall in the three transmitters operating simultaneously category with a score of 3,204 points from 320 contacts. They were using the VE1FO/1 call sign and had 15 operators operating in the class "A" category – up to and including 30 watts.

1951– Fritz Webb's cottage, Queensland – June 23rd and 24th – HARC with the VE1FO/1 call sign had a score of 2,439 points in the three transmitters operating simultaneously category. They had 271 contacts in the class "A" category – up to and including 30 watts. There were 15 operators operating the three transmitters. The 15 operators are shown in the photograph below. The scores were normally listed in QST starting with the highest to lowest scores. Whoever recorded this Field Day listed them alphabetically by call sign. It would take one a lot of time to figure out the actual HARC VE1FO overall standing. HARC VE1FO was the third Canadian in the three transmitters operating simultaneously category. Lakeshore ARC VE2NI had 4,293 points and Quinte ARC VE3NW had 2,502 points, only 66 points higher than HARC VE1FO.

H.A.R.C. FIELD DAY JUNE 1951
BACK ROW L-R RALPH VE1WD, BRIT VE1FQ, HOWIE VE1HC, DICK VE1HT DOUG VE1OM, ROY VE1WL, FRITZ VE1FB, UNKNOWN VE1, PAT VE1RP, DOUG VE1PQ.

H.A.R.C. FIELD DAY JUNE 1951
FRONT ROW L-R TOMMY VE1SF, WYLIE VE1KM NOW VY2YN, DON VE1LZ, GEORGE VE1PT, RALPH VE1HJ.

HARC Files

Fritz Webb was VE1DB and not VE1FB as noted on this photograph

1952 – CBC camp, Hubbards – June 21st and 22nd – HARC was the 3rd Canadian station and 41st overall in the three transmitters operating simultaneously category with a score of 2,367 points from 230 contacts. HARC had three gas driven plants in use by 14 operators in the class "A" category – up to and including 30 watts. 80-meter mobile was very popular back then and Brit Fader, VE1FQ, had to talk Art Crowell, VE1DQ, in to the site via their mobile rigs in their cars. Art did not know the location of the VE1FO/1 site. Club Radio Amateur de Hull, VE2BY/2, was the first Canadian station, 31st overall, in this category with 2,754 points and the London Amateur Radio Club, VE3AT/3, was the second Canadian station, 32nd overall, with 2,700 points.

1953 – CBC camp, Hubbards – June 20th and 21st – HARC was the 5th Canadian station and 55th overall in the three transmitters operating simultaneously category with a score of 2,007 points. VE1FO/1 had 181 contacts with 15 operators in the class "A" category – up to and including 30 watts. Yarmouth ARC VE1GM/1 had 2,672 points, Sackville ARC VE1GH/1 had 2,457 points, Calgary ARC VE6NQ/6 had 2,259 points and Verdun ARC VE2CB/2 had 2,022 points. They were the four stations ahead of HARC in this category.

1954 – Don Bain's cottage, Glen Margaret – June 19th and 20th – HARC was the 1st Canadian station and 26th overall in the three transmitters operating simultaneously category in class "A" up to and including 30 watts with a score of 3,321 points. 14 operators made 329 contacts for these points. This record rated a listing of the VE1FO/1 call at the beginning of the Field Day report as a top station in its area.

The caption of this photograph reads: "Halifax Amateur Radio Club, 3321 points in Class 3A with the club-assigned call VE1FO/1, posed with this display of equipment while breaking camp at Glen Margaret, N.S." It went on and gave the call signs only and I will try and include the names. Sitting left to right: Jack Mather, VE1LY, Brit Fader, VE1FQ, musician VE1SI and I do not have his name, Tommy Baker, VE1SF, and Wes Street, VE1EK. Standing left to right: Holland Sheppherd, VE1RR, and

Page 49, December 1954 QST

Binks Fisher, VE1AFN but listed here simply as ex VE3EA, Doug Johnson, VE1OM, Ray Wilson, VE1WL, Fritz Webb, VE1DB, Ralph Beach, VE1WD, Ralph Fraser, VE1HJ, and Harry McLean, VE1ED. Harry was from the old school and wore a neck tie at field day. Don Bain, VE1LZ, took and submitted the photograph. This is 12 of the 14 operators and counting Don one of the operators is missing. Note the Vibroplex bug sitting on top of the old BC-348 receiver. That is a good pick-up load of equipment because that old gear could be heavy.

The editor of QST stated in the January, 1955, issue that there were 8,384 people operating 2,026 amateur radio stations for solid 24-hours in the 1954 ARRL Field Day activity etching the Kennedy-Heawiside layer indelibly with 'CQ FD'. They all were having the time of their lives and that sounds like a good description to me. That guitar at the VE1FO/1 site would really loosen things up.

1955 – Don Bain's cottage, Glen Margaret – June 25th and 26th – HARC was the 4th Canadian station and 58th overall in the three transmitters operating simultaneously category with a score of 3,339 points. This record was from 334 contacts with 12 operators in class "A" up to and including 30 watts. The London ARC VE3YJ/3 was the first Canadian station and 28th overall with 522 contacts for 4,425 points. The Scarborough ARC VE3BXT/3 was the 2nd Canadian station and 53rd overall with 368 contacts and 3,537 points. The Vancouver ARC VE7ARV/7 was the 3rd Canadian station and 55th overall with 400 contacts and 3,420 points. The author of the 1955 Field Day report stated that 10,190 amateur radio operators participated in this annual test of emergency powered portables.

1956 – Swan Lake, Windsor Highway – June 23rd and 24th – HARC was the 1st Canadian station and 34th overall in the three transmitters operating simultaneously category with a score of 3,132 points. There were 15 operators operating class "A" – up to and including 30 watts. VE1FO/1 was the top Canadian station in Class "A" Call Area Leaders and therefore was listed within the square listing the leaders on the first page of the Field Day record. The ARRL stated that not only was the weather horrible in most areas but radio conditions were very poor as well.

The ARRL stated in the area of the instructions for this field day, that a moored balloon for an antenna over 6 feet in diameter with a capacity greater than 116 cubic feet would require permission from the Civil Aeronautics Authority. It also stated to make certain that it was not released or it would create an international incident. The cold war was on and civil defense was very big. Think of the "brownie points" some civil defense leader could get for shooting that "sucker" down!

1957 – Swan Lake, Windsor Highway – June 22nd and 23rd – HARC was the 2nd Canadian station and 42nd overall in the three transmitters operating simultaneously category with a score of 3,195 points. 15 operators made 330 contacts in class "A" – up to and including 30 watts. The Vancouver ARC, VE7ARV/7 was the 1st Canadian station and 27th overall in this category with 20 operators to make 566 contacts for 4,140 points.

1958 – Glen Margaret at Merriman's – June 28th and 29th – HARC was the 2nd Canadian station and 120th overall in the three transmitters operating simultaneously category with a score of 1,593 points. HARC had 14 operators that made 150 contacts in class "A" – up to and including 30 watts. The Vancouver ARC, VE7ARV/7 was the 1st Canadian station again but 51st overall this year. They had 13 operators, 478 contacts for 3,375 points.

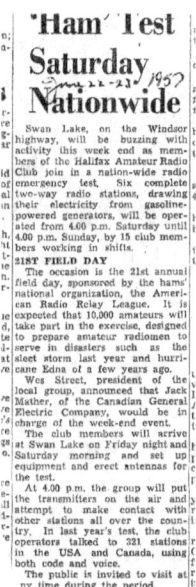

HARC Files

One could find a note like this in the newspapers on nearly every field day.

No doubt Cyril, VE1RJ now VE1AFB, had something to do with that back then.

Halifax Amateur Radio Club, VE1FO/1 (VE1IF keying)

QST 1958

This is Bob Shultz, VE1IF, operating the HARC field day station, VE1FO/1.

In 1958 Bob had his complete station operating from a Hudson Bomber.

Bob moved his shack to the top of a hill in early 1961 with a 70' tower.

Bob was also one of the first to operate amateur TV in this region.

Bob had the TV transmitter at this location on the hill and it was operated by VE1AFQ.

In late 1961 or early 1962 Bob had added colour to his TV transmitter and they tried DX; a signal from Rawdon to Moncton.

Bob lives in Mount Uniacke in 2008 and was HARC president in 1970.

ARRL stated there were 11,316 amateur operators that participated in this 1958 Field Day exercise. The Sydney ARC, VE1AEP, was the top score in the Maritimes with 2,250 points in the **four** transmitters operating simultaneously category.

This note appeared in QST: HARC took along the Halifax Civil Defence 40-foot trailer, 10-kw gas plant, and 6-foot portable tower, as well as club gear, for a full workout. This was a 100 per cent trailer operation, quite a **change** from the cosy cabins and cottages of former years – Halifax ARC, VE1FO.

Wes Street, VE1EK, stated that the Coordinator, Major J. Vickery accompanied the operators and watched them do their stuff in spite of the unfavourable radio conditions they experienced. Don Weeks, VE1WB, stated in his monthly report for September that conditions on field day were a big disappointment.

1959 – Swan Lake, Windsor Highway – June 27th and 28th – HARC was the 8th Canadian station and 106th overall in the three transmitters operating simultaneously category with a score of 2,244 points. HARC made 271 contacts with 10 operators operating in both class "A" and class "B". Class "A" was up to and including 30 watts. Class "B" was over 30 watts and up to and including 150 watts. There were 13,137 amateur radio operators that participated in the 1959 Field Day activities.

1960 – Rawdon Hills – June 25th and 26th – HARC was the 8th Canadian station and 120th overall in the three transmitters operating simultaneously category with a score of 2,250 points. HARC operated with 14 operators in class "B" and made 375 contacts. Class "B" was over 30 watts and up to and including 150 watts. There were 13,488 amateur radio operators that participated in the 1960 Field Day activities.

1961 – Rawdon Hills – June 24th and 25th – HARC was the 6th Canadian station and 68th overall in the three transmitters operating simultaneously category with a total of 3,564 points. This record was accomplished with 15 operators making 568 contacts in class "B"; a power output of over 30 watts and up to and including 150 watts. There were 3,000 stations and a total of 13,750 people operating in this field day

Page 16 QST December, 1961

This is a direct quote from page 21 of the November, 1961, issue of QST. "Statistically, the Canadian picture is an interesting one, with VE activity simply soaring skyward this year. Numerically speaking, Canadian's account for only about 4% of the total W/VE ham population, yet the VE Class A entries account for 7.5% of the total . . . meaning that in proportion to the number of Canadian to U.S. amateurs, Canadians had nearly 100% more participation or did twice as well as U.S. hams. If the W/K entries equated the Canadian participation, we'd have roughly twice as many logs to score! Think of that a minute, will you? Congratulations to the VEs, quite a battle for top score in VE1-land engaged with [HARC] VE1FO/1, [Pictou County ARC] VE1JV/1, and [St. Croix Valley ARC] VE1PF/1 scoring 3564, 3546, and 3537 respectively. VE3DOH/3, the Windsor Amateur Radio Club, led the eight-transmitter category, while the Scarboro Amateur Radio Club, sporting a new abbreviated call, VE3WE/3, wrested top-VE laurels from the Nortown Amateur Radio Club, VE3NAR – both groups in the 10-transmitter class."

1962 – Rawdon Hills – June 23rd and 24th – HARC was the 5th Canadian station and 84th overall in the **two** transmitters operating simultaneously category with a total of 2,484 points. This was via 389 contacts in class "B" – a power output of over 30 watts up to and including 150 watts. There were 20 operators operating these two transmitters.

Ellen White, W1YYM, wrote the record of this field day in QST. She stated there were a record 15,000 amateur operators that reported this activity, they had 3,464 transmitters in use and submitted 1,450 logs. There must have been some rain in most areas of the field day activity because she made this statement, I find rather cute:

"If it did not rain where you were, where were you?"

1963 – Rawdon Hills – June 22nd and 23rd – HARC was the 3rd Canadian station and 83rd overall in the two transmitters operating simultaneously category with a total of 2,826 points. There were 446 contacts made by 15 operators in class "B" – more than 30 watts and up to and including 150 watts. There were 15,654 participants in this field day operating 3,815 transmitters.

1964 – Rawdon Hills – June 27th and 28th – HARC was the 1st Canadian station and 16th overall in the two transmitters operating simultaneously category with a total of 5,250 points. This record was from 850 contacts with 10 operators operating in class "B" – more than 30 watts and up to and including 150 watts.

There were 14,757 participants in the 1964 field day operating 3,454 stations that included 1,510 field day score listings. The weather was fairly good for all W/VE stations participating in this field day.

Murray Alary, VE1ALS
Field Day 1964

1965 – Rawdon Hills – June 26th and 27th – HARC was 96th overall in the two transmitters operating simultaneously category and the 7th Canadian station with a total of 3,120 points. This involved 493 contacts made by 15 operators in class "B" – more than 30 watts and up to and including 150 watts. 1,460 logs were submitted to ARRL for this field day. There were 3,400 stations and 14,200 participants. HARC was beaten by the Annapolis Valley participants. They were 80th overall and using Gordon Banks VE1IM call sign.

1966 – Rawdon Hills – June 25th and 26th – HARC was the 1st Canadian station and 43rd overall in the two transmitters operating simultaneously category with a total of 4,868 points. There were 18 HARC operators that made 728 contacts in class "B" – more than 30 watts and up to and including 150 watts. There were 1,339 reports submitted to ARRL, from 3,266 stations by 13,600 participants. One will note the numbers appear to be decreasing from those during the years previous to this.

Marty Raine   Jim Payne   Fred Hunt   Chuck Adams   Tom Wilcox   Murray Helpard
ex VE1OK   VE1QI   ex VE1AFB   ex VE1UB   VE1OI - SK
now VE6TS   now VE1RG   now VE1SDU

Spud VE1BC – Tom, VE1UB, mailed this card to me about 1978. Bill, VE1MR, put the caption on the photo March 17th, 2009.

The amateur radio operators at CBC Halifax formed their own club with VE1UB as president in 1966. The CBC Amateur Radio Club operated the VE1CBC Repeater in the Halifax area. This photo was taken on Friday June 24th, 1966, just before they set out for their field day site at Harrietsfield. They were using President Tom's VE1UB call sign for this field day at Harrietsfield. They made 384 contacts in class A and B; up to 30 watts and over 30 watts up to 150 watts. Their twelve operators earned 2,924 points. They were 133rd overall and the 6th Canadian station in the 2 transmitters operated simultaneously category; the same as HARC. One has to agree that it is not bad for a dozen broadcast technicians acting as radio operators.

1967 – Rawdon Hills – June 24th and 25th – HARC was the 8th Canadian station and 71st overall in the three transmitters operating simultaneously category with a total of 5,192 points. 18 operators in Class "B" – over 30 watts up to and including 150 watts – made 797 contacts to obtain this record. There were almost 1,300 entries submitted to ARRL from about 3,050 stations that included 15,100 participants in the 1967 Field Day.

The Dartmouth ARC operated in the two transmitters operating simultaneously category and managed 570 contacts creating 3,840 points with 14 operators in Class "B" using call sign VE1HE/1.

1968 – Patton Road, Lower Sackville – June 22nd and 23rd – HARC was 50th overall and the 4th Canadian station in the two transmitters operating simultaneously category with a total of 5,616 points. HARC made 836 contacts with 18 operators operating in class "C" – over 50 watts up to and including 200 watts. There were 3,117 stations in operation this field day and 12,200 people participated that submitted 1,227 logs.

Gerry Harris, VE1AAC, does not recall the year but he remembers the HARC field day held on Patton Road, out in Lower Sackville very well. It was a terrible night. There was lots of heavy rain, thunder, and lightning flying off the antennas all night. There were lots of nervous guys around but Brit Fader, VE1FQ, was as happy as a pig in a mud puddle. He was the only one who would go out in the lightning to check the antennas. He was laughing the whole night. Thought it was great. It turned out to be one of the best years they ever had.

Don Bower, VE1AMC, remembers that field day event very well. He was driving a 66 Plymouth at the time so it was between 1966 and 1974. As St Elmo's fire danced all over the place and water was running over and through the power cables and coax on the floor of the EMO Volkswagen Camper. He remembers Dick Grantham, VE1AI, sending code by using a wooden lead pencil to push an old-fashioned straight key for fear of getting electrocuted from either above or below. He recalled that just about everyone got stuck trying to get out of a very soggy hayfield. He pulled several people out with his car. C. G. "Binx" Fisher, VE1AFN, was there as the generator quit in the middle of the night. Murray Alary, VE1ALS, and Don, VE1AMC, went and fixed it. Binx was sleeping on the floor and Don stepped on him in the dark going through the old EMO trailer. Binx was not impressed.

Dick Grantham, VE1AI, remembers that Sackville field day and not only using a pencil but keeping his feet off the steel floor and not touching anything. They had a lightning strike right in the field where they were operating. *But all in all, a really good time was had by all.*

1969 – Rawdon Hills – June 28th and 29th – HARC was the 2nd Canadian station and 29th overall in the two transmitters operating simultaneously category with a total of 7,074 points. 16 operators made 1,069 contacts in order to accumulate those points. They did this in both class "B" – over 10 watts and up to and including 50 watts – and in class "C" – over 50 watts up to and including 200 watts.

The set-up time period for this field day was to be recorded by each group participating in this field day and forwarded to ARRL. This could provide additional points by setting up quickly. Of course, the main object of any field day is to ensure the amateur radio equipment can be set-up quickly in case it is required for any emergency.

The 1st Canadian station in 1969, the Ottawa Valley Mobile Radio Club Inc. with call sign VE3RAM/3 was just ahead of HARC. Dick, VE1AI, stated at the monthly HARC meeting held on November 21st, 1969, that the VE1FO/1 score was 200 points short. The Secretary was requested to write ARRL and have this noted. A letter from ARRL was read at the January 16th, 1970, monthly meeting re this disputed field day score. "Evidently we did not make out the preamble to our field day message properly so lost the 200 points we claimed". HARC would have been ahead of the Ottawa Valley Mobile Radio Club Inc. There were

1,255 logs submitted for this field day that had 11,400 participants. There was no mention of the total number of stations involved.

Maritime News for July, 1969, claimed that HARC and the Dartmouth ARC provided a Field Day Trophy. This is another trophy that should be held in the HARC Club Room today unless the Dartmouth club is holding it in their club room.

Dartmouth ARC was assigned call sign VE1CG for the period of the Canada Games held in Halifax and Dartmouth in 1969.

1970 – Rawdon Hills – June 27th and 28th – HARC was the 21st station overall and the 2nd Canadian station operating in the two transmitters operating simultaneously category with a total of 8,687 points. VE1FO/1 accumulated these points with 16 operators and 1,314 contacts in class B; over 10 watts and up to and including 50 watts – and in class C; over 50 watts up to and including 200 watts. This field day submitted to ARRL a total of 1,313 logs from 11,762 participants using 3,259 stations making 753,765 contacts.

VE2ND/2 the Montreal Field Day Association was the 5th station overall in this two transmitters category with 14,211 points from 1,733 contacts by 7 operators only in the same two B and C categories.

The Dartmouth Amateur Radio Club with call sign VE1HE/1 was not that far behind HARC in 38th overall with 7,142 points from 1,057 contacts and 22 operators operating in class C only.

1971 – Rawdon Hills –– June 26th and 27th – HARC was 21st overall and the 1st Canadian station operating in section 2A with a total of 2,558 points. This record made the top ten listing within its area at the beginning of the Field Day records in QST. HARC with station VE1FO/1 managed this record with 18 operators making 1,154 contacts in class B; over 10 watts and up to and including 200 watts. There were 1,116 logs submitted to ARRL. 11,908 participants participated in this field day with 2,780 transmitters. This group made over 800,000 contacts.

1972 – Rawdon Hills – June 24th and 25th – HARC was 25th overall and the 1st Canadian station in the 2A category with a total of 2,764 points. This put the VE1FO/1 call in the class A Call Area Leaders Box at the beginning of the 1972 Field Day description. 15 operators managed 1,232 contacts in class B; over 10 watts up to and including 200 watts. There were only 1,007 logs submitted and no statistics were given for the number of transmitters or the number of participants. Hurricane Agnes was the big news for field day 1972. There was some consideration given to canceling or postponing field day. Agnes went up through the Eastern United States dumping a lot of rain on everyone and everything in its path. A number of field day participants terminated field day and went over providing emergency communications for flooding from this hurricane. Some of those who left field day to provide emergency communications came back for the last part of field day. In other words; Agnes made a mess of the field day record for 1972. There is a detailed description of

Agnes titled "Agnes has a Field Day" starting on page 68 and terminating on page 77 of the November, 1972, issue of QST.

1973 – Rawdon Hills – June 23rd and 24th – HARC was 28th overall and the 1st Canadian station with a total of 2,675 points in the 2A category. This record was created by 16 operators operating in class "B" – over ten watts and up to and including 200 watts – making 1,185 contacts. This also put the HARC station VE1FO/1 at the top for 2A in the Class A Call-Area Leaders Box at the beginning of the field day report. The Maritime News for September, 1973, stated that field day did not seem to have the participation that it had in previous years, yet in the field day report in November it claimed a 20% increase over 1972. There was a total of 12,221 participants, 1,132 logs submitted and 2,805 transmitters in operation.

1974 – Rawdon Hills – June 22nd and 23rd – HARC was the 2nd Canadian station and 30th overall in Section 2A with a total of 2,696 points right behind VE5NN, the Regina Amateur Radio Association at 29th overall. The VE1FO/1 station had 18 operators operating in class B; over 10 watts up to and including 200 watts. They made 1,198 contacts to give them the 2,696 points. VE5NN/5 was just ahead of VE1FO/1. They made 2,724 points in class B with 20 operators making 1,237 contacts. It took 2 more operators to bring Regina just above HARC in 29th position overall.

Field day 1974 had 2,650 transmitters operated by 11,903 participants and 1,126 logs were submitted to ARRL. Category 2A had 211 entries and 71.73% of those entries were operating phone. They claimed it was easier to get points via phone as opposed to cw.

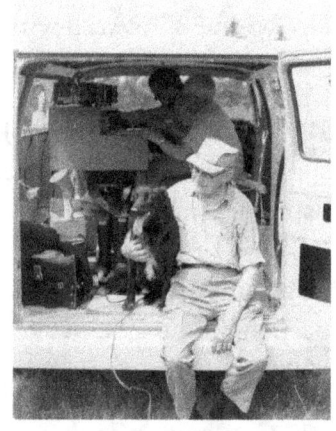

HARC Files

Charles G. "Binks" Fisher,
VE1AFN, and Mandy

Field Day 1974

1975 – Rawdon Hills – June 28th and 29th – HARC was the 1st Canadian station and 50th overall in the 2A category with a total of 2,890 points. 15 operators made 845 contacts in class B; over 10 watts and up to and including 200 watts. There was a change in the rules this year where one counted double points for each and every cw contact. This change may continue and be included in 1976. 10,406 operators submitted 1,107 log entries from 2,733 transmitters for the 1975 field day. VE1FO/1 was at the top of the Class A Call-Area Leaders Box at the beginning of the field day record.

1976 – Rawdon Hills – The 4th Saturday in June – June 26th and 27th – HARC was the 1st Canadian station and 22nd overall in the 2A category with a total of 4,496 points. 23 operators made 1,481 contacts in class B; over 10 watts and up to and including 200 watts. There were 1,323 reports filed with ARRL consisting of 3,372 transmitters from 16,120 participants. VE1FO/1 was again listed as the top station in the Class "A" Call Area Leaders block at the beginning of the records. Not only has the date been standardized to the 4th Saturday in June but the double points for CW contacts has been made a permanent score as well.

1977 – Rawdon Hills – The 4th Saturday in June – the 25th and 26th – HARC was the 1st Canadian station and 14th overall in the 2A category with a total of 5,248 points. HARC earned these points from 1,627 contacts with 30 operators in class B; over 10 watts and up to and including 200 watts. There was a big participation in this field day; 21,590 operators were operating 1,592 stations that included 4,194 transmitters. There were 307 entries in the HARC 2A category and since VE1FO/1 was the 1st Canadian the station was listed in the 2A section of the Class "A" Call-Area Leaders block at the beginning of the article on field day in QST for December, 1977.

1978 – Rawdon Hills – The 4th Saturday in June – the 24th and 25th – HARC was the 1st Canadian station and 6th overall in the 2A category with a total of 5,860 points. The 20 HARC operators made 1,743 contacts in class B. This was more than 10 but less than or equal to 200 watts. There were 22,319 ARRL operators operating 4,391 transmitters during the 1978 field day. There were 327 entries in the 2A category. VE1FO/1 was included in the box under 2A for the Class "A" Call-Area Leaders. Wes Street, VE1EK, had presented a meter to HARC that was to be used as an award for the best CW operator at field day. This meter had been mounted, a log was made up for it and it would have been available for this field day. A van was rented by HARC for the first time for this field day to help with the transportation.

1979 – Rawdon Hills –The 4th Saturday in June – the 23rd and 24th – HARC was the 1st Canadian station and 27th overall in the 2A category with a total of 5,144 points. 25 HARC operators had made 1,500 contacts in class B; greater than 10 watts and less than or equal to 200 watts to earn these points. There were 362 entries for the 2A section. There were 23,612 operators in the 1979 field day that operated 4,552 transmitters. VE1FO/1 was in the 2A section of the Class A Call-Area Leaders box at the beginning of the record.

1980 – The Ovens, Lunenburg – we found no record in any category for HARC – Apparently a few members of HARC simply participated in this Field Day at the Ovens but left no record of this activity. Brit, VE1FQ, advertised this field day in his Bulletin for June, 1980, as a one station set-up working presumably all bands. This was to be a sort of family outing as well. Bob Brown, VE1BFX, was to be in charge of this field day. Brit went on to state in his September Bulletin that field day had been observed in a low-profile manner. It took the form of a family camp out at The Ovens near Lunenburg and was spearheaded by Cliff McMullen, VE1BNK. One transmitter was operated on 10 through 80 meters and some 25 to 30 contacts were made. This field day was held on the fourth full weekend in June and fell on the 28th and 29th in 1980. There were 25,451 operators participating in this field day with 1,731 stations.

1981 – HARC did not participate in Field Day – This field day was held on June 27th and 28th the fourth full weekend in June. 23,816 operators participated in this field day with 4,559 stations. They sent ARRL 1,760 reports.

1982 – HARC did not participate in this Field Day – This field day was held on June 26th and 27th the fourth full weekend in June. This field day had 29,890 operators who sent in 1,735 reports but no mention was made of the number of stations in operation. Cliff McMullen, VE1BNK, was president of HARC and Dave Oldridge, VE1EI, was SEC during the 1981 and 1982 field days. I do not know if this had any effect on HARC not participating in field day. When Dave told me he had taken on the SEC position I wondered when he would find the time. We were practically living at Halifax Coast Guard Radio and the other members of HARC may have been as busy with their occupations. Dave was SEC from October 1st, 1980, until May, 1983.

1983 – This field day was held on the fourth full weekend in June the 25th and 26th. HARC did not participate in this field day. There were 25,552 amateur operators that participated in this field day with 4,217 transmitters that had 1,746 sites.

1984 – Hartlen Point – The fourth full weekend in June – the 23rd and 24th – HARC was the 2nd Canadian station and 73rd overall in the 2A category with a total of 3,688 points. HARC had 982 contacts with 25 operators in class "B" – less than 150 watts – to make these points. There were 443 entries in the 2A category and a total of 1,441 log entries were sent to ARRL. The record in QST does not list the number of stations or the number of participants. The one Canadian station that beat out HARC was the Ottawa ARC VE3RC and they did it with 40 operators in class "B" that made 4,440 points.

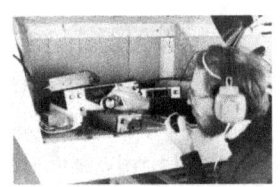

The six photographs above of the 1984 Field Day are from the HARC files.

1985 – Hartlen Point – The fourth full weekend in June – the 22nd and 23rd – HARC was the 1st Canadian station and 61st overall in the 2A category with a total of 4,266 points. 15 HARC operators made 1,294 contacts in class "B" – more than 10 watts and less than 150 watts – to earn these points. ARRL received 1,483 entries, 42 more than in 1984, and of this number 1,131 were operating with generators. The record keeper for this report was surprised there were so many generators and not more batteries. There were 457 stations operating in class 2A in 1985 and one will note 14 more than in 1984. There is usually someone different each year that records these field day reports and some list the total participation and some do not. This was a do not list year.

1986 – Hartlen Point – we found no history of any HARC participation on this Field Day in QST. Bill, VE1MR, claims the entry was mailed too late but feels this is accurate.

HARC was 1st in Canada and 31st overall in the 2A class with around 5,000 points. ASM Aaron Solomon, VE1OC, stated HARC had 15 operators in this field day and that it was at Eastern Passage. Hartlen Point is located in Eastern Passage. He recorded this record in his Maritime News for October, 1986. ARRL received 1,697 entries for this field day from 25,449 participants that had made 1,146,881 contacts for this 27-hour field day period. There were over four million points earned on this field day. 2A was again the most popular event at 515 or 30% of all the entries. 73% of the entries were using generators for power.

1987 – Hartlen Point – The fourth full weekend in June – the 27th and 28th – HARC was 39th overall and the 1st Canadian Station in the 2A section with a total of 5,014 points. HARC made 1,462 contacts with 20 operators in class 2 – less than 150 watts – to earn these points. The record for this field day in the November issue of QST gave a wealth of detail on this field day. The most popular entry in this field day was the 2A section, as usual, that had 544 entries. There were a total of 26,358 participants, 1,718 sites and 4,020 transmitters that participated in this field day. 1,451 sites were on complete emergency power for lights, radio equipment, etc. There were 1,196 sites using generators and 255 sites using batteries. Solar power was another popular source of power to run some of the sites. This field day produced 771,366 phone contacts and 446,772 CW contacts for a total of 1.2 million contacts. There were 758,000 bonus points for a grand total of over 4 million points. It sounds like a lot of fun was had by one and all. No doubt the mosquitos and bugs enjoyed this annual event immensely as well.

COWICHAN VALLEY AMATEUR RADIO CLUB

******** VE 7 CVA ********

P.O.Box 806,
DUNCAN, B.C.
V9L 3Y2

15th November, 1987

Halifax Amateur Radio Club,
VE1FO
Box 663,
Halifax, NS.
B3J 2T3

Dear Fellow Hams:

The members of the Cowichan Valley Amateur Radio Club offer our congratulations to the Halifax Radio Club for your outstanding performance in Field Day '87. Being the Canadian leader in your class (2A) is indeed an achievement worthy of pride.

We are newcomers to the 2A Class, but not to Field Day competition. We did faily well this year but look forward to a much better effort in 1988.

We'd like to challenge your Club: we think we can take over "top spot" in 1988! In a spirit of friendly "Coast-vs-Coast" Competition, we are serving notice that we will be gunning for the top honours next year.

Regardless of who "wins", we propose to send your club a small photo album of West Coast Field Day 88 snapshots for your collective amazement (amusement?) shortly after the last gasp of FD 88.

Will you join with us in this friendly effort? It should be fun!

Yours fraternally,

Ron Nicholson VE7DEV
for
Cowichan Valley Amateur Radio Club

Consrab
the video

THE HALIFAX AMATEUR RADIO CLUB

P. O. BOX 663
HALIFAX, NOVA SCOTIA

Comp. 5 Site 103
RR 1 Bedford. N.S.
B4A 2W9
March 17, 1988

Cowichan Valley ARC. VE7CVA
P.O. Box 806
Duncan, B.C.
V9L 3Y2

Dear Fellow Amateurs,

The members of the Halifax Amateur Radio Club thank the Cowichan Valley Amateur Radio Club for your letter of congratulations on our performance at Field Day '87. We are pleased to accept your challenge in the 2A Class for Field Day '88. HARC takes great pleasure in offering the CVARC our congratulations on your performance in achieving second place 2A Class in Canada on Field Day '87.

We are most pleased with the prospect of a West Coast challenge for we have become rather complacent with our wins in the 2A Class in Canada. However, we are certain that the "top honours" in 2A Class in Canada will remain with the HARC.

Your suggestion of a friendly "Coast-vs-Coast" competition has given us a further idea. We propose that a suitable trophy be made available for the winners of the 2A Class Club Group in Canada. We also propose that this trophy (plaque) be designed and co-sponsored by the CVARC and the HARC. The inscription could possibly read:

FIELD DAY #1 in CANADA
FOR
CLUB GROUP 2A CLASS

CO-SPONSORED BY THE COWICHAN VALLEY
AND HALIFAX ARCs'

Plates with the appropriate name would be installed for each year's winner. This trophy would then be a challenge trophy for the Canadian Amateur Radio Club Group competitors in the 2A Class for years to come.

If this meets with your approval we will proceed with the selection and design of the plaque. If you have additional suggestions let us hear from you.

Competitively yours, _VE1BLM_

Bernie A. Conrad, VE1BLM
Bernie A. Conrad. VE1BLM
Secretary. HARC

Box 228
Youbou, B.C. VOR 3EO

16 May 88

Dear Bernie;

Thanks very much for your letter of concurrance on the establishment
of the challenge for 2A activity on field day. We heartily agree with all
you suggestions.

Please proceed with the design and selection of the plaque. Upon
production kindly advise us of the costs and we will forward our share to you.

As to the possibility of "top honours" remaining with the HARC I
am afraid I must advise you that this is rather remote. The past couple of
field days the CVARC has been loafing along but now that there is a challenge
for us to do better, then we will gurd our loins and get on with it.

It would seem that Canada Post has been playing around again Bernie.
You letter dated 17 March did not arrive on this coast until early May and we
just processed it at our monthly meeting. Very sorry for the delay.

sincerely,

(K. E. Nicholson)President VE7DEY
C.V.A.R.C.

HARC Files

The plaque was purchased and each club paid 50% of the price. HARC managed to keep it and keep it until 2003! HARC entered the 2F section in 2003 and apparently everyone lost interest in the plaque.

1988 – Hartlen Point – This field day was held on the fourth full weekend in June – the 25th and 26th – HARC was 43rd overall and the **1st Canadian Station in the 2A** section with a total of 5,348 points in class 2 – less than 150 watts. (+ VE1VCU Jack Kiuru) was added to this score. 2A was again the most popular entry with 542 entries submitted. HARC with the VE1FO call sign participated with 35 operators and made 1,620 contacts to earn those 5,348 points. The Cowichan Valley ARC, VE7CVA, was so far behind VE1FO that I did not bother to count the number of their standing. It makes one wonder what happened to all their good intentions because they only had 587 contacts in class 2 with 18 operators and made 2,286 points. There were 27,471 participants in this field day operating 1,758 stations that made a total of 1,332,710 contacts. There were 891,000 SSB phone contacts, 442,000 CW contacts and 808,000 bonus points earned. This field day for the first time gave bonus points for working all 11 Canadian amateur radio stations with the suffix QST in the call sign. Only five of those stations reported to ARRL. The five were with the number of contacts: VE1QST – 636, VE2QST – 63, VE3QST – 1,262, VE5QST – 1,021 and VE7QST – 1,061.

1989 – Hartlen Point – the fourth full weekend in June – the 24th and 25th – HARC was the 1st Canadian station and 24th overall in the 2A category with a total of 6,506 points. (+ VE1MGT George Baker) was added to this score. VE1FO the HARC station had 30 operators in class 2 – less than 150 watts – on this field day that made 1,896 contacts. 2A was again the most popular event this field day with a total of 532 stations participating in this class. There were 28,701 participants at 1,742 sites who made 905,936 phone contacts and 465,922 CW contacts. This gave a total of 1.3 million contacts and 847,000 bonus points were earned. The total score earned by all entries was 4.7 million points.

1990 – Department of Highways site in Tantallon – the fourth full weekend in June – the 23rd and 24th – HARC (+ VE1RGG – Rick Gardiner) was the 1st Canadian station and 28th overall in the 2A category with a total of 6,264 points. The HARC station VE1FO had 50 operators in class 2 – less than 200 watts – and made 1,872 contacts. 2A was again the most popular section with 556 stations. One could get bonus points on this field day for making 10 contacts with slow speed novice CW stations. There were 30,861 participants in this field day with 1,828 stations. There were 883,687 phone contacts and 468,275 CW contacts making 1.35 million contacts. There were 903,000 bonus points making the total of more than 4.7 million points. 1,597 of the 1,828 stations were using 100% emergency power. 1,450 stations were using generators and 147 were on battery power.

1991 – Department of Highways site in Tantallon – the fourth full weekend in June – the 22nd and 23rd – HARC was the 1st Canadian station and 11th overall in the 2A category with a total of 7,288 points. The VE1FO station (+ VE1RIK Richard Atkinson) made 2,200 contacts in class 2 – less than 150 watts – to earn those points. Most of the HARC members must have participated in this field day because the record indicates 100 operators. The 2A category was again the most popular with a total of 546 entries. There were a reported 32,416 participants, at 1,835 sites operating 4,428 transmitters. There were 817,726 phone and 483,502 CW contacts. 1,772 of the sites were 100% emergency power, 1,630 generators were in use and 132 battery powered sites. It sounds like a good time was had by one and all.

1992 – Department of Highways site in Tantallon – the fourth full weekend in June – the 27th and 28th – HARC was 23rd overall and the 1st Canadian station in the 2A section with a total of 6,862 points. The most popular class again was the 2A class with 484 participants. VE1FO (+ VE1JMM Unknown) had 60 operators participate in class 2 – less than 150 watts – and they made 2,089 contacts to earn those 6,862 points. There were 37,278 participants in the 1992 field day with 1,940 sites running 4,541 transmitters. They made 826,403 phone and 512,049 CW contacts earning more than 1.33 million points. Of the 1,940 sites 1,424 were on complete emergency power, 1,470 were run by generators and 352 by batteries. This was another good field day and an excellent showing again by HARC.

1993 – Department of Highways site in Tantallon – The fourth full weekend in June – the 26th and 27th – HARC was the 1st Canadian station and 25th overall in the 2A category with a total of 7,618 points. VE1FO (+ VE1RUR Unknown) with 75 operators made 2,338 contacts in class 2 – less than 150 watts – for these 7,618 points. The 2A category was the most popular again with 636 entries in this category. There were 39,552 participants in this field day operating 2,086 stations at 1,620 sites. 1,503 stations were operated by generators, 518 by batteries and 542 were operated from natural power. A total of 962,580 phone contacts were made and 529,288 CW contacts were recorded for a total of 1.49 million contacts.

1994 – Department of Highways site in Tantallon – the fourth full weekend in June – the 25th and 26th – HARC was the 1st Canadian station and 26th overall in the 2A category with a total of 6,820 points. VE1FO (+ VE1LIA Lyman Allen) made 1,948 contacts in 2A with 75 operators in class 2 – less than 150 watts – for these 6,820 points. There were 36,208 participants that sent ARRL 2,058 entries for this field day. There is no record of the number of sites or the number of transmitters. 1.46 million QSO's were made on this field day and 65% were phone and the rest CW.

HARC Files

VE1FO Field Day 1994

HARC Files

VE1FO Field Day 1994

HARC Files

VE1FO Field Day 1994

HARC Files

VE1FO Field Day 1994

1995 – Department of Highways site in Tantallon – the fourth full weekend in June – the 24th and 25th – VE1FO (+ Dave McHattie, VE1RCN) was 15th overall and the 1st Canadian station in the 2A section with a total of 6,944 points. 32 operators made 2,098 contacts in order to earn those 6,944 points in class 2 – less than 150 watts. 36,000 attended this field day and 1.37 million contacts were made. The most popular section was the 2A section again with 644 stations in that section. VE1FO at the 15th spot was an excellent showing.

1996 – Department of Highways site in Tantallon – the fourth full weekend in June – the 22nd and 23rd – HARC was 91st overall and 1st in Canada in the 2A section with a total of 4,342 points. The 2A class was again the most popular with exactly 600 entries for that class, so VE1FO at 91st is one heck of a good standing. 32,043 amateurs participated in this field day. There were 1,928 sites operating 4,579 transmitters that made 761,609 phone contacts and 471,476 CW contacts. 1,528 stations were running everything including the coffee pot on emergency power. 1,309 were using generators for the power and there were 373 stations operating on battery power. VE1FO made 1,413 contacts with less than 150 watts of power and they had 60 operators to make those 4,342 points.

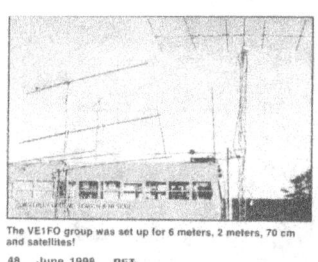

The VE1FO group was set up for 6 meters, 2 meters, 70 cm and satellites!

48   June 1998   QST.

1997

1997 – Department of Highways site in Tantallon – June 28th and 29th – the fourth full weekend in June – HARC was 23rd overall and 1st in Canada in the 2A section with a total of 6,658 points. The 2A section was the most popular again with 594 entries so HARC was right up next to the top. VE1FO (+ Mike Cousineau, VE1SCM) had 31 operators in class 2 – less than 150 watts, that made 1,837 contacts to earn those 6,658 points. There were more than 33,285 people that participated in this field day with 2,038 sites. They made 782,943 phone contacts and 496,350 CW contacts for a total of 1.2 million contacts. 1,622 sites were run completely on emergency power. 1,390 sites were using generators and 458 were using batteries.

1998 – Department of Highways site in Tantallon – June 27th and 28th the fourth full weekend in June – HARC was 30th overall and the 1st Canadian station in the 2A section with a total of 6,766 points. 2A was the most popular again with 606 entered. There were almost 31,000 people participating in this field day and 90% were using non-commercial power; generators, batteries, or alternate sources to operate their stations. There were 1.35 million contacts; over 490,000 CW and over 853,000 phone contacts. VE1FO made 2,278 contacts with less than 150 watts via 48 operators to earn those 6,766 points.

1999 – Department of Highways site in Tantallon – June 26th and 27th – The fourth full weekend in June – HARC was 68th overall and the 1st Canadian station in the 2A section with a total of 4,994 points. VE1FO was down a bit in points but so was the 2A section in events with 585 only this year. There were 31,266 participants this year that submitted 2,108 logs and made 1,470,218 contacts. There were 538,936 CW contacts and 926,872 phone contacts. More than 90% of the sites were on full emergency power. VE1FO (+ Moe Cyr VE1MPC) made 1,504 contacts with less than 150 watts with 50 operators for the 4,994 points.

2000 – Department of Highways site in Tantallon – The fourth full weekend in June – June 24ᵗʰ and 25ᵗʰ – HARC was 36ᵗʰ overall and the 1ˢᵗ in Canada in the 2A class with a total of 6,080 points. (+VE1INN Al Harris) VE1FO was in the class 2 (less than 150 watts) category and had 70 operators to earn those 6,080 points. See the chart at the end of the field day entry for 2005. The 2A class was the most popular again with 517 entries. A total of 2,043 logs were submitted from 30,151 participants. The total contacts were 1,421,816 and they were down 3.3% from 1999. The CW contacts had decreased 5.5% to 509,562 contacts. The emergency power and portable stations were 90% of the total stations. This is a direct quote from the field day report in QST: "Field Day proved to be a popular follow-up exercise for the hundreds of stations and groups that had participated in the nationwide Y2K standby on December 31st and January 1ˢᵗ".

2001 – Department of Highways site in Tantallon – The fourth full weekend in June – June 23ʳᵈ and 24ᵗʰ – HARC was 18ᵗʰ overall and the 1ˢᵗ Canadian in the 2A class with a total of 7,770 points. (+VE1DOH Darryl Williams) The 2A class was again the most popular with 498 entries. 2,062 entries were received representing 31,486 participants and ARRL stated this was an increase of 4.3% over 2000 and among the top five of all time participants. See the chart at the end of the field day entry for 2005. HARC earned these 7,770 points from 10 operators in the class 2 – less than 150 watts category.

2002 – Department of Highways site in Tantallon – fourth full weekend in June – June 22ⁿᵈ and 23ʳᵈ – HARC was 9ᵗʰ overall and the 1ˢᵗ Canadian station in the 2A section with a total of 9,832 points. VE1FO (+ Bill Elliott VE1MR) made 3,224 contacts with 44 operators in the less than 150 watts category to earn those 9,832 points. 2A was the most popular category with 518 participating in that category. ARRL received 2,093 entries from 34,498 participants that made a total of 1,424,222 contacts. HARC is in the Maritime Section. There were 8 stations that reported to ARRL from that section. See the chart at the end of the field day entry for 2005.

2003 – EMO, Dartmouth – fourth full weekend in June – 28ᵗʰ and 29ᵗʰ – HARC was 3ʳᵈ overall and the 1ˢᵗ Canadian station in the 2F section with a total of 5,408 points. The class F section was a new section that first appeared in 2003. There were a total of 62 entries submitted to ARRL for the 2F section. ARRL received a total of 2,080 entries for this field day. Total contacts were down 21% to 1.12 million only. ARRL blamed this decline in poor propagation. There was a total of 29 entries received from the Maritime Section. VE1FO was the only Maritime Station in 2F. There was one other Canadian station entered in 2F the Oakville ARC VE3HB with 1,057 contacts with less than 150 watts, 33 operators for 3,060 points. VE1FO (+ VE1YO the Dartmouth Club) made 1,347 contacts with less than 150 watts, 38 operators for the 5,408 points. See the chart at the end of the field day entry for 2005.

2004 – Department of Highways site in Tantallon – the fourth full weekend in June – the 26ᵗʰ and 27ᵗʰ – HARC was the 1ˢᵗ Canadian station and 14ᵗʰ overall in the 2A category with a total of 8,888 points. There were 33,002 participants in the 2004 field day and they were divided up per the chart below the 2008 record. VE1FO (+ VE1JMA Craig MacKinnon) made 2,471 contacts with less than 150 watts and 40 operators to make those 8,888 points. There were 79 ARRL sections in this field day and 494 entries in the 2A section. HARC at 14th out of 494 entries is a very good showing. 82% of the entries used some form of emergency power – generators, batteries, solar, wind, or some other. See the chart at the end of the field day entry for 2005 and 2008.

2005 –EMO, Dartmouth – June the 25th and 26th – HARC was 1st overall in the 2F section with a total of 7,200 points. There were 33,078 participants in the 2005 field day and they were divided up per the chart below and the chart below the 2008 record.

Recent Field Day Stats

	2005	2004	2003	2002	2001	2000
CW QSOs	503,205	517,738	67,748	537,130	536,072	511,422
Digital QSOs	21,765	20,940	12,525	17,170	14,283	10,376
Phone QSOs	692,722	787,444	646,564	869,922	868,174	906,226
Total QSOs	1,217,693	1,326,122	1,126,837	1,424,222	1,418,529	1,428,024
Total Entries	2212	2241	2080	2110	2062	2054
Novice/GOTA	396	436	346	428	208	240
Participants	33,078	33,002	32,100	34,398	31,530	30,342

VE1FO (+ the Dartmouth Club VE1YO) was the top station in this section. There was a total of 59 entries in the 2F section. VE1FO (+ VE1YO) made 1,778 contacts with less than 150 watts to obtain these 7,200 points by 47 operators.

2006 – York Redoubt – the fourth full weekend in June – the 24th and 25th – HARC VE1FO (+VE1YVN Sam Semple) was the 2nd Canadian station and 22nd overall in the 2A category with a total score of 7,744 points. VE3ZI was the Canadian station ahead of HARC and was 16th overall. There were 32,506 participants in the 2006 field day and they were divided up per the chart below the 2008 record. VE1FO + VE1YVN made 2,166 contacts with less than 150 watts and 48 operators to earn the 7,744 points. 2A was the most popular section with 484 entries from that section. ARRL received the 2,169 entries from all 80 ARRL/RAC sections.

Page 74 QST December 1967

Sam, VE1YVN, is waiting for Wayne, VE1TTT and water.

Thanks, guys, for all the help and especially the many laughs.

2007 – EMO, Dartmouth – the fourth full weekend in June – the 23rd and 24th – HARC VE1FO (+VE1YO the Dartmouth Club) was 3rd overall and the 1st Canadian station in the 2F section with a total of 8,482 points. They made 2,252 contacts with less than 150 watts by 38 operators to earn those 8,482 points. There were 69 entries received for the 2F section and as one can see HARC was 3rd of the 69. This was another very good record. There were 34,833 participants in the 2007 field day and they were divided up per the chart below the 2008 record.

2008 – York Redoubt – The fourth full weekend in June – the 28th and 29th – HARC VE1FO (+ VE1TRI Barry Diggins) was the 1st Canadian station in the 2A category and 11th overall with a total of 9,420 points. HARC earned these points from 2,785 contacts with a power output of less than 150 watts and with a total of 37 operators. There were 35,798 participants in

General Field Day Stats

	2008	2007	2006	2005	2004
CW QSOs	506,139	511,580	518,799	503,205	517,738
Digital QSOs	27,869	22,112	21,459	21,766	20,940
Phone QSOs	702,847	679,240	696,567	692,722	787,444
Total QSOs	1,236,855	1,212,932	1,236,825	1,217,693	1,326,122
Total Entries	2409	2331	2199	2212	2241
Novice/GOTA	447	467	432	396	436
Participants	35,798	34,833	32,506	33,078	33,002

Page 70 QST December, 2008

the 2008 field day and they were divided up per the chart below. There were 10 entries from the Maritime Section of ARRL.

2009 – HARC club station at St. Andrews school – the fourth full weekend in June the 27th and 28th – HARC VE1FO (+ VE1QD – Scott Wood) was 1st overall of the 78 entries in the 2F category with a total of 8,390 points. These points were earned from 2,110 contacts, with a power output of less than 200 watts by 50 operators. There were 11 entries from the Maritime Section of ARRL.

2010 – The HARC club station VE1FO (+ VE1QD – Scott Wood) at York Redoubt on June 27th and 28th earned 4,674 points as the 1st Canadian and 76th station overall in the 2A category. These points were earned by 1,413 contacts made with less than 200 watts by 42 operators. There were 11 entries from the Maritime Section of ARRL and a total of 474 entries in the 2A category.

2011 – HARC club station at St. Andrews school – the fourth full weekend in June the 25th and 26th – HARC VE1FO (+ VE1TRI Barry Diggins) was the 1st Canadian Station in the 2F category and 7th overall with a total of 1,001 contacts with a power output of less than 150 watts. 30 operators made these contacts for a score of 3,080 points. There were 11 entries in the Maritime Section and 74 entries within the 2F class.

2012 – HARC club station at York Redoubt – the fourth full weekend in June the 23rd and 24th – HARC VE1FO (+ VE1TRI Barry Diggins) was the 2nd Canadian Station in the 2A category and 91st overall with a total of 1,149 contacts with a power output of less than 150 watts. 40 operators made these contacts for a score of 3,980 points. The top Canadian Station was VE3NCR in Ottawa, just above VE1FO in the 90th overall position.

There were 407 entries in the 2A.

Radio club field day today

Like the sound of your own voice?

Well, this weekend's your chance to get it on the air when members of the Halifax Amateur Radio Club share their mikes while participating in their yearly field day.

Members of the club will also be chatting with the 35,000 other radio enthusiasts taking part across North America.

The group will be meeting at their headquarters at 3380 Barnstead Lane.

Festivities kick off at 3 p.m. today and run for 24 hours.

The 80-year-old club is one of the oldest in Canada and provides the backup communications centre for the municipality's emergency measures system, according to a news release.

The Chronicle Herald Saturday, June 22, 2013

2013 – HARC club station at the Emergency Operations Centre in the club house at St. Andrew's Centre, 3380 Barnstead Lane, Halifax – the fourth full weekend in June the 22nd and 23rd – in the 2F category – HARC VE1FO (+ VE1QD Scott Wood) was the 3rd station in the 2F category and the only Canadian Station – with 1,873 contacts, under 150 watts, with 30 operators and 7,396 points. There were 72 entries in the 2F category.

2014 – HARC field day was held in the 2A category at York Redoubt in the 81st Field Day on June 28th and 29th. 47,428 operators participated in this Field Day. HARC VE1FO (VE1QD) was the 24th station and the first Canadian station in the 2A category, with 1,997 contacts, under 150 watts, with 37 operators and 7,218 points. 392 stations participated in this 2A category.

2015 – HARC field day was held in the 2F category from the St. Andrews Community Centre on June 27th and 28th. VE1FO (VE1QD) was the 3rd station and the first Canadian station in the 2F category with 1,262 contacts, under 150 watts, with 73 operators and 5,884 points. 66 stations participated in this 2F category. There is more detail on the VE1FO field day results in the November 2015 issue of the HARC Reflector.

2016 – HARC field day was held in the 2A category on June 25th and 26th. VE1FO (VE1QD) was the 2nd Canadian station and the 45th overall with 1,177 contacts, under 150 watts with 62 operators and 5,106 points. There were 376 entries in the 2A category.

2017 – HARC field day was held on June 24th and 25th, 2017, in the 2F Class at the St. Andrew's Community Centre, the home of the club station VE1FO. HARC was the fourth highest station in this class and one of only three Canadian stations to participate in this class. VE3SOO was in the 23rd place and VY2CRC was in the 44th place. There was a total of 69 stations in the 2F class. VE1FO (+VE1FQ) had 1,193 contacts, fewer than 150 watts, with 43 operators and 5,546 points. The HARC Club Station at St. Andrew's Community Centre is the backup site for HRM-EMO communications. Working from the backup EMO site is extremely important to HRM-EMO who pays for the Club Station and facilities. HRM-EMO is an acronym for Halifax Regional Municipality Emergency Measures Organization.

2018 – Field Day was on June 23rd and 24th. HARC did not participate in this field day. The club had moved to Hammonds Plains on June 20th and the equipment was not available to participate. The Kings County Amateur Radio Club invited anyone that wanted to participate in field day to join them at Port Williams. A few of the HARC members joined them and it was a combination of the Kings County and Annapolis Valley Amateur Radio Clubs. They participated in the 2A category; VE1LD (+ VA1AVR), they were the first Canadian station and in the 23rd place of the 2A category. They made 1,892 contacts with less than 150 watts and with 15 operators for 6,448 points.

2019 – Field Day was on June 22nd and 23rd. HARC was the first Canadian station and the eighth overall in the 2F category. VE1FO (VE1FQ) 949 contacts with less than 150 watts and with 45 operators for 4,585 points.

2020 – Field Day was on June 27th and 28th and because of the COVID pandemic I had my own field day in my 5th wheel trailer at Sambro, Nova Scotia and gave the log to HARC. My record was recorded on page 80 in the December 2020 issue of QST. I had 84 CW contacts with less than 150 watts (50 watts) one operator, little ole me and made 478 points.

All Field Day records are now on the internet. Simply search ARRL Field Day and the year of interest.

One can see from these records that HARC has certainly done very well on Field Day.

POEMS

These poems were found in the Halifax Amateur Radio Club files. All that is known about them is recorded here. Some if not all were used in club newsletters over the years.

A HAM RADIO TOWER

Why do people get upset when one puts up a radio tower?
A few things in its favour:

IT DOESN'T:
Squeal its brakes
Screech its tires
Blow its horn
Roar it's motor
Slam its doors at ungodly hours
Shine its headlights in your bedroom window
Nor does it backfire.

IT DOESN'T:
Bite you
Bark or meow
Leave deposits on your property
Dig up your garden
Scratch on your door
Widdle on your trees
Nor does it dig into and scatter your garbage.

IT DOESN'T:
Drop leaves that you have to clean up
Grow branches over your house
Drop fruit or nuts which block your down pipes
Block your view like a tree or a building

Grow roots that damage your walk or driveway
Nor do its roots plug your drains.

IT DOESN'T:
Have boisterous parties
Or play loud music
Or have swimming parties through the night
It doesn't ring your phone (accidently?)
Nor does it ride bikes across your lawn.

IT'S JUST QUIET, AND HAS NOTHING TO SAY.

VE7BJ

* * *

"A MIGHTY MAN, "

The Net Control, on Traffic Net, -
A mighty man is he,
He steps into his awful job
As bravely as can be.
It seems as though no living being
Could face it, and stay sane,
Yet, on he pounds his harried way,
With little pain or strain.

The ITV, the yells of "break"
The code and QRN
With other odds and ends thrown in,
Would kill some weaker men,
But, does he shrink from all the din,
Or cower behind his mike?
No, he just bows his aching head
To each cruel hammer strike.

* * *

AN AMATEUR'S DREAM

by Spud Murphy VE3BGI

'twas the night before Xmas and all through the town
All the hams were asleep even ole Gordie Brown,
The boys were all snuggled down deep in their beds,
Receivers were off, transmitters were dead.

At this time of nite for year after year
Old Santa with reindeer would always appear.
Coming laden with gifts and presents galore
He never had missed the hams here before.

The radio ham nearly always did well.
But upon this fact we will not dwell,
You see something happened to alter the scene
For 3BGI was deep in a dream.
Please don't ever tell your children of this
Do not ever shatter their idea of bliss.
Oh no, it would be anything but right
To tell them what happened this horrible nite.

Spud went to bed early – one thirty or two.
To rest up for Xmas as good people do.
As he went to sleep little did it seem
He would ever have a more appropriate dream.

As Rudolph the red-nose came into sight
The dream came more clear on that long winter night.
Rudolph's Red Nose was used as a marker
Instead of ole Santa the driver was BARKER.

Sue enough there was Teddie, the angel himself,
And he'd brought along as a helper, an elf.
His bag was just filled with lots of good joys
A bag full of radio parts for the boys.

To Gordon – a key, with instructions to use,
To Harold a convertor, himself to amuse.
To Horace a transformer, 5000 volts,
To send to DX land many more jolts.

To Nelson a book, "Emergency Coordinator",
To Alex for mobile, a new generator.

SPURGEON "SPUD" ROSCOE VE1BC

To Dawson a map, and it was quite big,
To mark on the DX, he'd worked with his rig.

To Archie instruction, "How to get on Phone."
Like us, Archie just can't leave it alone.
To Charlie a Key – it is automatic,
Just some little thing Ted found in the attic.

For three Toms, Hal, Jim, Tim and Gil,
Ole Teddie dispensed with gifts until,
For Spud he had that rod to fishing,
The very darn thing for which he was wishing

After dispensing this goodwill, Ted says, "Rudy let's go,
These boys will be happy with those things I know".
And laying both hands on Rudy's Red Nose,
As quick as a flash up the chimney he rose.

Now, you can see why we ask you to hold,
The silence that is more precious than gold.
Children would lose all their faith in St. Nick,
No foolin they would, and pretty darn quick.

To end this tale I have to tell,
The dream was spoiled by an awful yell,
Patsy and Terry came in with a bound,
To tell Daddy of things on the tree they had found.

After breakfast that day we called on the phone,
To see how things were in Teddie's ole home,
Says Ella to us "Things aren't quite right,
Ole Teddie hasn't been home all night".

* * *

In this poem I have made my stand
Against AM phone and single sideband,
The reason is simple, the moral on hand
Never again will we hear heterodynes
Ranting and raving, screaming and screeching
Blasting away, the end unreaching.
Nor will we hear monkeys, calling each other
Day after day in the twenty-meter smother.
Calling CQ 'till their voices wear out
Turning their beams up, down, sideways, and inside out.

Yes, yes, a better thing could never take place
To the whole of the radio race,
Than for all of use, tired and true
To pound a key 'till our faces turn blue,
And make the move to CW.
(Thanks to Mrs. S. Butcher for contributing)

* * *

AR – SK

A ham knocked at the heavenly gate
He felt both tired and old,
He stood before the man of fate
For admission to the fold.
"What have you done?" St. Peter asked,
"To gain admission here?"
I have been a 75 fone operator, Sir
For many and many a year.
The pearly gates swung open wide
As St. Peter touched the bell,
"Come in and choose your harp", he said,
"You've had your share of h…"

Groundwave, Ottawa Amateur Radio Club, VE3RC. (Reprinted from the Bison, Indiana
R.C. Council)

* * *

"DEPRESSION IN DUCKLAND"

By Tom Maguire (1865-94)

With the development of British Socialism in the 1880's, a number of writers began to contribute satirical and polemical poems to Socialist periodicals and journals. Tom Maguire was active in William Morris' Socialist League, and later founder-member of the Independent Labour Party. "Depression in Duckland," published in 1892, is an Aesop-like tale satirizing capitalist economics.

357

A silly, self-sufficient goose
Laid golden eggs for an old duck's use;
And the old duck lived on the golden eggs,
While the goose ate worms and marsh-bank dregs.
But the duck had title deeds to show
That the marsh-bank dregs and the worms below
Were his sole, exclusive propertee,
On which the goose might fatten free;
To yield a regular egg supply
Was the one condition he bound her by.

So the goose had plenty of worms and dregs,
And the duck had plenty of golden eggs;
And the duck waxed fat and round and sleek,
While the goose waned wiry, worn and meek.
But on Sunday mornings the goose would hie
Regularly to the pond close by,
Where the duck would hold a service of prayer
For the good of the goose attending there,
And the sinful goose cried "Alas!" and "Alack!"
Whenever she heard the good duck quack.

For the duck would speak of the wicked ways
Of geese beginning and ending their days
A thriftless, shiftless, lazy lot,
Who didn't thank God for the worms they got.
He exhorted the goose to labour and lay
An extra golden egg per day,
To enable him to spread the light
Of his teachings and law in the lands of night,
For heathen turkeys and heathen "chucks"
Might all be geese, though they couldn't be ducks.

So the simple goose laid eggs galore,
And the artful duck still called for more,
Till at length so great was the egg supply,
That the duck complained of their quality.
"Supply" exceeding his sister "Demand"
The duck brought thing to a sudden stand,
Declared a stop to the laying of eggs,
The killing of worms or the drinking of dregs;
Saying out to the goose, "You must now make shift,
As I shall do, on the savings of thrift."

"Alas!" the goose cried out in her wore,
"May I lay for myself?" but the duck said "No!"
"The on!" she exclaimed in wild dismay,
"May I drink the dregs?" but the duck said "Nay!
The dregs are mine, and mine are the worms,
And did you not agree to my terms?"
"But," argued the goose, "I have changed your dregs,
By labour and skill into golden eggs!
What is the remedy for my lack?"
The duck's laconic reply was "Quack!"

* * *

You say
The surgeon was drunk
when he put in the stitches.
And they gave you shots
in the seat of your britches.
Your gown exposes
your derriere.
And as it turns out
your ailment ain't rare.
The bedpans cold
and the nurse ain't cute.
You're bored as hell
with flowers and fruit.
You ask for booze
and they bring you pills.
You're long on miseries
short on thrills.
They ban midnight parties
and sex and such
and hospital social life
just ain't much.
Kildare and Casey
never appear.
And you're scared
of sterile atmosphere.
You're bathed in a way
that doesn't please
and you suffer
great indignities.
Sounds to me

Like the going's mighty thick.
Only one thing to do, pal.
GET WELL QUICK

* * *

THE HEAVEN-BOUND C.Q.

In frigid, rugged North there is a well-known mining town,
And in that place there is a church, with a person of renown.

In this church one Sunday morn, with the congregation there,
The parson said, "Good people", God has answered all our prayers.

I hope that all you folks, like me, too-day are in good voice,
CJKL this Sunday morn have made our church their choice."

Gargled the choir and tried the note, the organ gave the tone,
The parson with the Good Book took his stand at the microphone.

The congregation took their books and settled in the pew,
While the parson with 5000 watts threw out the long C.Q.

At last, long last, the parson signed, - - the choir prepared to sing,
But in the lull there came a voice that made their eardrums ring:

"This is Peter Henry calling, " - - the parson stood aghast,
"Of all the times I've called upstairs an answer's come at last!"

His mind worked fast, himself bethought "I'll be a man of fame.
In two thousand year I'm the first to know St. Peter's second name!"

The Reverend gent was very sad when later on he found –
Instead of Heaven calling, it was "VE3PH - - Peter Henry, Gordon Browne."

Tom, VE3DAD

* * *

THE RADIO HAM

If you should see upon the street
A man who walks with dipole feet,

With a train of little pips behind,
He's a radio ham with a micro-mind.

With micro-seconds and micro-waves
And micro-volts he fills his days.
And thus in the course of time we find
His brain has shrunk to a micro-mind.

His eyes give out a neon gleam
His ears fan out like a radio beam,
As he chews, his molars oscillate
And his heart pumps blood at a video-rate.

This man obtains with passing years,
Infinite impedance between his ears,
And at last he succumbs to a heavy jolt
When he gets what he thinks is a micro-volt.

The doc looks up from his microscope
And says to his nurse; "Behold this dope,
No trace of a brain cell can I find,
He's a Radio Ham with a micro-mind".

73
Don VE1AJA

* * *

THE RAG-CHEW BLUES

Zeke's wife had fixed the lovely meal,
('Twas fit for any King),
And didn't want to have it spoiled
By any little thing.
So, just before she served it up,
She told Zeke, "Come and eat,"
And set it well back on the stove,
Where it would hold the heat.

Now, Zeke was talking on the air,
Just waiting for her call.
So, now he'd sign and QRT,
To get right on the ball.

But though he told the other guy,
"I've simply got to run",
And told him more than twenty times,
The good it did was none.

At last, when Zeke could break away,
The meal was ruined forever.
His wife was blazing mad, and said,
"That fool is far from clever,
He heard you say you'd had the call,
And that you had to run.
I guess the stupid boob was sure
That you were having fun.
The meal is ruined; my work for naught.
Of Hams I've had my fill.
Why can't they use the Golden Rule?
Their brains add up to nil."

Now Zeke is scared, when mealtime comes,
To go near his old rig.
For QSO's, that time of day,
He doesn't care to fig,
For, now he feels the other guy
Won't try on other's shoes,
And thus, will cause a man and wife
To suffer Rag-Chew Blues.

VE1RT, with apologies to all XYLs

* * *

THIS HAPPY BREED

THE BRITISH ISLES CONSIST OF FOUR RACES: THE SCOTS, WHO KEEP THE SABATH AND EVERYTHING ELSE THEY CAN LAY THEIR HANDS ON; THE WELSH, WHO PRAY UPON THEIR KNEES AND UPON THEIR NEIGHBOURS; THE IRISH, WHO DON'T KNOW WHAT THE DEVIL THEY WANT BUT ARE PREPARED TO DIE FOR IT; AND THE ENGLISH, WHO BELIEVE THEM SELVES TO BE A RACE OF SELF-MADE MEN, THEREBY RELIEVING THE ALMIGHTY OF A DREADFUL RESPONSIBILITY.

88 Helen VP9JD

* * *

IN MEMORY OF VE1HC, THE HAPPY CANADIAN

As Remembered by W1BRI

One of the saddest days for me
Was when we lost VE1HC.
Dear Howie, How I loved that name
The band somehow is not the same.

I recall, as I write line after line,
How he taught me to make his Digby wine,
And how he laughed when I told the gang
How the wine blew up with a tremendous bang.

I also recall that at Saunders Bay
He came running down the parking way
And grabbed me with much a vigorous squeeze
My body tensed; my mind in a freeze
Until at last I could plainly see
The tag he wore said VE1HC

I remember he would tell with joy
Of his menu planned – Oh Boy, Oh Boy,
And upon my face came a great big smile
As he told how the goodies made his day worthwhile.

I know that I shall never see
Another ham like VE1HC
And we who knew this Canadian friend
Have memories that will never end.

Some nights as I lie awake
I seem to hear him say "Break, Break,"
This is VE1HC,
"Can anyone down there still hear me?"
And then sign off from his distant star
With his 73 and God Bless America

The Happy Canadian or Hotel Charley
Is forever, as is Scrooge and Marley
A radio club now has his call
As memoriam for one and all.

* * *

WHAT HAMS ARE MADE OF - - -
(W9GJS REPORTS AUTHOR UNKNOWN)

I want to tell a story,
A story I have heard;
You may think it all a fable,
But it's gospel, every word.

The Good Lord took a blacksmith,
A tinkerer and a Bo;
Who used to hold a trick job,
Train dispatching, don't you know.

A phone lineman and a "central,"
An "electric" engineer;
A radio announcer,
And a man who couldn't hear.

He mixed 'em all together,
With a grain of salt or two;
And set 'em in the shadows,
To let the mixture brew.

Then he ran it through a filter,
And he screened it through a grid;
He seasoned it with pepper,
And lifted off the lid.

He moulded it and twisted it,
And shaped it here and there;
Then took it out and looked at it,
And sat it on a chair.

He put some earphones upon it's head,
And shot it full of juice;
But when it didn't seem to care,
The Lord said, "What's the use?"

"I've tried to make a genius,
But it won't work worth a damn,
So I'll simply let it go at that,
And called the thing a "Ham."

* * *

WE FOUND THIS ONE IN QST.

I care not where they put me
When with this life I'm through
As long as from my resting place
I can still call "CQ"

K0DCC
Page 34, January 1958, QST

* * *

*This exercise has been a lot of fun and I enjoyed it very much.
I trust it has been of some interest.*

73

Spurgeon G. "Spud" Roscoe

Radioman Special Royal Canadian Navy 1956-1961
Graduate Radio College of Canada, Toronto, Ontario
Graduate Air Services Training School, Ottawa
Graduate National Radio Institute, Washington, D.C.
First Class Certificate of Proficiency in Radio # 6-108
Coast Guard Radiotelegraph Operators Certificate # 054
Amateur Radio Station VE1BC

www.ingramcontent.com/pod-product-compliance
Lightning Source LLC
Chambersburg PA
CBHW080903170526
45158CB00008B/1970

9781039133204